Mushroom Biotechnology for Improved Agriculture and Human Health

Scrivener Publishing
100 Cummings Center, Suite 541J
Beverly, MA 01915-6106

Publishers at Scrivener
Martin Scrivener (martin@scrivenerpublishing.com)
Phillip Carmical (pcarmical@scrivenerpublishing.com)

Mushroom Biotechnology for Improved Agriculture and Human Health

Edited by

Charles Oluwaseun Adetunji
Microbiology Department, Faculty of Science, Edo State University Uzairue, Edo State, Nigeria

Chukwuebuka Egbuna
Nutritional Biochemistry and Toxicology Unit, University of Port-Harcourt, Rivers State, Nigeria

Oluwatosin Ademola Ijabadeniyi
Dept. of Biotechnology and Food Science, Durban University of Technology, Durban, South Africa

and

Samantha C. Karunarathna
Qujing Normal University China, Key Laboratory for Plant Diversity and Biogeography of East Asia, Kunming Institute of Botany, Chinese Academy of Sciences, Kunming, Yunnan, China

Scrivener Publishing

WILEY

This edition first published 2025 by John Wiley & Sons, Inc., 111 River Street, Hoboken, NJ 07030, USA and Scrivener Publishing LLC, 100 Cummings Center, Suite 541J, Beverly, MA 01915, USA

For more information about Scrivener publications please visit www.scrivenerpublishing.com.

Wiley Global Headquarters
111 River Street, Hoboken, NJ 07030, USA

For details of our global editorial offices, customer services, and more information about Wiley products visit us at www.wiley.com.

Library of Congress Cataloging-in-Publication Data

ISBN 978-1-394-21263-7

Front cover image courtesy of Wikimedia Commons
Cover design by Russell Richardson

Set in size of 11pt and Minion Pro by Manila Typesetting Company, Makati, Philippines

Printed in the USA

10 9 8 7 6 5 4 3 2 1

Contents

Preface

The ever-increasing global population has posed significant food insecurity, nutrition, and health challenges. Consequently, there is a pressing need for sustainable solutions to address these issues. The intervention of microbial biotechnology in various sectors (food, agriculture, and health) has shown remarkable growth in recent years, driven by sustainable innovations and the biotechnological utilization of beneficial microorganisms, such as mushrooms, for the optimal benefit of humanity. Recent advancements in mushroom biotechnology hold great promise as a natural problem solver for achieving sustainable solutions. These include enhanced nutritional values in agricultural crops, sustained health benefits from pharmacologically active substances used in managing human diseases, and improved crop production. This book is timely and will greatly benefit individuals in both industry and academia.

This book is authored by renowned scientists who have acquired extensive knowledge, expertise, vision, and dedication in their scientific careers. They have earned international recognition, particularly for their work on novel mushroom strains in solving human problems through biotechnology.

The book is intended for a diverse audience, including global leaders, industrialists, professionals in the food industry, agriculturists, agricultural microbiologists, plant pathologists, botanists, and experts in agriculture, microbiology, biotechnology, nanotechnology, environmental microbiology, and microbial biotechnology. It caters to investors, innovators, farmers, policymakers, extension workers, educators, researchers, and individuals in other interdisciplinary scientific fields. Additionally, this book serves as an educational resource and project guide for undergraduate and postgraduate students, as well as educational institutions conducting research in agriculture and nanotechnology. This book is highly recommended for professionals, scientists, environmentalists, industrialists, researchers, higher education students, innovators, and policymakers interested in agriculture.

I want to express my deepest appreciation to all the contributors who have dedicated their time and efforts to making this book a success. Furthermore, I sincerely thank my co-editors for their hard work and dedication throughout this project. I also wish to gratefully acknowledge the suggestions, assistance, and support of Martin Scrivener and the team at Scrivener Publishing.

Professor Charles Oluwaseun Adetunji
(Ph.D, AAS affiliate, FRSB (UK) FNYA; FBSN; FNSM, MNBGN)
Director of Research and Innovation, Edo State University, Uzairue, Nigeria
December 2024

1

Application of Mushrooms in the Bioremediation of Environmental Pollutants

Isibor Patrick Omoregie[1*], Oluwafemi Adebayo Oyewole[2], Kayode-Edwards Ihotu[1], Agbontaen Osagie David[3], Konjerimam Ishaku Chimbekujwo[4], Simon Sunday Ameh[5], Samuel Adeniyi Oyegbade[1] and Charles Oluwaseun Adetunji[6]

[1]*Department of Biological Sciences, Covenant University, Ota, Ogun State, Nigeria*
[2]*Department of Microbiology, Federal University of Technology, Minna, Niger State, Nigeria*
[3]*Department of Public Health, University of South Wales, Pontypridd, UK*
[4]*Department of Microbiology, Modibbo Adama University, Yola, Adamawa State, Nigeria*
[5]*Department of Biochemistry, Federal University of Technology, Minna, Niger State, Nigeria*
[6]*Department of Microbiology, Edo State University, Uzairue, Edo State, Nigeria*

Abstract

The application of mushrooms in bioremediation involves utilizing certain species of mushrooms to remediate or clean up contaminated environments. Mushrooms possess unique properties that make them effective in breaking down and absorbing various pollutants, including heavy metals, organic compounds, and even oil spills. This process, known as mycoremediation, leverages the abilities of mushrooms to degrade, detoxify, and accumulate harmful substances, such as heavy metals, polyaromatic hydrocarbons, solid wastes, and agricultural wastes, in their tissues. The mycelium (the underground network of fungal threads) plays a pivotal role in this process, as it has the ability to secrete enzymes that break down these pollutants, and it can also absorb and concentrate contaminants. Mycoremediation has been applied to diverse environments, including soil, water

**Corresponding author*: patrickisibor007@gmail.com

Charles Oluwaseun Adetunji, Chukwuebuka Egbuna, Oluwatosin Ademola Ijabadeniyi and Samantha C. Karunarathna (eds.) Mushroom Biotechnology for Improved Agriculture and Human Health, (1–28) © 2025 Scrivener Publishing LLC

bodies, and industrial sites, as a more sustainable and natural approach to remediation compared to traditional methods. However, the efficacy of mycoremediation depends on factors such as mushroom species selection, environmental conditions, and the specific contaminants present. The potential of mushrooms in bioremediation is still being explored, and techniques are being improved for optimal outcomes.

***Keywords*:** Mycoremediation, contaminants, pollutants, fungi, mycelium, environmental cleanup

Introduction

Bioremediation involves the utilization of microorganisms like bacteria and fungi, and plants, along with their enzymes, to restore polluted environments to their natural state. Bacteria that can break down specific soil pollutants, such as chlorinated hydrocarbons, through the process of microbial biodegradation are harnessed within the bioremediation process [1].

Furthermore, bioremediation involves harnessing biological processes to mitigate, and ideally eliminate, the harmful impacts caused by pollutants in specific locations. "*In situ*" bioremediation refers to when these biological processes are applied at the polluted site itself. On the other hand, if the contaminated material (such as soil and water) is intentionally moved to a different location to enhance biocatalysis, it becomes an "*ex situ*" case [2]. Although bacteria play a primary role in bioremediation, fungi and their potent oxidative enzymes are crucial for breaking down stubborn polymers and foreign chemicals. Additionally, various plants, whether natural, genetically modified, or in conjunction with rhizosphere microorganisms, are exceptional at eradicating or immobilizing pollutants. Nonetheless, this article solely focuses on bacteria due to the wealth of available genomes, enabling a systems biology approach to address significant environmental challenges [3].

Bioremediation primarily centers on intervention to mitigate pollution, placing it within the realm of biotechnology. It should not be confused with biodegradation, which explores the biological underpinnings of the metabolism of unusual or resistant compounds, mainly by bacteria. Large fungi, known as macrofungi, possess the capacity to gather and disintegrate a diverse range of harmful metals, proving to be an extremely efficient method for rejuvenating compromised environments. Typically, mushrooms employ three effective techniques—biodegradation, bioconversion, and biosorption—to successfully rehabilitate tainted or polluted soils [4].

White-rot fungi, as named due to their distinctive degradation process that causes wood substrates to bleach, employ enzyme secretion to digest wood lignin, resulting in a bleached wood appearance. The white-rot fungi process stands apart from established bioremediation methods like bacterial systems due to its unique natural mechanisms, conferring several advantages in pollutant degradation [5]. Notably, these fungi hold an edge over bacterial systems as they do not necessitate preconditioning to specific pollutants. Unlike bacteria, which require pre-exposure to induce pollutant-degrading enzymes, the degradation potential of white-rot fungi remains uninhibited by this prerequisite in addition to not being limited by concentration thresholds. Numerous strains of white-rot fungi with the capacity to degrade aromatic compounds have been identified. The versatile application of white-rot fungi encompasses bioremediation for polluted soils, heavy metal accumulation, mineralization, bio-deterioration, biodegradation, transformation, and co-metabolism [6].

Phanerochaete chrysosporium, a distinctive fungus, is emerging as the preeminent model for bioremediation. This organism possesses the capacity to degrade an array of substances including lignin macro molecules and various organopollutants such as polycyclic aromatic hydrocarbons (PAHs), polychlorinated biphenyls, dioxins, chlorophenols, chlorolignins, nitroaromatics, synthetic dyes, and diverse pesticides. The study selected robust degraders, including *Phanerochaete sordida, Pleurotus ostreatus, Trametes versicolor, Nematolana frowardii*, and *Irpex lacteus*, for examination [7].

P. chrysosporium has been observed to influence the bioleaching of organic dyes. This organism marked the initial discovery of extracellular ligninase enzymes capable of depolymerizing lignin and related compounds *in vitro*. Notably, *P. chrysosporium* effectively degrades toxic xenobiotics, including aromatic hydrocarbons, chlorinated organics, insecticides, pesticides, and nitrogen aromatics, with the degradation process involving laccases, polyphenol oxidases, and lignin peroxidases [8].

Trametes versicolor has demonstrated the production of three effective lignolytic enzymes capable of efficiently degrading lignin, polycyclic aromatic hydrocarbons, a mixture of polychlorinated biphenyls, and various synthetic dyes [9]. *T. versicolor*, along with its enzymes, has exhibited the ability to delignify and bleach kraft pulp [10], as well as proficiently dechlorinate and decolorize effluents from bleached kraft pulp [11]. This promising capability establishes the potential for innovative and environmentally friendly technologies within the pulp and paper industry. Amaral *et al.* [12] also documented the utilization of *T. versicolor* as biocatalysts for the decolorization of various industrial dyes and wastewater treatment.

Recent investigations have indicated that *P. ostreatus* possesses the capacity to break down a diverse range of polycyclic aromatic hydrocarbons (PAHs). This ability extends to PAH degradation in non-sterile soil, irrespective of the presence of cadmium and mercury. *P. ostreatus* has been observed to catalyze the humification process for anthracene, benzo(a) pyrene, and flora in two PAH-contaminated soils originating from a manufactured gas facility and an abandoned electric cooping plant [13].

Lentinus squarrosulus has demonstrated the ability to mineralize soil contaminated with varying concentrations of crude oil, yielding elevated nutrient levels in the treated soil [14].

Unique Characteristics of Fungi

Fungi constitute a diverse group of eukaryotic organisms that play crucial roles in various ecosystems and have significant impacts on human life. They are distinct from plants, animals, and bacteria.

Fungi are made up of eukaryotic cells with well-defined nucleus and organelles. They lack chlorophyll and do not perform photosynthesis like plants. Instead, they are heterotrophic, obtaining nutrients by absorbing organic matter from their environment. Most fungi have a filamentous body structure composed of thread-like structures called hyphae. These hyphae collectively form a mass called a mycelium [15].

Some fungi, like yeast, are unicellular and do not form mycelium. Fungi reproduce both sexually and asexually. Asexual reproduction often involves the production of spores through processes like budding, fragmentation, or sporulation. Sexual reproduction involves the fusion of specialized sexual structures to form spores that carry genetic variation. Fungi are decomposers and play a vital role in nutrient cycling by breaking down complex organic matter into simpler compounds.

They form symbiotic relationships with other organisms, such as mycorrhizal associations with plants, where they help plants absorb nutrients from the soil. Fungi can be pathogenic, causing diseases in plants, animals, and humans. Examples include athlete's foot, ringworm, and fungal infections in crops. Fungi are used in various industries. Yeast is used in baking and brewing also certain fungi are used to produce antibiotics like penicillin [13].

Fungi are also important in bioremediation, helping to break down pollutants in the environment. Fungi are found in a wide range of environments, including terrestrial and aquatic habitats. They thrive in moist conditions and are often found in damp areas. Fungi are classified into several major groups, including chytrids, zygomycetes, ascomycetes, basidiomycetes, and microsporidia. Each group has unique characteristics and life cycles. Fungi exhibit a vast diversity of forms, sizes, and colors. They can range from microscopic molds to large mushrooms. The cell walls of fungi are composed of chitin, a tough and rigid carbohydrate similar to the exoskeletons of insects and other arthropods. Fungi can cause a range of diseases in humans and other organisms. Some infections are superficial (e.g., skin infections), while others can be systemic and life-threatening. Fungi have also contributed to the development of various medications, such as antibiotics and immunosuppressants. They are fascinating organisms with diverse ecological roles and significant impacts on various aspects of life on Earth [16].

Fungi can be used in wastewater treatment systems to help break down organic matter and pollutants present in wastewater, contributing to the purification of water before it is released back into the environment. Mycorrhizal fungi play a crucial role in soil health and nutrient cycling. When reintroduced to degraded or contaminated soils, they can help restore soil structure, improve plant growth, and enhance the overall health of ecosystems. It is important to note that while mushrooms and fungi have demonstrated promising capabilities in environmental cleanup, the implementation of these techniques can be complex and site-specific [17]. Factors like mushroom species selection, environmental conditions, and the type of pollutants present need to be carefully considered. Ongoing research is exploring ways to harness the potential of fungi for sustainable and effective bioremediation strategies [18].

Mycoremediation is generally perceived as a more "natural" and less invasive method of environmental cleanup, which may lead to greater public acceptance and support for remediation efforts. The harvested biomass of mushrooms grown during mycoremediation can be repurposed for various applications, such as compost or animal feed, reducing waste and maximizing the benefits of the remediation process [19]. Mycoremediation is a growing field with ongoing research, which means there is potential for continued innovation and optimization of techniques for different types of pollutants and environments. While mycoremediation offers these advantages, it is not a one-size-fits-all solution. Each site and type of contamination may have unique considerations, and the success of mycoremediation depends on factors such as mushroom species selection, environmental conditions, and pollutant characteristics. Additionally, mycoremediation may not always replace traditional methods entirely but can complement them as part of a comprehensive remediation strategy [20].

Specific Contaminants Targeted by Mushrooms

Mushrooms and fungi have been studied for their ability to target and remediate a variety of contaminants in the environment. Different mushroom species exhibit varying degrees of effectiveness against specific types of pollutants.

Oyster mushrooms (*Pleurotus* spp.) and white-rot fungi are known for their ability to degrade hydrocarbons found in oil spills and contaminated soil. Certain mushrooms, such as oyster mushrooms, can accumulate and tolerate heavy metals like cadmium, lead, and mercury. They can be used for mycoremediation in metal-contaminated sites [21].

Enzymes produced by various mushrooms can break down and detoxify certain pesticides and herbicides, contributing to the degradation of these contaminants. Some fungi, particularly white-rot fungi, have been shown to degrade chlorinated compounds like polychlorinated biphenyls (PCBs) and dioxins [22]. Certain mushrooms have demonstrated the ability to metabolize endocrine-disrupting compounds, which can have harmful effects on hormonal systems in humans and wildlife. The mycelium of certain mushrooms has been shown to absorb and degrade pharmaceuticals and personal care products that enter water bodies through wastewater discharge. Some fungi have been investigated for their potential to absorb and accumulate radioactive elements, contributing to the cleanup of areas contaminated by nuclear accidents or waste. White-rot fungi and other specialized species have been used to degrade and detoxify various industrial chemicals, including textile dyes [22].

Many mushroom species, particularly those that belong to the Agaricales order, have shown the ability to break down and metabolize PAHs, which are harmful organic pollutants. It is important to note that the effectiveness of mushrooms in remediating specific contaminants can vary based on factors such as mushroom species, environmental conditions, and the type and concentration of the pollutants. Mycoremediation is a promising field of research, and ongoing studies continue to explore the potential of various mushroom species for targeted environmental cleanup. However, while mushrooms offer valuable tools for bioremediation, they are not a panacea and may not be suitable for all types of contaminants or environmental conditions. Mycoremediation is often best used as part of a comprehensive remediation strategy that may also involve other techniques and approaches.

The mycelium of certain mushrooms has been shown to absorb and degrade pharmaceuticals and personal care products that enter water bodies through wastewater discharge. Some fungi have been investigated for their potential to absorb and accumulate radioactive elements, contributing to the cleanup of areas contaminated by nuclear accidents or waste. White-rot fungi and other specialized species have been used to degrade and detoxify various industrial chemicals, including textile dyes. Many mushroom species, particularly those that belong to the Agaricales order, have shown the ability to break down and metabolize PAHs, which are harmful organic pollutants.

The effectiveness of mushrooms in remediating specific contaminants can vary based on factors such as mushroom species, environmental conditions, and the type and concentration of the pollutants. Mycoremediation is a promising field of research, and ongoing studies continue to explore the potential of various mushroom species for targeted environmental cleanup.

However, while mushrooms offer valuable tools for bioremediation, they are not a panacea and may not be suitable for all types of contaminants or environmental conditions. Mycoremediation is often best used as part of a comprehensive remediation strategy that may also involve other techniques and approaches.

Mechanisms of Mushroom Bioremediation

The enzymes produced by the mycelium of certain mushrooms can break down complex pollutants into simpler, less harmful compounds. This process often involves enzymatic reactions that convert pollutants into water, carbon dioxide, and other harmless substances. Depending on the mushroom species, the mycelium may eventually produce fruiting bodies (mushrooms) that contain spores. These spores are then released into the environment, potentially aiding in the further dispersion and colonization of the mycelium to continue the remediation process.

Mycoremediation projects require ongoing monitoring to assess the progress of pollutant reduction and the health of the ecosystem. Adjustments to the process may be necessary based on the site's conditions and the performance of the mushrooms. Over time, as mycoremediation progresses, the pollutants are broken down and the environment is restored. The success of mycoremediation can be measured by improvements in soil quality, reductions in contaminant concentrations, and the recovery of native plant and animal species.

It is important to note that mycoremediation is a complex and site-specific process that requires careful consideration of factors such as mushroom species selection, environmental conditions, pollutant characteristics, and regulatory guidelines. While mycoremediation holds promise as an eco-friendly and sustainable cleanup method, it is often used in combination with other remediation techniques to achieve the most effective and comprehensive results. Mycoremediation is a form of bioremediation that utilizes various species of mushrooms and their associated mycelium to clean up contaminated environments. This process takes advantage of the unique abilities of mushrooms to absorb, accumulate, and break down a wide range of pollutants, thereby contributing to the restoration of ecosystems and the removal of harmful substances from the environment.

Site Assessment and Selection: The first step in mycoremediation involves assessing the contaminated site to determine the type and extent of pollution. Different mushroom species have varying abilities to target specific pollutants, so the selection of appropriate mushroom species is crucial [20].
Mushroom Cultivation: Mushroom cultivation involves growing the selected mushroom species in a controlled environment, such as a laboratory or greenhouse, to ensure optimal growth conditions. This process can involve preparing a suitable growth medium (substrate) and providing the necessary temperature, humidity, and light conditions.
Inoculation of Contaminated Site: Once the mushrooms mature and have developed mycelium (the thread-like network of hyphae), they are introduced to the contaminated site. This can be done by applying mushroom spawn, mycelium-infused substrates, or mature fruiting bodies directly to the polluted area.
Mycelium Growth and Pollutant Absorption: The mycelium of the mushrooms begins to grow and spread throughout the contaminated soil or substrate. As it grows, the mycelium absorbs and accumulates pollutants through various processes, such as physical adsorption, absorption, and enzymatic degradation.
Pollutant Breakdown and Transformation: The enzymes produced by the mycelium of certain mushrooms can break down complex pollutants into simpler, less harmful compounds. This process often involves enzymatic reactions that convert pollutants into water, carbon dioxide, and other harmless substances.
Fruiting and Spore Production: Depending on the mushroom species, the mycelium may eventually produce fruiting bodies (mushrooms) that contain spores. These spores are then released into the environment,

potentially aiding in the further dispersion and colonization of the mycelium to continue the remediation process.
Monitoring and Management: Mycoremediation projects require ongoing monitoring to assess the progress of pollutant reduction and the health of the ecosystem. Adjustments to the process may be necessary based on the site's conditions and the performance of the mushrooms.
Completion and Ecosystem Recovery: Over time, as mycoremediation progresses, the pollutants are broken down and the environment is restored. The success of mycoremediation can be measured by improvements in soil quality, reductions in contaminant concentrations, and the recovery of native plant and animal species.

Mycoremediation is a complex and site-specific process that requires careful consideration of factors such as mushroom species selection, environmental conditions, pollutant characteristics, and regulatory guidelines. While mycoremediation holds promise as an eco-friendly and sustainable cleanup method, it is often used in combination with other remediation techniques to achieve the most effective and comprehensive results.

Absorption and Accumulation of Contaminants by Mushrooms

Mushrooms, often overlooked for their ecological significance, have demonstrated a remarkable ability to absorb and accumulate various contaminants from their surrounding environment. This phenomenon has led to the exploration of mushrooms as natural agents for environmental remediation, a process known as mycoremediation. The absorption and accumulation of contaminants by mushrooms is driven by their unique biology and physiology, making them promising candidates for cleaning up polluted ecosystems.
Absorption Mechanisms: Mushrooms possess intricate mycelial networks composed of hyphae, thread-like structures that extend through the substrate. These hyphae play a crucial role in absorbing contaminants. Substances present in the environment, such as heavy metals, organic pollutants, and even radioactive elements, are absorbed by the mycelium through processes like physical adsorption, absorption, and chelation [23]. This absorption occurs due to the high surface area-to-volume ratio of the mycelium and the presence of specialized binding sites.
Accumulation and Sequestration: Once absorbed, contaminants can be stored within the mycelium or translocated to the fruiting bodies (mushrooms)

through a process called accumulation. Some mushrooms have a higher affinity for specific contaminants due to the chemical properties of their mycelium. The contaminants can be sequestered within the cells, effectively immobilizing them and preventing their movement within the ecosystem.
Hyperaccumulation and Tolerance: Certain mushroom species exhibit a unique trait known as hyperaccumulation, where they can accumulate contaminants to levels significantly higher than those found in the surrounding environment. This ability is linked to the adaptive mechanisms developed by mushrooms to tolerate or detoxify pollutants. Researchers have identified several detoxification mechanisms, such as enzymatic breakdown, complexation, and compartmentalization of contaminants within the mushroom cells.
Potential Applications: The absorption and accumulation capabilities of mushrooms have spurred interest in using them for mycoremediation projects. By cultivating specific mushroom species in contaminated environments, pollutants can be effectively removed or reduced from the ecosystem. This approach is particularly valuable for sites contaminated with heavy metals, organic pollutants, and even complex compounds like polycyclic aromatic hydrocarbons (PAHs).
Challenges and Considerations: While mycoremediation shows promise, its success depends on factors such as mushroom species selection, environmental conditions, and the types of contaminants present. It is essential to carefully choose mushrooms that can tolerate and accumulate specific pollutants while ensuring that the process does not lead to the dispersion of contaminants to new areas.

The absorption and accumulation of contaminants by mushrooms present a natural and sustainable solution for environmental remediation. The ability of mushrooms to sequester pollutants within their biomass offers an innovative approach to reducing contamination levels and restoring ecosystems. As researchers continue to explore the potential of mycoremediation, this natural process has the potential to complement traditional remediation methods and contribute to a cleaner, healthier environment.

Transformation and Degradation of Pollutants

Mushrooms, beyond their culinary and ecological significance, exhibit a fascinating capacity to transform and degrade a wide range of pollutants, contributing to the natural processes of environmental detoxification. This phenomenon has spurred research into the potential use of mushrooms as agents for bioremediation, a field that explores their ability to break down and transform various contaminants, including organic pollutants, pesticides, and complex chemicals.

Mushrooms possess enzymatic machinery that enables them to degrade contaminants through various metabolic pathways. These enzymes, including ligninolytic enzymes like peroxidases and laccases, play a crucial role in breaking down complex molecules into simpler compounds. The mycelium and fruiting bodies work in tandem to metabolize contaminants, utilizing them as a potential carbon and energy source for growth and development.

Ligninolytic enzymes are particularly essential in the degradation of recalcitrant pollutants, often termed xenobiotics, which are resistant to conventional degradation processes. These enzymes initiate oxidative reactions that dismantle complex chemical structures, making them more accessible for microbial degradation or mineralization. This ability to transform xenobiotics makes mushrooms powerful allies in the cleanup of contaminated environments.

Mushroom-mediated bioremediation holds promise for remediating sites contaminated with persistent organic pollutants (POPs), polycyclic aromatic hydrocarbons (PAHs), and other hazardous substances. By introducing selected mushroom species into polluted environments, contaminants can be transformed into less toxic or even harmless compounds. This process reduces the bioavailability of pollutants, limiting their potential impact on ecosystems and human health.

Mushroom-mediated bioremediation often occurs in conjunction with other microorganisms, creating synergistic interactions within the soil or substrate. Bacterial and fungal communities collaborate to break down contaminants more efficiently, resulting in a combined effect that enhances the overall degradation process. This demonstrates the complexity and interconnectedness of natural remediation mechanisms.

The efficacy of mushroom-mediated bioremediation depends on factors like mushroom species, substrate composition, environmental conditions, and the types of pollutants present. Proper species selection and understanding of the biochemical pathways involved are crucial to ensuring successful biodegradation outcomes. The transformation and degradation of pollutants by mushrooms offer a fascinating example of nature's capacity for environmental detoxification. By harnessing their enzymatic prowess and metabolic pathways, researchers are exploring innovative ways to enhance bioremediation efforts. Mushroom-based bioremediation holds the potential to not only detoxify contaminated sites but also inspire sustainable solutions for pollution management and ecosystem restoration.

As our understanding deepens, these fungal allies may play an increasingly pivotal role in addressing some of the most challenging pollution issues of our time.

Role of Enzymes and Metabolic Processes

Enzymes and metabolic processes are fundamental components of nature's toolbox for addressing pollution and contaminants in the environment. These biological mechanisms play a pivotal role in breaking down, transforming, and detoxifying various pollutants, contributing to the self-regulation and restoration of ecosystems. Understanding the intricate interplay between enzymes, metabolic pathways, and pollutant transformation is essential for harnessing these natural processes in environmental remediation.

Enzymes are biocatalysts that accelerate chemical reactions within living organisms. They facilitate the conversion of substrates into products through specific binding sites, lowering the activation energy required for reactions to occur. In the context of environmental remediation, enzymes act as molecular tools that initiate the breakdown of complex pollutants into simpler, less toxic compounds. Metabolic pathways are sequences of enzyme-catalyzed reactions that occur within cells. These pathways are central to the transformation of pollutants, especially xenobiotics-foreign compounds that resist natural degradation. Enzymes, such as oxidases and dehydrogenases, participate in these pathways, leading to the breakdown and mineralization of pollutants into less harmful forms.

Biodegradation, a process facilitated by enzymes, involves the conversion of pollutants into metabolites that can be incorporated into microbial biomass or assimilated by other organisms. This process is often linked to detoxification, as the breakdown of complex contaminants reduces their toxicity and bioavailability. In polluted environments, microbial communities utilize enzymatic reactions to co-metabolize pollutants, resulting in a net reduction of their concentration. Enzymatic processes rarely operate in isolation. Microbial communities within ecosystems engage in synergistic interactions, where different microorganisms contribute complementary enzymatic activities to efficiently degrade pollutants. This collaborative approach enhances the overall degradation rate, demonstrating the complexity and adaptability of natural remediation mechanisms.

Enzymes and metabolic pathways are subject to evolutionary pressures that shape their efficiency and specificity. Microorganisms exposed to pollutants can undergo genetic adaptations, resulting in the production of novel enzymes optimized for breaking down specific contaminants. This enzymatic diversity highlights the dynamic response of biological systems to environmental challenges. Enzymes and metabolic processes are increasingly harnessed for bioremediation strategies. *In situ*, bioremediation involves introducing microorganisms that possess the necessary enzymes to degrade pollutants. By providing optimal environmental conditions, these microorganisms accelerate the natural degradation process, leading to the eventual transformation of contaminants into harmless substances.

While enzymes and metabolic processes offer powerful tools for environmental remediation, challenges such as substrate availability, enzyme activity, and environmental conditions can impact their efficiency. Selecting appropriate microbial strains, optimizing nutrient availability, and monitoring degradation progress are critical aspects of successful bioremediation endeavors.

Enzymes and metabolic processes form the foundation of nature's remediation toolkit, enabling the transformation and detoxification of pollutants. Harnessing these biological mechanisms offers innovative and sustainable solutions for addressing environmental contamination. As we deepen our understanding of enzymatic reactions and metabolic pathways, we unlock new possibilities for leveraging the power of nature to restore and protect ecosystems from the impacts of pollution.

Advancements and Research in Mushroom Bioremediation

Fungal bioremediation, harnessing the unique abilities of fungi to degrade and transform pollutants, is undergoing rapid advancements, driven by innovative research and technological breakthroughs. As we strive for more effective and sustainable solutions to environmental contamination, several emerging trends are shaping the field of fungal bioremediation:

Researchers are exploring the integration of nanoparticles into fungal bioremediation strategies to enhance the efficiency and specificity of pollutant degradation. Nanoparticles can serve as carriers for enzymes or as catalysts to initiate and accelerate enzymatic reactions, amplifying the remediation capabilities of fungi. This approach offers a targeted and controlled means of addressing pollution hotspots in contaminated sites.

Advances in genetic engineering allow scientists to modify fungal strains to enhance their bioremediation potential. Genetic manipulation can introduce genes encoding specialized enzymes or pathways, enabling fungi to break down specific pollutants more efficiently. Engineered fungi also have the potential to adapt to challenging environments and pollutants that naturally occurring strains might struggle with. Synthetic biology techniques enable the design and construction of novel biological systems with tailored functions. In fungal bioremediation, synthetic biology is being explored to create custom-made fungal strains optimized for targeting specific contaminants. This approach allows researchers to engineer fungal consortia with complementary metabolic pathways for efficient pollutant degradation.

High-throughput omics technologies, such as genomics, transcriptomics, proteomics, and metabolomics, are providing unprecedented insights into the molecular mechanisms underlying fungal bioremediation. These technologies enable a holistic understanding of fungal responses to pollutants and environmental conditions, guiding the optimization of bioremediation strategies. Combining different fungal species or integrating fungi with other microorganisms forms hybrid systems that leverage the strengths of each component. Fungal-bacterial consortia, for example, enhance pollutant degradation by capitalizing on the complementary enzymatic activities of different microorganisms. These hybrid approaches provide versatile solutions for tackling complex and recalcitrant contaminants.

Fungi have shown remarkable versatility in utilizing a wide range of substrates, including unconventional ones like waste materials, agricultural residues, and industrial byproducts. Researchers are exploring ways to optimize fungal growth and metabolism using these substrates, promoting both sustainable bioremediation and waste valorization. As laboratory successes translate to real-world applications, fungal bioremediation is transitioning from bench-scale experiments to larger field trials. Researchers are working on scaling up fungal bioremediation processes, addressing logistical challenges, and demonstrating the feasibility of using fungi for large-scale environmental cleanup. Collaboration between fungal bioremediation and nanotechnology is on the rise. Nanomaterials can enhance fungal growth, enzyme activity, and pollutant uptake, leading to more efficient and controlled bioremediation processes. This interdisciplinary approach paves the way for synergistic solutions that capitalize on the strengths of both fields.

As fungal bioremediation continues to evolve, these emerging trends hold the promise of revolutionizing the field, enabling more targeted, efficient, and sustainable strategies for remediating contaminated environments. By embracing innovative approaches and cross-disciplinary collaborations, researchers are expanding the horizons of fungal bioremediation, driving us closer to a cleaner and healthier planet.

Emerging Trends in Fungal Bioremediation

Fungal bioremediation, an eco-friendly approach to decontaminating polluted environments, is witnessing exciting advancements as scientific understanding deepens and technological innovations accelerate. Several emerging trends are shaping the field of fungal bioremediation, offering novel strategies and enhanced efficiency for addressing environmental challenges. High-throughput omics technologies, such as genomics, transcriptomics, proteomics, and metabolomics, are providing comprehensive insights into fungal responses to pollutants. This data-rich approach allows researchers to understand the molecular mechanisms behind fungal bioremediation, leading to more informed strain selection, optimization, and monitoring.

Synthetic biology empowers scientists to engineer fungal strains with enhanced bioremediation capabilities. By introducing genetic modifications, researchers can tailor fungi to efficiently metabolize specific contaminants or tolerate harsh conditions. Engineered fungal strains are opening new avenues for targeted and specialized remediation solutions. Fungi often work in tandem with other microorganisms, such as bacteria, to enhance bioremediation efficiency. Microbial consortia and co-culture systems leverage the synergistic interactions between different organisms, enabling complementary degradation pathways and improved pollutant removal. Nanotechnology is merging with fungal bioremediation to create hybrid approaches with enhanced capabilities. Nanoparticles can serve as carriers for enzymes or be engineered to facilitate pollutant binding and uptake by fungi. This integration enhances pollutant degradation efficiency and offers controlled and targeted remediation strategies.

Extremophilic fungi, which thrive in extreme environments, are gaining attention for their potential in remediating contaminated sites with challenging conditions. These fungi possess unique metabolic pathways and enzyme systems that enable them to break down pollutants in harsh environments, expanding the scope of fungal bioremediation. Fungal bioaugmentation involves introducing selected fungal strains to polluted sites to enhance their natural remediation capabilities. *In situ*, bioremediation strategies leverage indigenous fungi already present in contaminated environments. Both approaches capitalize on fungi's adaptability and intrinsic bioremediation potential. Fungi are being explored for their ability to degrade microplastics, addressing the growing concern of plastic pollution. Certain fungal species have shown promise in breaking down microplastics into non-harmful byproducts, offering a potential solution to this global environmental challenge.

Digital technologies such as machine learning and modeling are being integrated to predict fungal behavior, optimize bioremediation conditions, and monitor remediation progress in real time. These tools aid in decision-making and enhance the precision and efficiency of fungal bioremediation processes. Advancements in fungal bioremediation are transitioning from laboratory settings to field applications. Researchers are developing strategies to scale up fungal remediation processes, addressing logistical challenges and demonstrating the feasibility of large-scale environmental cleanup.

Leveraging cutting-edge technologies, interdisciplinary collaborations, and innovative strategies, are important for researchers to expand the potential of fungi to serve as effective, sustainable, and versatile tools for restoring polluted ecosystems and safeguarding environmental health. As fungal bioremediation gains traction, increased public awareness and regulatory considerations are essential. Ethical concerns, safety protocols, and guidelines for releasing genetically modified fungi into the environment are becoming critical aspects of the field's development.

Genetic Modification of Mushrooms for Enhanced Bioremediation

Genetic modification, a powerful tool in modern biotechnology, offers a pathway to enhance the bioremediation capabilities of mushrooms. By introducing specific genes or modifying existing ones, scientists can optimize mushrooms for efficient pollutant degradation and environmental detoxification. This approach holds immense promise for addressing pollution challenges in novel and effective ways.

Genetic modification enables the introduction of genes encoding specialized enzymes into mushrooms. These enzymes, capable of breaking down specific pollutants, can be sourced from other organisms or designed for enhanced activity. This customization empowers mushrooms to target a wide range of contaminants, including complex and recalcitrant pollutants.

Genetic modification is a means through which researchers can enhance the metabolic capabilities of mushrooms, enabling them to metabolize pollutants more efficiently. This includes increasing the expression of key metabolic enzymes involved in degradation pathways. Modified mushrooms can display improved substrate utilization, tolerance to pollutants, and overall bioremediation performance. Genetic modification can equip mushrooms with stress response genes that help them thrive in contaminated environments. By bolstering stress tolerance mechanisms, such as oxidative stress defenses, modified mushrooms can endure harsh conditions while actively engaging in pollutant degradation. This adaptation enhances their bioremediation potential.

Modified mushrooms can be engineered to interact synergistically with other microorganisms, forming consortia that efficiently degrade pollutants. This collaborative approach capitalizes on the strengths of different organisms, such as bacteria and fungi, working together to enhance pollutant removal and transformation. Genetic modification enables mushrooms to target specific pollutants by optimizing their enzymatic toolkit. This specificity reduces the risk of unintended ecological consequences, as the modified mushrooms focus solely on the intended contaminants without disrupting non-targeted species or compounds.

Mushrooms can be engineered to enhance their nutrient uptake and transport systems, facilitating the efficient acquisition of essential elements required for pollutant degradation. This optimization ensures that modified mushrooms have the resources needed to sustain their bioremediation activities. While genetic modification holds tremendous potential, ethical and safety considerations are paramount. Ensuring that modified mushrooms do not pose risks to native ecosystems, non-target organisms, or human health is crucial. Careful containment and risk assessment protocols are essential before releasing genetically modified organisms into the environment.

Genetic modification of mushrooms for enhanced bioremediation represents a frontier where biotechnology meets environmental stewardship. By precisely tailoring mushrooms' genetic makeup, we can unleash their potential as efficient and specialized agents for pollution mitigation. As research advances, the responsible application of genetic modification in fungal bioremediation could revolutionize our approach to environmental cleanup, offering sustainable solutions to some of our most pressing pollution challenges. Mushroom bioremediation, also known as mycoremediation, offers a host of environmental benefits that make it a promising and sustainable approach for cleaning up contaminated sites. By harnessing the natural abilities of mushrooms, this method provides a range of advantages that contribute to environmental restoration, ecosystem health, and overall sustainability:

Mushroom bioremediation relies on naturally occurring organisms and processes. It avoids the use of harsh chemicals and disruptive mechanical methods, minimizing the ecological footprint and ensuring a non-invasive approach to pollution cleanup. Different mushroom species exhibit preferences for specific contaminants. This selectivity allows for the targeted removal of pollutants from contaminated environments. By choosing the right mushroom species, researchers can effectively address a wide range of pollutants, from heavy metals to organic compounds [23].

Traditional remediation methods often involve soil excavation and disruption of ecosystems. Mushroom bioremediation takes place within the existing soil matrix, minimizing disturbance to habitats and preventing erosion, which helps preserve local biodiversity. Mushrooms enhance soil health by breaking down pollutants and promoting nutrient cycling. As they degrade contaminants, they release nutrients that enrich the soil, fostering a healthier and more productive environment for plant growth. Mushrooms used in bioremediation are generally safe for humans and wildlife. Unlike chemical treatments, which can have unintended consequences for non-target species, mushroom bioremediation poses minimal risks to the broader ecosystem [24].

Mushroom bioremediation is energy-efficient compared to traditional methods. It does not require the energy-intensive processes of excavation, transport, and treatment that some conventional methods entail.

Mushrooms capture carbon dioxide during their growth, contributing to carbon sequestration. Moreover, their ability to remediate contaminated sites reduces the need for energy-intensive practices that release greenhouse gasses, thus helping mitigate climate change.

Mushrooms thrive in diverse environments, including polluted ones. This adaptability allows for the application of bioremediation in a wide range of settings, from industrial brownfields to urban areas and natural ecosystems. Mushrooms often work in collaboration with indigenous microbial communities. This partnership enhances the breakdown of contaminants and contributes to the overall ecological balance of the site. Mushroom bioremediation requires minimal infrastructure and can be integrated into various landscapes without significant visual impact. It offers an aesthetically pleasing and sustainable way to address contamination issues. In a world increasingly concerned about environmental health and sustainability, mushroom bioremediation stands out as a holistic and ecologically sound solution. By leveraging the natural abilities of fungi, we have the opportunity to restore polluted environments, preserve biodiversity, and pave the way for a greener and cleaner planet [25].

Benefits of Mushroom Bioremediation

Mushroom bioremediation is a sustainable and ecologically friendly method for cleaning up contaminated environments and offers numerous environmental benefits that contribute to the restoration of ecosystems and the promotion of environmental health. Here are some key advantages of using mushrooms for bioremediation:

Mushroom bioremediation harnesses the natural ability of fungi to break down and transform pollutants. This approach relies on existing biological processes, avoiding the need for introducing synthetic chemicals or disrupting the environment. Unlike traditional remediation methods that often involve heavy machinery, excavation, and chemical treatments, mushroom bioremediation operates with a minimal environmental footprint. It reduces soil disturbance and prevents further damage to already vulnerable ecosystems [26].

Different mushroom species have specific affinities for certain contaminants. This selectivity allows for targeted remediation, addressing specific pollutants without affecting non-target organisms or compounds. As mushrooms degrade contaminants, they release nutrients into the soil. This process contributes to soil restoration, improving its fertility, structure, and overall health. The enhanced soil quality can support the growth of native vegetation and restore ecosystem balance.

Mushrooms play a role in carbon sequestration by capturing carbon dioxide during their growth. This helps mitigate climate change by reducing the concentration of greenhouse gasses in the atmosphere. Mushroom bioremediation requires less energy compared to traditional methods like incineration or chemical treatments. This reduces the carbon footprint associated with pollution cleanup efforts. Mushroom bioremediation promotes biodiversity by restoring contaminated sites to a healthier state. As the ecosystem recovers, it provides habitat and resources for diverse plant and animal species. Mushroom bioremediation is non-toxic and safe for both the environment and human health. It eliminates the risks associated with the release of harmful chemicals into the ecosystem [27].

Mushrooms often form symbiotic relationships with other microorganisms in the soil. These partnerships can enhance the overall effectiveness of bioremediation by promoting collaborative pollutant degradation. By converting polluted sites into healthy ecosystems, mushroom bioremediation supports sustainable land use practices. These remediated areas can eventually be reintegrated into productive land for agriculture, forestry, or recreational purposes. Mushroom bioremediation is an environmentally sound strategy that aligns with the principles of sustainability and conservation. By harnessing the natural processes of fungi, we can restore contaminated sites, enhance soil health, and contribute to the overall well-being of our planet's delicate ecosystems. Bioremediation, a sustainable approach to cleaning up contaminated environments using natural processes, offers several economic advantages over traditional methods that involve mechanical, chemical, or physical interventions [28].

Bioremediation relies on natural biological agents, such as microorganisms or plants, which are often less expensive to procure than the chemicals or machinery required for traditional methods. This reduces the overall cost of treatment. Bioremediation typically requires minimal infrastructure, as it utilizes existing environmental conditions for pollutant degradation. In contrast, traditional methods often involve constructing and maintaining complex systems like incinerators, landfills, or treatment facilities. Bioremediation processes generally have lower energy requirements compared to energy-intensive traditional methods like incineration or thermal treatment. This translates to reduced energy costs and a smaller carbon footprint.

Bioremediation offers the potential for long-term cost savings due to its ability to address contamination at its source. By promoting natural degradation processes, ongoing maintenance and monitoring costs can be significantly reduced over time. Traditional methods often generate substantial amounts of waste, requiring additional costs for disposal and management. Bioremediation processes generate fewer secondary waste streams, minimizing the need for waste treatment and disposal [21].

Bioremediation restores contaminated sites to a functional state, allowing them to be repurposed for agriculture, construction, or other productive uses. This contributes to economic revitalization and value creation. Bioremediation takes place within the existing environment, minimizing the need for disruptive activities like excavation and site preparation. This reduces project timelines and associated costs. Bioremediation aligns with environmental regulations and sustainable practices, reducing potential legal and compliance costs associated with traditional methods that may involve the use of chemicals or hazardous materials [4].

Bioremediation often enjoys greater public acceptance due to its eco-friendly and non-intrusive nature. Positive public perception can lead to smoother project execution and fewer delays. Bioremediation can be adapted to various site-specific conditions and pollutants, offering scalability for projects of different sizes and complexities. This adaptability reduces the need for custom engineering and design, streamlining project costs.

The economic advantages of bioremediation, encompassing reduced costs, lower energy consumption, and sustainable site restoration, position it as a viable and cost-effective alternative to traditional pollution cleanup methods. As environmental and economic considerations become increasingly intertwined, bioremediation offers a balanced solution that benefits both the environment and the bottom line [15].

Challenges and Limitations of Using Mushrooms

Although mushrooms hold great potential for bioremediation, there are several challenges and limitations that must be considered when using them as a remediation tool, because some factors mitigate the effectiveness, scalability, and practicality of mushroom-based bioremediation strategies.

Mushroom species have varying affinities for specific contaminants. Finding the right match between mushroom species and pollutants is essential for effective bioremediation. Not all mushrooms can degrade all types of pollutants, which limits the versatility of this approach. Mushroom bioremediation is influenced by factors such as soil type, pH, moisture levels, and temperature. Unsuitable environmental conditions can hinder the growth and activity of mushrooms, affecting their ability to remediate pollutants effectively. Mushroom-based bioremediation often proceeds at a slower rate compared to some traditional methods. Mushrooms require time to grow, establish mycelial networks, and degrade pollutants. This slower pace may not be suitable for sites with urgent remediation needs.

Mushroom growth is influenced by seasonal changes and climatic conditions. Some mushroom species are more active during specific times of the year. This seasonality can affect the timing and consistency of bioremediation efforts. Mushrooms may struggle to degrade complex pollutants or those with intricate chemical structures. Certain persistent organic pollutants (POPs) or heavily chlorinated compounds may not be easily broken down by mushroom enzymes. Continuous monitoring and maintenance are required for successful mushroom bioremediation projects. Ensuring optimal environmental conditions, nutrient availability, and proper mushroom growth can be resource-intensive and require ongoing commitment [25].

Mushrooms have the potential to accumulate contaminants within their fruiting bodies. If not managed carefully, harvested mushrooms might release pollutants back into the environment or introduce them into the food chain. Introducing non-native mushroom species into an ecosystem for bioremediation could have unintended consequences, disrupting native fungal communities or interacting with other organisms in unexpected ways. The public perception of using mushrooms for bioremediation might not be universally positive. Concerns about introducing genetically modified organisms or unfamiliar species into the environment could hinder project support.

Regulations for mushroom bioremediation are not as well-established as those for traditional methods. Developing appropriate regulatory frameworks for the use of modified mushrooms and assessing their long-term impacts is a challenge. While mushroom bioremediation offers a natural and sustainable solution to pollution challenges, it is not without its limitations and challenges. Careful consideration of site-specific conditions, mushroom species selection, ethical considerations, and project timelines is essential to ensure the successful implementation of mushroom-based bioremediation strategies.

Future Prospects and Research Opportunities

Mushroom-based bioremediation holds immense potential for growth and expansion as a viable and sustainable solution to environmental pollution. With ongoing research, technological advancements, and increasing awareness of its benefits, this approach is poised to play a significant role in shaping the future of pollution cleanup and ecosystem restoration:

Ongoing exploration of the fungal kingdom may uncover new mushroom species with unique bioremediation capabilities. Identifying fungi that efficiently target specific pollutants can expand the range of contaminants that can be addressed through this method. Advances in genetic engineering offer the potential to modify mushrooms for enhanced bioremediation. Genetic manipulation can enable mushrooms to metabolize a wider variety of contaminants, accelerate degradation rates, and adapt to diverse environmental conditions [14].

Combining mushroom-based bioremediation with other remediation methods, such as phytoremediation or bacterial degradation, can create hybrid approaches that synergistically tackle pollution challenges. These collaborative strategies offer comprehensive solutions for complex contamination scenarios. As research progresses, a deeper understanding of mushroom ecology and its interaction with contaminants will allow for the development of tailored solutions for specific sites. Customized fungal consortia and management strategies can optimize bioremediation outcomes.

Mushroom-based bioremediation is moving from laboratory-scale studies to field applications. With successful field trials, this method could be scaled up to address contaminated sites of varying sizes, from industrial brownfields to urban areas and natural ecosystems. Collaboration between mushroom bioremediation and nanotechnology offers innovative possibilities. Nanoparticles can enhance fungal growth, enzyme activity, and pollutant uptake, leading to more efficient and controlled bioremediation processes.

Mushroom-based bioremediation can be integrated into circular economy principles, where waste materials are used as substrates for mushroom growth. This approach not only remediates pollution but also contributes to waste reduction and resource utilization. Collaborative efforts among researchers, industries, governments, and environmental organizations can drive the expansion of mushroom bioremediation globally. Sharing knowledge, best practices, and technological advancements can accelerate its adoption on a broader scale [19].

Increased public awareness and understanding of the advantages fungal bioremediation has for the environment can lead to better acceptance and support for this technique. Transparent communication about its safety, efficacy, and long-term impacts is crucial for building trust. As mushroom bioremediation gains traction, regulatory frameworks will likely evolve to address ethical, safety, and environmental concerns. Proper guidelines and regulations will facilitate its responsible implementation and prevent potential risks.

The potential growth and expansion of mushroom-based bioremediation align with global efforts to transition toward sustainable practices. As technological capabilities improve and our understanding of fungi's role in environmental cleanup deepens, this method has the potential to become a key tool in the arsenal of solutions aimed at restoring polluted environments and fostering a healthier planet. Continued research and innovation in mushroom bioremediation are pivotal to unlocking the full potential of this environmentally friendly approach. As we confront increasingly complex pollution challenges, advancing our understanding of fungi's capabilities and developing novel strategies become increasingly essential for realizing the benefits of mushroom-based bioremediation [6].

Continued research allows us to explore how different mushroom species interact with emerging contaminants, including pharmaceuticals, microplastics, and novel pollutants. Understanding how mushrooms respond to these new challenges will enable us to expand their bioremediation repertoire. Research can help us better understand the mechanisms underlying the interaction between mushrooms and heavy metals. By identifying key genes, enzymes, and pathways involved in metal uptake and detoxification, we can optimize mushrooms for efficient metal removal. Innovation in genetic engineering techniques can lead to the development of genetically modified mushrooms with enhanced bioremediation capabilities. By introducing specific genes or pathways, we can tailor mushrooms to degrade specific pollutants more effectively.

Research into the synergistic interactions between mushrooms and other microorganisms can lead to the design of highly efficient bioremediation consortia. Understanding how fungi collaborate with bacteria and other organisms can help us develop strategies that capitalize on their combined strengths. Research-driven insights into mushroom ecology and adaptation can enable us to design bioremediation strategies that are tailored to the unique conditions of specific contaminated sites. This site-specific approach increases the likelihood of successful remediation outcomes. Continued research is essential for assessing the long-term environmental fate of remediated pollutants and the potential ecological impacts of using mushrooms for bioremediation. This knowledge ensures that our solutions do not inadvertently create new problems [16].

The discovery of new and understudied mushroom species with unique bioremediation capabilities holds great potential. Ongoing research can uncover previously unknown fungi that can contribute to pollution cleanup in innovative ways. Innovation can lead to the development of cost-effective cultivation techniques, substrates, and growth conditions for mushrooms used in bioremediation. This reduces the overall cost and makes the approach more accessible for various applications. Research can provide insights into the challenges and strategies for scaling up mushroom bioremediation from controlled laboratory settings to real-world field applications. Understanding how mushrooms perform under diverse conditions is crucial for successful implementation. Innovative research can contribute to public education and engagement efforts. Sharing breakthroughs, success stories, and safety assurances can help build public trust and acceptance of mushroom-based bioremediation. In a world grappling with environmental degradation, research and innovation are the driving forces that enable us to develop effective and sustainable solutions. By continuing to push the boundaries of knowledge and technology, we can harness the remarkable potential of mushrooms and propel bioremediation to the forefront of pollution cleanup strategies.

Conclusion and Recommendations

Mushrooms, often underestimated, hold immense significance in sustainable environmental management due to their unique capabilities and contributions across various aspects of ecological health, pollution mitigation, and ecosystem restoration. Their role extends beyond just bioremediation, encompassing several areas that collectively contribute to a more sustainable future. Mushrooms have the exceptional ability to absorb, accumulate, and transform a wide range of pollutants, including heavy metals, organic compounds, and even radioactive elements. Their natural remediation potential reduces the need for energy-intensive and invasive cleanup methods, aligning with sustainable practices. Mushrooms play a pivotal role in enhancing soil health by breaking down organic matter, cycling nutrients, and promoting microbial diversity. They contribute to soil structure and fertility, aiding in the restoration of degraded land and fostering sustainable agriculture [20].

Mushrooms capture and sequester carbon dioxide during their growth, effectively contributing to carbon storage. This process helps mitigate climate change by reducing atmospheric greenhouse gas concentrations. Mushrooms can be cultivated on organic waste materials, converting waste into valuable resources. This aligns with the principles of the circular economy, reducing waste generation and contributing to sustainable resource management. Mushrooms are integral components of ecosystems, playing roles in nutrient cycling, symbiotic relationships, and decomposition. By fostering healthy fungal communities, we contribute to overall ecosystem health and biodiversity conservation [29].

Certain mushroom species are cultivated for food, providing a nutritious protein source while requiring fewer resources compared to traditional livestock production. Mushroom farming can also be integrated with agroforestry and sustainable land use practices. Some mushrooms produce natural compounds with pesticidal properties that can be used as alternatives to synthetic chemical pesticides. This reduces chemical pollution and supports ecologically balanced pest management. Research on mushrooms continually unveils new insights into their biology, ecology, and applications. These discoveries fuel innovation in various fields, including biotechnology, medicine, and sustainable land management. The unique and fascinating nature of mushrooms can spark public interest and engagement in environmental issues. Educating people about mushrooms' ecological roles and potential benefits encourages a deeper understanding of sustainable practices. Mushrooms have cultural significance in many societies and indigenous communities. Preserving and respecting traditional knowledge about mushrooms can contribute to sustainable resource management and cultural heritage conservation [30].

Mushrooms embody the concept of symbiosis between humans and the environment. Their versatility, from bioremediation to soil restoration to food production, positions them as essential allies in sustainable environmental management. Embracing mushrooms' potential and integrating them into holistic approaches can help us move toward a more balanced, resilient, and ecologically sound future.

References

1. Doukani, K., Boukirat, D., Boumezrag, A., Bouhenni, H., Bounouira, Y., Biodegradation of Pollutants, in: *Handbook of Biodegradable Materials*, pp. 899–925, Springer International Publishing, Cham, 2023.
2. Lin, L., Li, L., Tao, M., Wu, Q., Zhou, L., Wang, B., Wang, L., Shao, X., Zhong, C., Qian, G., Assembly of an active microbial consortium by engineering compatible combinations containing foreign and native biocontrol bacteria of kiwifruit. *Comput. Struct. Biotechnol. J.*, *21*, 3672–3679, 2023.
3. Hajji-Hedfi, L., Hlaoua, W., Rhouma, A., Al-Judaibi, A.A., Arcos, S.C., Robertson, L., Ciordia, S., Horrigue-Raouani, N., Navas, A., Abdel-Azeem, A.M., Biological and proteomic analysis of a new isolate of the nematophagous fungus *lecanicillium sp. BMC Microbiol.*, *23*, 1, 1–10, 2023.
4. Orozco-Mosqueda, M.D.C., Kumar, A., Fadiji, A.E., Babalola, O.O., Puopolo, G., Santoyo, G., Agroecological management of the grey mould fungus *Botrytis cinerea* by plant growth-promoting bacteria. *Plants*, *12*, 3, 637, 2023.
5. Cho, S.S. and Park, S.H., Biocontrol and bioremediation by the rhizosphere microflora of *Tricholoma matsutake*. *Appl. Microbiol. Biotechnol.*, *104*, 6, 2391–2402, 2020.
6. Manjareeka, M. and Das, N., Fungi mediated biosynthesis of nanoparticles and their applications in various sectors: a review. *Front. Microbiol.*, *8*, 2558, 2017.
7. Das, N. and Das, R.K., Biological and environmental aspects of mushroom, in: *Mycoremediation and Environmental Sustainability*, pp. 39–61, Springer, Cham, Switzerland, 2019.
8. Agrawal, N., Verma, P., Shahi, S.K., Degradation of polycyclic aromatic hydrocarbons (phenanthrene and pyrene) by the ligninolytic fungi *Ganoderma lucidum* isolated from the hardwood stump. *Bioresour. Bioprocess.*, *5*, 1, 1–9, 2018.
9. Novotny, L., Dvorska, L., Lorencova, A., Beran, V., Pavlik, I., Fish: a potential source of bacterial pathogens for human beings. *Vet. Med.*, *49*, 9, 343–358, 2004.
10. Gamelas, J.A., Tavares, A.P., Evtuguin, D.V., Xavier, A.M., Oxygen bleaching of kraft pulp with polyoxometalates and laccase applying a novel multi-stage process. *J. Mol. Catal. B: Enzym.*, *33*, 3-6, 57–64, 2005.
11. Akhtar, N. and Mannan, M.A.U., Mycoremediation: expunging environmental pollutants. *Biotechnol. Rep.*, *26*, e00452, 2020.

12. Amaral, L., Martins, A., Spengler, G., Molnar, J., Efflux pumps of Gram-negative bacteria: what they do, how they do it, with what and how to deal with them. *Front. Pharmacol.*, 4, 168, 3 Jan 2014.
13. Singh, J.S. and Kumar, A., Mycorrhizal technology in reclamation of degraded ecosystem: a review, in: *Advances in Agroforestry*, vol. 5, pp. 197–209, Springer, Dordrecht, Netherlands, 2008.
14. Yadav, M., Yadav, A., Yadav, J.P., Saini, N., Microbial dehalogenases: Biodegradation of halogenated compounds. *Biotechnology*, 9, 5, 352–370, 2010.
15. Pandey, V.C. and Singh, K., Impact of biological amendments of soil on plant parasitic nematodes and soil nematode community. *Eur. J. Soil Biol.*, *46*, 2, 138–146, 2010.
16. Sannino, F., Gianfreda, L., Gioffrè, A.M., Bioremediation and functional recovery of hydrocarbon-polluted soils using the bioaugmentation strategy of two differently acclimated microbial consortia. *Chemosphere*, *179*, 77–83, 2017.
17. Geethanjali, P.A. and Jayashankar, M., Efficiency of Paddy Residues as Substrates for Fungal Laccase Production. *Int. J. Theor. Appl. Sci.*, *10*, 1, 190–193, 2018.
18. Uddin, M., Zhang, D., Proshad, R., Haque, M.K., Role of mushrooms in soil mycoremediation: a review. *Chin. J. Appl. Environ. Biol.*, *26*, 2, 460–468, 2020.
19. Al-Qodah, Z., Yahya, M.A., Al-Shannag, M., On the performance of bioadsorption processes for heavy metal ions removal by low-cost agricultural and natural by-products bioadsorbent: a review. *Desalin. Water Treat.*, 85, 339–57, 2017 Aug 1.
20. Kang, M.Y., Rico, C.W., Lee, S.C., *In vitro* antioxidative and antimutagenic activities of oak mushroom (*Lentinus edodes*) and king oyster mushroom (*Pleurotus eryngii*) byproducts. *Food Sci. Biotechnol.*, *21*, 167–173, 2012.
21. Khatoon, H., Rai, J.P.N., Jillani, A., Role of fungi in bioremediation of contaminated soil, in: *Fungi Bio-Prospects in Sustainable Agriculture, Environment and Nano-technology*, pp. 121–156, Academic Press, Cambridge, MA, USA, 2021.
22. da Luz, J.M.R., Paes, S.A., Ribeiro, K.V.G., Mendes, I.R., Kasuya, M.C.M., Degradation of green polyethylene by *Pleurotus ostreatus*. *PLoS One*, *10*, 6, e0126047, 2015.
23. Chaudhary, K., Agarwal, S., Khan, S., Role of phytochelatins (PCs), metallothioneins (MTs), and heavy metal ATPase (HMA) genes in heavy metal tolerance, in: *Mycoremediation and Environmental Sustainability: Volume 2*, pp. 39–60, 2018.
24. Bilal, M. and Asgher, M., Biodegradation of agro-wastes by the lignocellulolytic activity of an oyster mushroom, *Pleurotus Sapidus*. *J. Natl. Sci. Found. Sri Lanka*, *44*, 4, 399–407, 2016.

25. Gupta, S., Annepu, S.K., Summuna, B., Gupta, M., Nair, S.A., Role of mushroom fungi in decolourization of industrial dyes and degradation of agrochemicals, in: *Biology of Macrofungi*, pp. 177–190, 2018.
26. Park, J., Seo, Y., Kim, Y.H., Biological control of *Meloidogyne hapla* using an antagonistic bacterium. *Plant Pathol. J.*, *30*, 3, 288, 2014.
27. Nicola, L., Tosi, S., Savini, D., *In vitro* evaluation of nematophagous activity of fungal isolates. *J. Basic Microbiol.*, *54*, 1, 1–5, 2014.
28. Lin, L., Xu, K., Shen, D., Chou, S.H., Gomelsky, M., Qian, G., Antifungal weapons of Lysobacter, a mighty biocontrol agent. *Environ. Microbiol.*, *23*, 10, 5704–5715, 2021.
29. Hussain, M., Zouhar, M., Rysanek, P., Parasitic activity of the root knot nematode Meloidogyne incognita in okra in a naturally infested field and its suppression by *Lecanicillium muscarium*. *Pak. J. Zool.*, *50*, 6, Dec 1 2018.
30. Mohammed, A.A., Kadhim, J.H., Kamaluddin, Z.N., Selection of highly virulent entomopathogenic fungal isolates to control the greenhouse aphid species in Iraq. *Egypt. J. Biol. Pest Control*, *28*, 1–7, 2018.

2

Application of Mushroom in the Management of Pest and Diseases Affecting Agricultural Crops

Ruth Ebunoluwa Bodunrinde[1], Konjerimam Ishaku Chimbekujwo[2], Oluwafemi Adebayo Oyewole[3], Charles Oluwaseun Adetunji[4*], Oluwabukola Atinuke Popoola[4], John Tsado Mathew[5] and Adeyomoye Olorunsola Israel[6,7]

[1]*Department of Microbiology, Federal University of Technology Akure, Ondo State, Nigeria*
[2]*Department of Microbiology, Modibbo Adama University, Yola, Adamawa State, Nigeria*
[3]*Department of Microbiology, Federal University of Technology, Minna, Niger State, Nigeria*
[4]*Department of Microbiology, Edo State University Uzairue, Iyamho, Edo State, Nigeria*
[5]*Genetics, Genomics and Bioinformatics Department, National Biotechnology Development Agency, Abuja, FCT, Nigeria*
[6]*Department of Chemistry, Ibrahim Badamasi Babangida University, Lapai, Niger State, Nigeria*
[7]*Department of Physiology, Ondo Medical University, Ondo State, Nigeria*

Abstract
Pathogenic microbe is one major element that leads to agricultural production loss, which can reduce crop output and quality. Synthetic pesticides are necessary for farmers to hinder crop diseases and pests. Ecological and surrounding issues like soil compression, resistant weed growth, and pesticide use have become significant, leading to serious negative impacts on agriculture's continual advancement. Biocontrol of plant pathogen using fungi (mushroom) is a substitute for

**Corresponding author*: adetunjicharles@gmail.com;
ORCID: http://orcid.org/0000-0003-3524-6441

Charles Oluwaseun Adetunji, Chukwuebuka Egbuna, Oluwatosin Ademola Ijabadeniyi and Samantha C. Karunarathna (eds.) Mushroom Biotechnology for Improved Agriculture and Human Health, (29–44) © 2025 Scrivener Publishing LLC

synthetic chemicals. The utilization of plant suppressors inhibits disease-causing microbes in plants. Pathogens and plants relate with vase number of microbes, and this interplay directly or indirectly affects plants' health. Various mechanisms have been employed for biocontrol like competition, parasitism, antibiosis, and induced resistance in plants. Moreover, different properties of mushrooms as biocontrol agents have also been identified. Secondary metabolites secreted by some species of basidiomycetes have been shown to have antiviral, antifungal, antibacterial, and nematocidal activity agents plant pathogens. Biocontrol microbes have also been treated with sprinkling methods on foliar diseases like leaf spots. This chapter has explored mushrooms as a means of controlling pests and diseases on various crops. Among the various types of fungi, basidiomycetes species in particular is recognized as an organism that functions as microbes and insect benefactor and a biological regulator using different mechanisms for many fungi and plant-pathogenic bacteria.

Keywords: Biological control, diseases, pests, crops, mechanism, beneficial microorganism

2.1 Introduction

Studies have revealed that the substrates of mushrooms are not ideal to be left without almost attention, since the compost is close to 75% inorganic matter. The studies at the Stroud Water Research Center suggested that the impact of the mushroom substrate on the underground water table is reduced to the minimum [1]. Post-mushroom substrate is crucial as a source of nutrients in plants and animals. These spent mushroom substrates were found to add to the nitrate level in the soil. Apple scab and blight conditions in tomatoes have been effectively controlled by the use of post-mushroom substrates. The researchers reported that when thorough pasteurization is done, the substrates serve as disease-control agents. The damage and destruction of agricultural produce such as grains both at pre-harvest and post-harvest is an alarming situation worldwide. Some crops were affected by insects, pests, and fungi [2, 3]. When there is a loss of crops both in quality and quantity, there will be food insecurity and a reduction in public health in such a community [4]. The cost of production, purchase, and some adverse effects in the application of synthetic pesticides leads researchers' and farmers' interest in the development of other alternatives that are eco-friendly and safe when crops are consumed by humans [5, 6]. In the quest for alternatives to synthetic pesticides, researchers worked on mushroom and it exhibited antimicrobial, antifungal, and other cidal potentials [7–9]. They also found that mushroom possesses insecticidal properties against *Drosophila melanogaster* [10].

Fungi are a very important part of the soil biodiversity; these groups of organisms have the ability to tackle longtime global challenges, which are plant repellence and food security. Mattila *et al.* [11] and Zhang *et al.* [12], also reported that some bioactive compounds found in the mushroom extracts could be the reason for the effectiveness and antifeedant effects against maize pests, thereby rendering the grains both safe for consumption and storage for a period of time.

Valverde *et al.* [13], at the end of their study, concluded that contact toxicity, feeding deterrence, and the ability of the grains to repel infection by pests with the use of mushroom extracts. *Cantharellus cibarius* can serve a dual purpose such as food and also as biopesticides. These biopesticides can confidently be utilized by farmers owing to its affordability, stability, and eco-friendliness [12].

2.2 Properties of Mushroom as Biocontrol Agents (Basidiomycetes)

Antifungal Activity

In vitro studies using Grifoline, which was obtained from *Albatrellus dispansus*, successfully eradicated many plant fungi [14]. The isolation of Phellinsin A occurred in *Phellinus* sp. The development of fungi was prevented by this substance [15]. The pine cone fungus, *Strobilurus tenacellus* [16], was the source of a notable agricultural composite. The *S. tenacellus* mycelia are the source of stabulins, which are fungal toxic substances. Strobilurins are greatly functional because they prevent yeast and other molds from breathing [17]. The biological and chemical action of strobilurins includes ubihydroquinone cytochrome reductase, which is involved in breathing [18] although the presence of (E)-β-methoxyacrylate moiety [19] determines their activity. The disease of plant management has made strobilurin fungus killers a practical solution [20]. Basidiomycota, Ascomycota, and Deuteromycota fungi are controlled by strobilurins, which are structural composites that inhibit mitochondrial inhilation and have distinct effects. An antimycotic peptide called pleurostrin, which has a bisection dimension of ganodermin, was from oyster Basidiomycetes (*Pleurotus ostreatus*). Plant pathogenic fungus *Botrytis cinerea* and others like *F. oxysporum*, and *Peyronellaea arachidicolla* were prevented from growing by ganodermin. *G. lucidum* was obtained from two other biological operational proteins, namely lectin and ribonuclease [21]. *Stereum ostrea* was treated against three different plant pathogenic fungi, i.e., *Botrytis cinerea*, *Colletotrichum miyabeanus*, and *Colletotrichum gloeosporioides*, using drains from culture, methanol, or concentrate of *Stereum ostrea* [22].

Antibacterial Activity

Several mushroom secondary products of metabolism were discovered to possess antibiotic activity [23]. Efficient and cost-effective products for human beings and crop disease management are mushrooms. Poriales, Agaricales, and Stereales exhibit antibiotic properties that could serve as an alternative to novel antibiotics [24]. Basidiomycetes' lower products of metabolism have extensively been studied for their impact on human beings and microorganisms associated with animals. Basidiomycetes draw out demonstrated averagely high bacteriocidal actions, while 3 others decreased the severity of wilt disease caused by *Ralstonia solanacearum* in and outside the cell [25]. The anti-fungi strobilurin F 500 was found to have an effect on tobacco's ability to resist wild fire-causing bacteria (*Pseudomonas syringae* pv. *tabaci*) [26], Coprinol, which is derived from the plant *Copartum* spp. The majority of plant microbes were resistant to it [27].

Antiviral Activity

The viricidal properties of mushrooms on crops were classified as biogenic reaction transformers and viral obstructors [28]. Basidiomycete COHs have been studied in both human beings and animal prototypes [29]. Basidiomycetes (*A. aegerita*) spore-producing structures which were drained to form Lectin called AAL (*Agrocybe aegerita* lectin) [30]. The aphid-borne mosaic virus was found to have viricidal action of the spore-producing structure of *A. brasiliensis* and *L. edodes* [31]. The anti-plant viral action of a normal carbohydrate and acid polymer varies. The inactive state of Glucuronoxylomannan was significant, as the midpoint between TMV and the complete assemble indicated a more potent activity [32]. β-L-glutaminyl-3,4-benzoquinone, which is found in sporophores of *A. bisporus*, can effectively prevent viruses that infect plants [33]. The inhibition of TMV disease on *Nicotiana glutinosa* was observed. The impact of Basidiomycete metabolic products on pathogenic plant viruses was minimally researched on [31].

Plant Nematocidal Activity

Scientists worldwide have become more and more attentive to reducing the use of chemical methods to fight parasites [34]. The present research has shown that some safe-to-eat Basidiomycetes (*Pleurotus* sp.) can exhibit nematocidal action by producing a poison [35]. Mushrooms may possess a resistant strategy that can defend it against myceliophagous nematodes [36].

Additional research indicates that *P. ostreatus* secretes a toxin for nematodes that mimics peroxides which prevent them from moving, leading to the degradation of its members, with a death rate of 95 percent in nematode, both pathogenic (*Bursaphelenchus xylophilus*) and *Panagrellus redivivus* (free-living) [37]. The different *P. eryngii* strains exhibit nematocidal properties generally which from 4. 8 percent to 99.6 percent. Several elements, such as temperature and growth phase, as well as the genetic properties of particular strains of Basidiomycetes, may affect the anti-nematoda impact of some species [38]. The fact that few deceased maggots had mycelium within them, indicated a possible correlation between fungal mycelia and the breakdown of the host's body [38]. Biological active composites from *Nenohopanus nambi* did suggest potential treatments for the root-knot nematode (*Meloidogyne incognita*). A concentrate of 500 milligrams per liter was found to be greatly poisonous to *M. Incognita*, inducing death in 30 minutes which is a fatal outcome [39]. Both (4-methoxyphenyl)-1,2-propandiol and p-anisaldehyde exhibit weak fungicidal and systemic characteristics. S-coriolic acid and linoleic acids are among the fatty acids obtained from *P. Pulmonarius* that had an anti-nematoda impact for putrefaction nematode (*Caenorhabditis elegans*), with LD50 worth of five and ten micrograms per milliliter, respectively [40]. The fatty acid's distance and rate of not saturated were crucial in the anti-nematode impact. Nematocidal monoterpene 1,2-dihydroxymintlactone was secreted from *Cheimonophyllum candidissimum*. It was used to detect the LD50 worth of twenty-five micrograms per milliliter.

C. candidissimum was responsible for the secretion of two compounds Cheimonophyllons and cheimonophyllal, which were found in the wood-based habitat [41]. The medium drain of *Amauroderma macer*, *Laccaria tortilis*, and *Tylopilus striatulus* demonstrated significant anti-nematode action which inhibits *Bursaphelenchus xylophilus* (pine wood nematode) [42]. Irpexlacteus was used to secrete Furaldehydes, 5-pentyl-2-furaldehyde, and 5(4-penteny)-2-furaldehyde that were nematicidal and attacked *Aphelencoides besseyi*. *Crinipellis stipitaria* demonstrated exceptional anti-nematode activity which inhibited *C. elegans* [43].

2.3 Mushroom Substrate as Biocontrol Agent for Plant

In agriculture, synthetic pesticides have a negative impact on farming and the need for using eco-friendly output to protect pests. Various biological manage control substance, such as living microbes and biological pest control agent, has been developed to manage plant infections. Unlike chemically synthesized pesticides, biological manage agents have little effect on untargeted microbes as they do not produce any enduring hazardous leachates which do not result in the growth of immune microorganisms or insects. Even so, their efficacy is often limited and their durability is short [44]. The antimicrobial characteristics of the spent mushroom substrate make it a potential solution for the treatment of microbes in plants. The effectiveness of mushroom and mycelia drains in eradicating plant microbe has been demonstrated by a study carried out within living cells [45]. The *in vivo* efficacy is not revealed. Plant disease incidence can be reduced through the use of spent mushroom substrate innovations. *L. edodes* has been the subject of several studies on spent mushroom substrate-based biological managed compounds on diseased plants. *L. edodes*. Spent mushroom substrate exhibits microbes inhibiting action *in vitro* [46]. It went ahead to utilize hot H_2O concentrate to suppress the growth of *Phytophthora capsici* and inhibit *Pyricularia oryzae* asextual reproductive body in *oryza sativa*. Also, diseases of crops can be subdued using *P. ostreatus* spent mushroom substrate as a substitute. Agricultural by-product, *P. ostreatus* spent mushroom substrate with *T. asperellum*, the severity level of *F. oxysporum* formed infection was dramatically reduced, and tomato vegetable development was increased [47]. A polymer distillate from *P. ostreatus* spent mushroom substrate and disposed mushroom (*L. edodes*) affect bacterial resulting from *Xanthomonas gardneri* in tomato embryonic leaf, leaflets, and 5 leaf crops was decreased to 50 percent gravity [48]. Aliquid distillate extract of *Ly. decastes* spent mushroom substrate was sprinkled on cucumber fruit, which provided protection against *Colletotrichum lagenarium*. The inclusion of *Ly. decastes* spent mushroom substrate in the soil obviated the injury resulted by *Al. brassicicola*. The vaporous constitute of spent mushroom substrate was found in leaves of Arabidopsis thaliana, which has been associated with the effect [49].

2.4 Mechanism of Action of Mushrooms in the Control of Pests and Diseases

Mushrooms combat pests and diseases through several mechanisms. One crucial mechanism is the production of natural antibiotics and antifungal compounds. These compounds inhibit the growth, proliferation, and development of harmful pests and pathogens. Mushrooms also have a symbiotic relationship with beneficial organisms like mycorrhizal fungi. These fungi form a mutually beneficial association with the mushroom's roots, helping to improve nutrient uptake and enhance the immune system of the plant. Mushrooms are known to have chitinase enzymes that break down the chitin found in the cell walls of many pests and pathogens. This enzyme effectively weakens the defense mechanisms of these organisms, making them more susceptible to attack [12]. The tool used by microorganisms that inhibit pests and infections in basidiomycete's genus has been both direct (parasitic, competition, antibiosis) and indirect (intense systemic resistance) mechanisms [50].

Parasitism

The ability of fungi (mushrooms) to parasitize nematodes, arthropods, and other microorganisms is well-known [51]. Following the identification and interconnection of microbes with a competitor, catalyst system biosynthesis and/or secretion can occur to promote the development of a competitor that harms its host [52]. Appressorium and catalytic release are two ways in which mechanical perforation can penetrate the infection process [53]. Hydrolytic enzyme (Chitinase) formation can inhibit mycelia, including *A. nidulans*, *C. gloeosporioides*, *R. solani*, and *S. rolfsii* [54]. Conversely, the most potent insect disease-inhibiting agent is believed to be enzymes (lipases, protease, and chitinase) [55]. As a result, these enzymes are secreted by *P. fumosoroseus* and have proven to be an efficient way of managing *T. molitor* [56]. In *P. lilacinus*, there are some species that have been identified. Through the sprouting of spores, diverting into filaments, and developing an adhesion structure produced by fungus in various nematodes, can percolate both systemically and via eggshells [57]. Excessive expression of genes that control catalytic synthesis in *Meloidogyne incognita*, *Panagrellus redivivus*, and *Caenorhabditis elegans* leads to increased parasitical action and severity [58].

Competition

Competitiveness for resources and area enables microbes to grow in the same place [59]. The formation of siderophore restricts the amount of iron that microbes can obtain [60]. In *P. lilacinus*, hydroxamate and carboxylate siderophores were synthesized primarily in the laboratory [61]. This technique has a straight impact on management, and the antagonist is often preceded by other tools [62]. Some microbes are inhibited from developing, due to the speedy development of *Paecilomyces* sp. [63]. For example, sprinkling *P. variotii* on sunflower seeds acts as a barrier between cells and microbe (*Macrophomina phaseolina*) [64]. However, this interaction can also cause harm to other advantageous microorganisms [65].

Antibiosis

Antimicrobial minor product of metabolism is produced by fungal sp. which have been extensively documented. Amidst them are the alkaloids, phenolic components, and biological vaporous constituents (steroids, flavonoids and peptides, polyketides, and terpenoids) [66]. These metabolic products play a crucial role in the biocontrol of pests and infected plants.

Induced Resistance in Plants

The article on activated recalcitrant after the colonization of a plant's root structure has limited information [67]. The effect of *P. variotii* infusion in tomato roots was observed. It completely prevents the symptoms of X. campestris' symptoms on leaves. Plant seed shad of cotton that was previously infused in *P. lilacinus* seed solution led to fewer *Aphis gossypii* population cotton plants than those otherwise healthy [68].

Infused immunity can also be formed by the distillate of *Paecilomyces* immunity to *X. oryzae* or *P. syringae* in *oryza sativa* may also be caused by Zhi Neng Cong (ZNC), an industrial distillate. A 500 nanogram per milliliters serving of ZNC did not prevent microbe development in the laboratory, but a smaller amount of 100 nanograms per milliliters produced resistance against certain microbes. In contrast, Arabidopsis plants exhibit an increase in reactive O_2 compound like superoxide and H_2O_2, which also stimulate the production of salicylic acids that are essential for their defense system [69].

2.5 Several Areas Where Mushrooms Can Be Applied

1. Biocontrol Agents: some examples of mushrooms, such as *Trichoderma* and *Beauveria* species, were found to have natural antagonistic potentials against some pests and pathogens. These mushrooms can be used as biocontrol agents to reduce pests and diseases in agricultural produce.
2. Fungal Extracts: Extracts from specific mushroom species, like *Ganoderma lucidum* and *Agaricus bisporus*, have shown potential as natural fungicides and insecticides. These extracts were formulated into biologically based pesticides that effectively target and control pests and diseases [6].
3. Mycofiltration: Studies have shown that mushrooms have the ability to break down and remove contaminants, such as pesticides, toxins, and heavy metals from the environment. This process, known as mycofiltration, can be utilized to manage pest and disease issues caused by chemical pollutants.
4. Mycoremediation: Certain mushrooms have been studied for their ability to remediate soil contaminated with various pathogens. They can help suppress disease-causing

organisms and restore the health of the soil, reducing the incidence of plant diseases. Mycoremediation depends on the effective enzymes, formed by mushroom, while degrading some types of substrates and contaminants. Studies showed that mycoremediation using mushroom farming will eradicate two of the world's major glitches, i.e., accumulation of wastes and production of proteinaceous food concurrently [70–73].

2.6 Mushrooms as Disease Control Agents

Mushrooms have been worked on by several researchers to check their ability to perform the following;

1. Antimicrobial Properties: Certain types of mushrooms, such as Shiitake (*Lentinula edodes*) and Maitake (*Grifola frondosa*), contain bioactive components that possess antimicrobial effects. These components can help in inhibiting the growth of harmful bacteria, fungi, and other microorganisms, potentially reducing the risk of infections.
2. Antiviral Properties: Some mushrooms, such as Reishi (*Ganoderma lucidum*) and Turkey Tail (*Trametes versicolor*), have been studied for their antiviral properties. They contain compounds that may have antiviral activity against various viruses, including influenza and herpes viruses.
3. Immunomodulatory Effects: Certain mushrooms, such as Reishi and Cordyceps (*Cordyceps militaris*), have been found to have immunomodulatory potentials, which can help regulate the immune system's response, potentially enhancing proper immune functioning and reducing the severity of immune-related disorders and/or diseases.
4. Disease Prevention in Plants: Mushrooms and their byproducts can also be used for disease control in plants. Some mushroom extracts and compounds have shown effectiveness against plant pathogens, including fungi and bacteria. They can be used as natural biopesticides or as preventive treatments to protect crops from diseases.

5. Animal Health: The antimicrobial and immunomodulatory properties of certain mushroom species can also benefit animal health. Adding these mushrooms or their extracts to animal feed might help enhance the animal's immune system and reduce the risk of infections or diseases. The application and dosage of mushroom-based disease control agents may vary depending on the target disease and the specific mushroom species used [74].

2.7 Conclusion

New methods of biocontrol are developed due to the loss of efficiency of pesticides on some microbes, constraints in production, surroundings, and human diseases as a result of these products, and the inefficiency of genetic immunity based on fast pathogenic changes. This also reduces the entire number of synthetic vital substances without affecting manufacturing damage is challenging at present, using biological stimulating agents that could help reduce soil contamination and enhance the utilization of synthetic products. This review highlights the distinct bio-functions of mushroom components based on their control over crop pathogens and nematodes. The identification of new innate products from Basidiomycetes for controlling crop nematodes and pathogens will result in more environmentally friendly methods of crop preservation. It is imperative to thoroughly investigate the interplay between Basidiomycetes' biological active metabolic products and plant cells and to quickly commercialize potential bioactivating molecule(s) derived from Basidiomycetes to mitigate undesirable effects of synthetic products.

References

1. Danho, M., Gaspar, C., Haubruge, E., The impact of grain quantity on the biology of Sitophilus zeamais Motschulsky (Coleoptera: Curculionidae): oviposition, distribution of eggs, adult emergence, body weight and sex ratio. *J. Stored Prod. Res.*, 38, 259–66, 2002.
2. Shiferaw, B., Prasanna, B.M., Hellin, J., Bänziger, M., Crops that feed the world 6. Past successes and future challenges to the role played by maize in global food security. *J. Food Secur.*, 3, 3, 307–27, 2011.
3. Tefera, T., Demissie, G., Mugo, S., Beyene, Y., Yield and agronomic performance of maize hybrids resistant to the maize weevil Sitophilus zeamais Motschulsky (Coleoptera: Curculionidae). *J. Crop Prot.*, 46, 94–9, 2013.

4. Ngowi, A.V.F., Mbise, T.J., Ijani, A.S.M., London, L., Ajayi, O.C., Smallholder vegetable farmers in Northern Tanzania: Pesticides use practices, perceptions, cost and health effects. *Crop Prot.*, 26, 11, 1617–24, 2007.
5. Spyrou, I.M., Karpouzas, D.G., Menkissoglu-Spiroudi, U., Do botanical pesticides alter the structure of the soil microbial community. *Microb. Ecol.*, 58, 4, 715–27, 2009.
6. Dubey, N.K., Shukla, R., Kumar, A., Singh, P., Prakash, B., Prospects of botanical pesticides in sustainable agriculture. *Curr. Sci.*, 98, 4, 479–80, 2010.
7. Watanabe, F., Schwarz, J., Takenaka, S., Miyamoto, E., Ohishi, N., Nelle, E., Hochstrasser, R., Yabuta, Y., Characterization of vitamin B12 compounds in the wild edible mushrooms black trumpet (Craterellus cornucopioides) and golden chanterelle (Cantharellus cibarius). *J. Nutr. Sci. Vitaminol. (Tokyo)*, 58, 6, 438–41, 2012.
8. Kuka, M., Cakste, I., Galoburd, R., Sabovics, M., Chemical composition of latvian wild edible mushroom Cantharellus cibarius, in: *9th Baltic conference on food science and technology "Food consum well-being"*, Jelgava, pp. 248–252, 2014.
9. Kozarski, M., Klaus, A., Vunduk, J., Zizak, Z., Niksic, M., Jakovljevic, D., Vrvic, M.M., Van Griensven, L.J., Nutraceutical properties of the methanolic extract of edible mushroom Cantharellus cibarius (Fries): primary mechanisms. *Food Funct.*, 6, 6, 1875–86, 2015.
10. Trigueros, V., Lougarre, A., Ali-Ahmed, D., Rahbé, Y., Guillot, J., Chavant, L., Fournier, D., Paquereau, L., Xerocomus chrysenteron lectin: Identification of a new pesticidal protein. *Biochim. Biophys. Acta Gen. Subj.*, 1621, 3, 292–8, 2003.
11. Mattila, P., Salo-Väänänen, P., Könkö, K., Aro, H., Jalava, T., Basic composition and amino acid contents of mushrooms cultivated in Finland. *J. Agric. Food. Chem.*, 50, 22, 6419–6422, 2002, doi: 10.1021/jf020608m.
12. Zhang, M., Cui, S.W., Cheung, P.C.K., Wang, Q., Antitumor polysaccharides from mushrooms: a review on their isolation process, structural characteristics and antitumor activity. *Trends Food Sci. Technol.*, 18, 1, 4–19, 2007, doi: 10.1016/j.tifs.2006.07.013.
13. Valverde, M.E., Hernández-Pérez, T., Paredes-López, O., E dible Mushrooms: Improving Human Health and Promoting Quality Life. *Int. J. Microbiol.*, 2015, 376387, 2015, doi: 10.1155/2015/376387.
14. Luo, D.Q., Shao, H.J., Zhu, H.J., Liu, J.K., Activity *in vitro* and *in vivo* against plant pathogenic fungi of grifolin isolated from the basidiomycete Albatrellus dispansus. *Z. Naturforsch. C J. Biosci.*, 60, 50–56, 2005.
15. Hwang, E.I., Yun, B.S., Kim, Y.K., Kwon, B.M., Kim, H.G., Lee, H.B., Jeong, W.J., Kim, S.U., Phellinsin A, a novel chitin synthases inhibitor produced by Phellinus sp. PL3. *J. Antibiot.*, 53, 903–911, 2000.
16. Robles-Hernández, L., Gonz/alez-Franco, A.C., Soto-Parra, J.M., Montes-Domínguez,.F., Review of agricultural and medicinal applications of basidiomycete mushrooms. *Tecnociencia Chihuah*, 2, 95–107, 2008.

17. Anke, T., Oberwinkler, F., Steglich, W., Schramm, G., The strobilurins-new antifungal antibiotics from the basidiomycete Strobilurus tenacellus. *J. Antibiot.*, 30, 806–810, 1977.
18. Weber, W., Anke, T., Steffan, B., Steglich, W., Antibiotics from basidiomycetes. XXXII. Strobilurin E: A new cytostatic and antifungal (E)-BETA-methoxyacrylate antibiotic from Crepidotusfulvotomentosus Peck. *J. Antibiot.*, 43, 207–212, 1990.
19. Fredenhagen, A., Kuhn, A., Peter, H.H., Cuomo, V., Giuliano, U., Strobilurins, F, G and H, three new antifungal metabolites from Bolinea lutea. I. Fermentation, isolation and biological activity. *J. Antibiot.*, 43, 655–660, 1990.
20. Gullino, M.L., Leroux, P., Smith, C.M., Uses and challenges of novel compounds for plant disease control. *Crop Prot.*, 19, 1–11, 2000.
21. Wang, H. and Ng, T., Ganodermin, an antifungal protein from fruiting bodies of the medicinal mushroom Ganoderma lucidum. *Peptides*, 27, 27–30, 2006.
22. Imtiaj, A., Jayasinghe, C., Lee, G.W., Lee, T.S., Antibacterial and antifungal activities of Stereum ostrea, an inedible wild mushroom. *Mycobiology*, 35, 210–214, 2007.
23. Barros, L., Baptista, P., Estevinho, L.M., Ferreira, I.C., Effect of fruiting body maturity stage on chemical composition and antimicrobial activity of Lactarius sp. mushrooms. *J. Agric. Food. Chem.*, 55, 8766–8771, 2007.
24. Alves, M.J., Ferreira, I.C., Dias, J., Teixeira, V., Martins, A., Pintado, M., A review on antimicrobial activity of mushroom (Basidiomycetes) extracts and isolated compounds. *Planta Med.*, 78, 1707–1718, 2012.
25. Erjavec, J., Ravnikar, M., Brzin, J., Grebenc, T., Blejec, A., Gosak, M.Ž., Sabotič, J., Kos, J., Dreo, T., Antibacterial activity of wild mushroom extracts on bacterial wilt pathogen Ralstonia solanacearum. *Plant Dis.*, 100, 453–464, 2016.
26. Herms, S., Seehaus, K., Koehle, H., Conrath, U., A strobilurin fungicide enhances the resistance of tobacco against tobacco mosaic virus and *Pseudomonas syringae* pv tabaci. *Plant Physiol.*, 130, 120–127, 2002.
27. Johansson, M., Sterner, O., Labischinski, H., Anke, T., Coprinol, a new antibiotic cuparane from a Coprinus species. *Z. Naturforsch. C J. Biosci.*, 56, 31–34, 2001.
28. Brandt, C.R. and Piraino, F., Mushroom antivirals. *Recent Res. Dev. Antimicrob. Agents Chemother.*, 4, 11–26, 72, 2000.
29. Gao, Y., Zhou, S., Huang, M., Xu, A., Antibacterial and antiviral value of the genus Ganoderma P. Karst. species (Aphyllophoromycetideae): A review. *Int. J. Med. Mushrooms*, 5, 1–12, 2003.
30. Sun, H., Zhao, C.G., Tong, X., Qi, Y.P., Alectin with mycelia differentiation and antiphytovirus activities from the edible mushroom Agrocybe aegerita. *J. Biochem. Mol. Biol.*, 36, 214–222, 2003.

31. Di Piero, R.M., Novaes, Q.S., Pascholati, S.F., Effect of Agaricus brasiliensis and Lentinulaedodes mushroom son the infection of passionflower with Cowpea aphid-borne mosaic virus. *Braz. Arch. Biol. Technol.*, 53, 269–278, 2010.
32. Kovalenko, O.G., Polishchuk, O.N., Wasser, S.P., Virus Resistance induced by glucuronoxylomannan isolated from submerged cultivated yeast-like cell biomass of medicinal yellow brain mushroom Tremella mesenterica Ritz.: Fr. (Heterobasidiomycetes) in hypersensitive host plants. *Int. J. Med. Mushrooms*, 11, 199–205, 2009.
33. Tavantzis, S. and Smith, S., Isolation and evaluation of a plant-virus-inhibiting quinone from sporophores of Agaricus bisporus. *Phytopathology*, 72, 619–621, 1982.
34. Haydock, P.P., Woods, S.R., Grove, I.G., Hare, M.C., Perry, R.N., Moens, M., Chemical control of nematodes, in: *Plant Nematology*, 2nd ed., pp. 459–479, CABI Publishing, Oxfordshire, UK, 2013.
35. Comans-Pérez, R., Aguilar-Marcelino, L., Mendoza de Gives, P., Sánchez, J., López-Arellano, M., Singh, M., *In vitro* lethal capability of ten strains of edible mushrooms against Haemonchus contortus (Nematoda) infective larvae [conference poster], in: *Proceedings of the 8th International Conference on Mushroom Biology and Mushroom Products (ICMBMP8)*, New Delhi, India, 19–22 November 2014, Volume I & II, pp. 557–562, 2014.
36. Luo, H., Li, X., Li, G., Pan, Y., Zhang, K., Acanthocytes of Stropharia rugosoannulata function as a nematode-attacking device. *Appl. Environ. Microbiol.*, 72, 2982–2987, 2006.
37. Kwok, O., Plattner, R., Weisleder, D., Wicklow, D., A nematicidal toxin from *Pleurotus ostreatus* NRRL 3526. *J. Chem. Ecol.*, 18, 127–136, 1992.
38. Palizi, P., Goltapeh, E., Pourjam, E., Safaie, N., Potential of oyster mushrooms for the biocontrol of sugar beet nematode (Heterodera schachtii). *J. Plant Prot. Res.*, 49, 27–34, 2009.
39. Bua-art, S., Saksirirat, W., Hiransalee, A., Kanokmedhaku, S., Lekphrom, R., Effect of bioactive compound from luminescent mushroom (Neonothopanus nambi Speg.) on root-knot nematode (Meloidogyne incognita Chitwood) and non-target organisms. *KKU Res. J.*, 16, 331–341, 2011.
40. Stadler, M., Mayer, A., Anke, H., Sterner, O., Fatty acids and other compounds with nematicidal activity from cultures of Basidiomycetes. *Planta Med.*, 60, 128–132, 1994.
41. Lorenzen, K. and Anke, T., Basidiomycetes as source for new bioactive natural products. *Curr. Org. Chem.*, 2, 329–364, 1998.
42. Dong, J.Y., Li, X.P., Li, L., Li, G.H., Liu, Y.J., Zhang, K.Q., Preliminary results on nematicidal activity from culture filtrates of Basidiomycetes against the pine wood nematode, Bursaphelenchus xylophilus (Aphelenchoididae). *Ann. Microbiol.*, 56, 163–166, 2006.
43. Lambert, M., Kremer, S., Anke, H., Antimicrobial, phytotoxic, nematicidal, cytotoxic, and mutagenic activities of 1-hydroxypyrene, the initial

metabolite in pyrene metabolism by the basidiomycete *Crinipellis stipitaria. Bull. Environ. Contam. Toxicol.*, 55, 251–257, 1995.
44. Stack, J.P., Kenerley, C.M., Pettit, R.E., Application of biological control agents, in: *Biocontrol of plant diseases*, pp. 43–54, CRC Press, England, United Kingdom, 2020.
45. Katas, H., Akhmar, M.A.M., Abdalla, S.S.I., Biosynthesized silver nanoparticles loaded in gelatine hydrogel for a natural antibacterial and antibiofilm wound dressing. *J. Bioact. Compat. Polym.*, 36, 2, 111–123, 2021, doi: 10.1177/0883911520988303.
46. Kaur, H., Nyochembeng, L.M., Mentreddy, S.R., *et al.*, Assessment of the antimicrobial activity of Lentinula edodes against *Xanthomonas campestris* pv. Vesicatoria. *Crop Prot.*, 89, 284–288, 2016, doi: 10.1016/j.cropro.2016.08.001.
47. Singh, G., Tiwari, A., Gupta, A., *et al.*, Bioformulation development via valorizing silica-rich spent mushroom substrate with Trichoderma asperellum for plant nutrient and disease management. *J. Environ. Manage.*, 297, 113278, 2021, doi: 10.1016/j.jenvman.2021.113278.
48. Aguiar, T., Luiz, C., Neto, A., *et al.*, Residual polysaccharides from fungi reduce the bacterial spot in tomato plants. *Bragantia*, 77, 2, 299–313, 2018, doi: 10.1590/1678-4499.2016514.
49. Fujita, R., Yokono, M., Ube, N., *et al.*, Suppression of *Alternaria brassicicola* infection by volatile compounds from spent mushroom substrates. *J. Biosci. Bioeng.*, 132, 1, 25–32, 2021, doi: 10.1016/j.jbiosc.2021.03.003.
50. O'Brien, P.A., Biological control of plant diseases. *Australas. Plant Pathol.*, 46, 293–304, 2017.
51. Munawar, M., Khan, S.A., Javed, N., Haq, I.U., Gondal, A.S., Bio-management of tomato wilt complex caused by Meloidogyne incognita and *Fusarium oxysporum* f.sp. lycopersici. *Nematology*, 17, 479–485, 2015.
52. Rabea, E.I., Badawy, M.E.T., Stevens, C.V., Smagghe, G., Steurbaut, W., Chitosan as antimicrobial agent: Applications and mode of action. *Biomacromolecules*, 4, 1457–1465, 2003.
53. Spadaro, D. and Droby, S., Development of biocontrol products for postharvest diseases of fruit: The importance of elucidating the mechanisms of action of yeast antagonists. *Trends Food Sci. Technol.*, 47, 39–49, 2016.
54. Chen, C.C., Kumar, H.A., Kumar, S., Tzean, S.S., Yeh, K.W., Molecular cloning, characterization, and expression of a chitinase from the entomopathogenic fungus Paecilomyces javanicus. *Curr. Microbiol.*, 55, 8–13, 2007.
55. Khan, A., Williams, K.L., Nevalainen, H.K., Effects of Paecilomyces lilacinus protease and chitinase on the egg shell structures and hatching of Meloidogynejavanicajuveniles. *Biol. Control*, 31, 346–352, 2004.
56. Gómez, C., Amaya, I., de la Cruz-Quiroz, R., Rodríguez-Herrera, R., Aguilar, C.N., Tenebrio molitor biomass as inducer of lipases and proteases produced by Paecilomyces fumosoroseus. *Mex. J. Biotechnol.*, 2, 142–150, 2017.

57. Khan, A., Williams, K.L., Nevalainen, H.K., Infection of plant-parasitic nematodes by Paecilomyceslilacinus and Monacrosporium lysipagum. *BioControl*, 51, 659–678, 2006.
58. Wang, J., Liu, F., Pan, C., Enhancing the virulence of Paecilomyce slilacinus against Meloidogyne incognita eggs by overexpression of a serine protease. *Biotechnol. Lett.*, 32, 1159–1166, 2010.
59. Dukare, A.S., Paul, S., Nambi, V.E., Gupta, R.K., Singh, R., Sharma, K., Vishwakarma, R.K., Exploitation of microbial antagonists for the control of postharvest diseases of fruits: A review. *Crit. Rev. Food Sci. Nutr.*, 59, 1498–1513, 2019.
60. Talibi, I., Boubaker, H., Boudyach, E.H., Ait Ben Aoumar, A., Alternative methods for the control of postharvest citrus diseases. *J. Appl. Microbiol.*, 117, 1–17, 2014.
61. Ruanpanun, P., Tangchitsomkid, N., Hyde, K.D., Lumyong, S., Actinomycetes and fungi isolated from plant-parasitic nematode infested soils: Screening of the effective biocontrol potential, indole-3-acetic acid and siderophore production. *World J. Microbiol. Biotechnol.*, 26, 1569–1578, 2010.
62. Latz, M.A., Jensen, B., Collinge, D.B., Jørgensen, H.J., Endophytic fungi as biocontrol agents: Elucidating mechanisms in disease suppression. *Plant Ecol. Divers.*, 11, 555–567, 2018.
63. Arora, K., Sharma, S., Krishna, S.B., Adam, J.K., Kumar, A., Non-edible Oil cakes as a novel substrate for DPA production and augmenting biocontrol activity of Paecilomyces variotii. *Front. Microbiol.*, 8, 753, 2017.
64. Anis, M., Abbasi, M.W., Zaki, M.J., Bioefficacy of microbial antagonists against *Macrophomina phaseolina* on sunflower. *Pak. J. Bot.*, 42, 2935–2940, 2010.
65. Yu, Z., Zhang, Y., Luo, W., Wang, Y., Root colonization and effect of biocontrol fungus Paecilomyces lilacinuson composition of ammonia-oxidizing bacteria, ammonia-oxidizing archaea and fungal populations of tomato rhizosphere. *Biol. Fertil. Soils*, 51, 343–351, 2015.
66. Mousa, W.K. and Raizada, M.N., The diversity of anti-microbial secondary metabolites produced by fungal endophytes: An interdisciplinary perspective. *Front. Microbiol.*, 4, 65, 2013.
67. Suárez-Estrella, F., Arcos-Nievas, M.A., López, M.J., Vargas-García, M.C., Moreno, J., Biological control of plant pathogens by microorganisms isolated from agro-industrial composts. *Biol. Control*, 67, 509–515, 2013.
68. Lopez, D.C., Zhu-Salzman, K., Ek-Ramos, M.J., Sword, G.A., The entomopathogenic fungal endophytes Purpureocillium lilacinum (formerly Paecilomyces lilacinus) and Beauveria bassiana negatively affect cotton aphid reproduction under both greenhouse and field conditions. *PLoS One*, 9, e103891, 2014.
69. Wang, X., Yao, Y., Chen, B., Zhang, M., Liu, Z., Wang, Q., Ma, J., Paecilomyces variotii extracts and controlled-release urea synergistically increased nitrogen use efficiency and rice yield. *ACS Omega*, 5, 13303–13311, 2020.

70. Akinyele, B.J. Olaniyi O.O. and Arotupin, D.J., Bioconversion of Selected Agricultural Wastes and Associated Enzymes by Volvariella volvacea: An Edible Mushroom. *Res. J. Microbiol.*, 6, 63–70, 2011.
71. Akinyele, B.J. and Akinkunmi, C.O. Fungi associated with the spoilage of berry and their reaction to electromagnetic field. *J. Yeast Fungal Res.*, 3, 4, 49–57, 2012.
72. Nagy, P., Bernard Faye, B.O., Marko, S., Thomas, U., Wernery, J J. Microbiological quality and somatic cell count in bulk milk of dromedary camels (Camelus dromedarius): Descriptive statistics, correlations, and factors of variation. *J. Dairy Sci.*, 96, 9, 5625–5640, 2013.
73. Oyetayo, O.V., A. Nieto-Camacho, T.M. Ramırez-Apana, R.E. Baldomero and M. Jimenez, 2013. Total phenol, antioxidant and cytotoxic properties of wild macrofungi collected from Akure Southwest Nigeria. *Jordan J. Biol. Sci.*, 6, 105–110, 2013.
74. Zhao S., Gao Q., Rong C., Wang S., Zhao Z., Liu Y., Xu J., Immunomodulatory effects of edible and medicinal mushrooms and their bioactive Immunoregulatory product. *J. Fungi.*, 6, 269, 2020. Doi: 10.3390/jof6040269.

3

Agricultural Applications of Novel Mushroom-Based Nanopesticide

Isibor Patrick Omoregie[1]*, Oluwafemi Adebayo Oyewole[2], Adamu Binta Buba[3], Alhassan Muhammad Alhassan[2] and Charles Oluwaseun Adetunji[4]

[1]*Department of Biological Sciences Covenant University, Ota, Nigeria*
[2]*Department of Microbiology, Federal University of Technology, Minna, Nigeria*
[3]*National Biotechnology Development Agency, Abuja, Nigeria*
[4]*Department of Microbiology, Edo State University, Uzairue, Nigeria*

Abstract

Nanobiopesticides derived from mushrooms offer a novel approach to tackling agricultural pests and diseases. These nanoparticles utilize bioactive compounds found in mushrooms to provide targeted pest and pathogen control. Their unique properties enhance penetration, allowing precise delivery to pests and pathogens while minimizing non-target effects. This technology reduces environmental impact and toxicity concerns associated with conventional pesticides. Synthesis methods involve green techniques, aligning with sustainable practices. Challenges include variability in effectiveness and regulatory considerations. Nevertheless, the potential benefits are vast, from effective pest management and disease suppression to improved crop yields and ecosystem health. Nanobiopesticides showcase the synergy of nature and technology, aimed at reshaping agricultural practices toward a more sustainable and eco-friendly future. This paper outlines the relevant literary discourse surrounding these concepts while emphasizing the significance of addressing agricultural challenges related to pests and diseases with this novel approach.

Keywords: Bioactive compounds, targeted delivery, pests and disease control, environmental sustainability, non-toxicity

**Corresponding author*: patrickisibor007@gmail.com

Charles Oluwaseun Adetunji, Chukwuebuka Egbuna, Oluwatosin Ademola Ijabadeniyi and Samantha C. Karunarathna (eds.) Mushroom Biotechnology for Improved Agriculture and Human Health, (45–70) © 2025 Scrivener Publishing LLC

3.1 Introduction

Nanobiopesticides are innovative and advanced pest management solutions that combine the principles of nanotechnology and biology. These formulations involve the use of nanomaterials, which are extremely small particles on the nanometer scale (1 nanometer = 1 billionth of a meter), in combination with bioactive compounds derived from natural sources. The goal of nanobiopesticides is to provide targeted and efficient control of agricultural pests and diseases while minimizing environmental impact. Agricultural pest and disease challenges refer to the various problems and threats that affect crops and livestock in agricultural systems. These challenges can have significant economic, environmental, and social impacts on food production, security, and livelihoods. Pests include insects, mites, rodents, and other organisms that feed on crops or livestock, causing damage and reducing yields. Pathogens encompass bacteria, viruses, fungi, and other microorganisms that cause diseases in plants and animals. Pest infestations and disease outbreaks can substantially reduce crop yields and livestock productivity. This can result in food shortages, increased costs for farmers, and higher prices for consumers. Agricultural pests and diseases cause billions of dollars in economic losses each year due to decreased yields, increased production costs (such as for pesticides and veterinary care), and trade restrictions on affected products [1].

The world's growing population relies on agricultural products for food security. Pest and disease outbreaks can disrupt this supply, leading to food shortages and malnutrition, particularly in vulnerable regions. Pesticides and other control methods can impact non-target species and disrupt ecosystems, hence, it remains a crucial challenge to maintain a balance between controlling pests and preserving biodiversity. Climate change can influence the distribution and behavior of pests and pathogens. Areas with warmer temperatures and changing precipitation patterns may allow pests and diseases to spread to new regions, affecting crops that were previously unaffected. Over time, pests and diseases can evolve resistance to the chemicals and methods used to control them, making management more challenging. This can lead to the development of superbugs or superweeds that are difficult to control. Integrated Pest Management (IPM) involves using a combination of strategies to manage pests and diseases, including cultural practices (crop rotation, planting resistant varieties), biological controls (predators, parasites), and judicious use of pesticides [2].

The global movement of agricultural products and people can facilitate the spread of pests and diseases across borders. Quarantine measures and trade regulations are used to prevent the introduction of new pests and diseases. Advancements in biotechnology, genomics, and data analytics are being used to develop more effective and sustainable methods for managing agricultural pests and diseases. This includes the development of genetically modified organisms (GMOs) and precision agriculture techniques. Training farmers, extension workers, and policymakers on pest and disease identification, prevention, and management is crucial for effective control strategies. Raising awareness among stakeholders about the importance of preventing and managing these challenges is also essential.

Addressing agricultural pest and disease challenges requires a multi-pronged approach that combines scientific research, sustainable agricultural practices, policy interventions, and international cooperation to ensure food security and the resilience of agricultural systems.

Nanobiopesticides, a cutting-edge innovation at the intersection of nanotechnology and agriculture, hold promise as a more efficient and environmentally friendly solution for managing agricultural pests and diseases. These nanoscale formulations combine the benefits of nanotechnology with biological compounds to create highly targeted and effective pest management tools. Nanobiopesticides enable precise targeting of pests and pathogens. Nanoparticles can be engineered to interact specifically with the pest's physiology or the pathogen's biology, minimizing harm to non-target organisms and reducing the overall environmental impact. The unique properties of nanoparticles can enhance the efficacy of bioactive compounds. By encapsulating biological agents such as plant extracts, enzymes, or microbial agents within nanoparticles, their stability, solubility, and longevity can be improved, leading to a more consistent and lasting pest control effect [1].

Conventional chemical pesticides often have negative effects on non-target organisms, soil health, and water quality. Nanobiopesticides offer a more environmentally friendly alternative, as they can be designed to degrade more rapidly, minimizing residual effects and reducing the overall ecological footprint. Nanobiopesticides have the potential to be effective at lower application rates compared to traditional pesticides. This means that farmers can achieve the desired pest control while using fewer resources, which can lead to cost savings and reduced exposure to harmful chemicals. Pests and pathogens can develop resistance to traditional pesticides over time. Nanobiopesticides, by employing multiple modes of action and utilizing natural biological agents, could help delay the development of resistance and extend the useful lifespan of pest management strategies. Nanobiopesticides can contribute to precision agriculture practices by allowing farmers to target specific areas or individual plants with precision. This can optimize resource use and reduce waste while maintaining effective pest control [3].

Nanobiopesticides align well with the principles of integrated pest management, which emphasizes a holistic approach to pest control that combines multiple strategies. Nanobiopesticides can complement other IPM components like cultural practices, biological control, and resistant crop varieties. The use of nanobiopesticides may lead to reduced chemical residues on crops, benefiting food safety and quality. This is particularly important for crops destined for export markets with strict residue limits. Nanotechnology enables the creation of innovative formulations, such as nanoparticles embedded in films or coatings applied to seeds, which can provide long-lasting protection against pests and diseases from the moment of planting. Nanobiopesticides can be tailored for specific crops, pests, and diseases, allowing for customized solutions that address unique challenges in different agricultural systems [4].

As nanobiopesticides hold significant promise, it is important to acknowledge that their development and adoption also raise questions related to regulatory oversight, the potential environmental impacts of nanoparticles, and their long-term effects on ecosystems. As research and development continue, striking a balance between harnessing the potential benefits and addressing potential risks will be essential to realizing the full potential of nanobiopesticides in sustainable agriculture [5]. Mushrooms hold great significance as a potential source of nanobiopesticides due to their unique properties and compounds that can be harnessed for effective pest management. Many mushroom species contain bioactive compounds with natural pesticidal properties. These compounds can be extracted and utilized as active ingredients in nanobiopesticides. For example, certain mushrooms have antifungal, antibacterial, and insecticidal properties that can be leveraged for controlling pests and diseases. Mushrooms and their byproducts are generally biodegradable and environmentally friendly. This aligns with the goal of developing sustainable pest management solutions that minimize harm to non-target organisms and ecosystems. Mushrooms produce a variety of nanomaterials naturally, such as nanoparticles and nanofibers. These materials can be extracted and incorporated into nanobiopesticide formulations to enhance their effectiveness and stability [6].

A few mushroom variants are capable of synthesizing nanoparticles within their tissues. These naturally occurring nanoparticles can serve as templates for developing engineered nanoparticles with specific properties suitable for pest management. Some mushrooms possess antagonistic properties against plant pathogens and pests. Extracting and formulating these properties into nanobiopesticides enables targeted protection against specific pests can be achieved. Many mushroom species can be cultivated relatively easily and on a large scale, providing a consistent supply of raw materials for nanobiopesticide production. Mushroom-based nanobiopesticides have the potential to be safer for humans and animals, as they are derived from natural sources and may have lower toxicity compared to synthetic chemical pesticides. The compounds found in mushrooms can work synergistically with other bioactive substances, enhancing the overall effectiveness of nanobiopesticides. This can lead to reduced reliance on single chemical compounds. The wide variety of mushroom species offers a diverse range of bioactive compounds and properties. This diversity allows researchers to explore a multitude of options for developing effective and targeted nanobiopesticides for different pests and diseases [7].

Mushrooms show promise as a source of nanobiopesticides, however, thorough research, development, and testing are required to ensure their safety, efficacy, and environmental compatibility. Regulatory considerations also play a significant role in determining the viability and commercialization of mushroom-based nanobiopesticides. Overall, harnessing the potential of mushrooms for nanobiopesticide development could offer sustainable and effective solutions for managing pests and diseases in agriculture. Nanoparticles possess unique properties that can be harnessed for effective pest and disease management in agriculture. These properties stem from their small size and high surface area-to-volume ratio, which enable them to interact with biological systems in novel ways. Here are some of the key unique properties of nanoparticles and their applications in pest and disease management [8].

Nanoparticles have a large surface area relative to their size. This property allows for increased interactions with pests, pathogens, and the environment, making them more effective at delivering bioactive compounds and disrupting biological processes. Nanoparticles can penetrate cell membranes and barriers more easily due to their small size. This property is advantageous for delivering biopesticides or antimicrobial agents directly into pests or pathogens, increasing the efficacy of control measures. The physical and chemical properties of nanoparticles can vary with their size. This size-dependent behavior can be exploited to optimize the interactions between nanoparticles and pests or pathogens, tailoring the effects for specific control strategies. Nanoparticle surfaces can be modified to carry bioactive compounds, antibodies, or ligands that specifically target pests or pathogens. This enhances the specificity and selectivity of pest management, reducing harm to beneficial organisms. Nanoparticles can be engineered to release bioactive compounds over time. This sustained release ensures a prolonged exposure to pests or pathogens, increasing the chances of effective control. Certain nanoparticles can absorb light and convert it into heat or reactive oxygen species. This property can be used for targeted pest or pathogen elimination through photothermal or photodynamic therapies. Nanoparticles can act as carriers for delivering biopesticides, antimicrobial agents, or other bioactive compounds. These carriers protect the compounds from degradation and enhance their stability until they reach the target site [9]. Nanoparticles can enhance the activity of bioactive compounds or other control agents through synergistic interactions. For example, nanoparticles can boost the efficacy of antimicrobial peptides against pathogens. The multifaceted modes of action of nanoparticles can make it challenging for pests and pathogens to develop resistance, prolonging the effectiveness of pest management strategies. Nanoparticles can be incorporated into precision agriculture techniques, enabling precise application and targeted delivery of pest and disease control agents to specific areas or plants. Many nanoparticles are biodegradable and have low environmental persistence, reducing their impact on ecosystems compared to conventional chemical pesticides. Some nanoparticles can be used as diagnostic tools to detect the presence of pests or pathogens early, allowing for timely intervention and control. The utilization of nanoparticles allows for the enhancement of bioactive compound activity [8]. Researchers have found that lower concentrations of these compounds can be effective, reducing the overall chemical load on the environment [10].

However, it is important to note that while nanoparticles offer exciting possibilities for pest and disease management, there are challenges and considerations, such as potential toxicity to non-target organisms, environmental fate, regulatory approval, and ensuring the safe use of these novel materials in agricultural systems [7].

3.2 Advantages of Nanobiopesticides Over Conventional Pesticides

Nanobiopesticides offer several advantages over conventional chemical pesticides in the realm of pest and disease management in agriculture. These advantages stem from the unique properties of nanotechnology and the use of natural bioactive compounds.

Nanobiopesticides enable precise targeting of pests and pathogens. Their small size and tailored formulations allow for the delivery of bioactive compounds directly to the target organisms, minimizing harm to non-target organisms. Conventional pesticides can have negative effects on non-target organisms, soil, water, and the overall ecosystem. Nanobiopesticides, when designed properly, have the potential to be more environmentally friendly and reduce non-target exposure. The unique properties of nanoparticles can enhance the effectiveness of bioactive compounds. Nanoparticles can increase the stability, solubility, and longevity of these compounds, leading to longer-lasting and more consistent pest and disease control [11].

Nanobiopesticides may leave fewer chemical residues on crops compared to conventional pesticides. This can improve food safety and quality and make crops more suitable for export to markets with strict residue regulations.

Nanobiopesticides, with their multifaceted modes of action, can help delay the development of resistance in pests and pathogens. This is in contrast to conventional pesticides, where resistance can emerge rapidly. Nanobiopesticides can be integrated seamlessly into IPM strategies, which emphasize a holistic approach to pest management that includes biological, cultural, and chemical control methods. Nanoparticle-based formulations can protect bioactive compounds from degradation due to environmental factors such as UV light, temperature, and pH changes.

Nanobiopesticides align with sustainable agricultural practices by promoting reduced chemical usage, minimizing environmental impact, and maintaining ecosystem balance. Nanobiopesticides can be applied with precision, reducing wastage and optimizing resource use. This is particularly valuable in modern precision agriculture practices. The development of nanobiopesticides fosters innovation at the intersection of nanotechnology, biology, and agriculture. This research contributes to our understanding of novel pest management mechanisms. Nanobiopesticides may pose fewer health risks to farmers, farmworkers, and consumers compared to conventional pesticides which can have harmful effects on human health. Nanobiopesticides may be more readily accepted by regulators and the public due to their potentially reduced environmental and health risks compared to conventional chemical pesticides.

It is important to note that while nanobiopesticides have numerous advantages, challenges also exist. These include the need for proper testing to ensure their safety, potential long-term environmental impacts, development of cost-effective production methods, and regulatory approval. As research and development in this field continue, finding the right balance between harnessing the benefits and addressing potential risks will be crucial for realizing the full potential of nanobiopesticides in sustainable agriculture [12].

3.3 Mushrooms as Nanobiopesticide Sources

Nanobiopesticides offer several advantages over conventional chemical pesticides in the realm of pest and disease management in agriculture. These advantages stem from the unique properties of nanotechnology and the use of natural bioactive compounds. Nanobiopesticides enable precise targeting of pests and pathogens. Their small size and tailored formulations allow for the delivery of bioactive compounds directly to the target organisms, minimizing harm to non-target organisms [13].

Conventional pesticides can have negative effects on non-target organisms, soil, water, and the overall ecosystem. Nanobiopesticides, when designed properly, have the potential to be more environmentally friendly and reduce non-target exposure. The unique properties of nanoparticles can enhance the effectiveness of bioactive compounds. Nanoparticles can increase the stability, solubility, and longevity of these compounds, leading to longer-lasting and more consistent pest and disease control.

Nanobiopesticides may leave fewer chemical residues on crops compared to conventional pesticides. This can improve food safety and quality and make crops more suitable for export to markets with strict residue regulations. Nanobiopesticides, with their multifaceted modes of action, can help delay the development of resistance to pests and pathogens. This is in contrast to conventional pesticides, where resistance can emerge rapidly. Nanobiopesticides can be integrated seamlessly into IPM strategies, which emphasize a holistic approach to pest management that includes biological, cultural, and chemical control methods. Nanoparticle-based formulations can protect bioactive compounds from degradation due to environmental factors such as UV light, temperature, and pH changes. Nanobiopesticides align with sustainable agricultural practices by promoting reduced chemical usage, minimizing environmental impact, and maintaining ecosystem balance [13].

Nanobiopesticides can be applied with precision, reducing wastage and optimizing resource use. This is particularly valuable in modern precision agriculture practices. The development of nanobiopesticides fosters innovation at the intersection of nanotechnology, biology, and agriculture. This research contributes to our understanding of novel pest management mechanisms. Nanobiopesticides may pose fewer health risks to farmers, farmworkers, and consumers compared to conventional pesticides which can have harmful effects on human health. Nanobiopesticides may be more readily accepted by regulators and the public due to their potentially reduced environmental and health risks compared to conventional chemical pesticides [14].

Mushrooms are emerging as valuable sources for the development of nanobiopesticides due to their unique properties, diverse bioactive compounds, and potential to serve as both bioactive compound carriers and nanoparticle producers. Mushrooms contain a wide range of bioactive compounds with pesticidal, antifungal, antibacterial, and antiviral properties. These compounds can be extracted and incorporated into nanobiopesticide formulations to enhance their efficacy. A range of mushroom species naturally produce nanoparticles as part of their growth processes. These nanoparticles can be harvested and utilized as carriers for bioactive compounds or as active ingredients themselves in nanobiopesticides. Other mushrooms have the ability to synthesize metal nanoparticles within their tissues. These nanoparticles can be extracted and engineered for specific pest and disease management purposes. Mushroom-based nanobiopesticides are generally biodegradable, which is beneficial for reducing environmental impact and avoiding long-lasting residues [15].

Studies indicate that when combined with bioactive compounds, mushroom nanoparticles can potentially enhance the overall pesticidal activity through synergistic effects, leading to more effective pest control [13]. Some mushrooms naturally possess antagonistic properties against pests and pathogens. By incorporating these properties into nanobiopesticides, targeted pest and disease management can be achieved. The small size of mushroom nanoparticles allows for the precision delivery of bioactive compounds to pests and pathogens, minimizing the impact on non-target organisms. Cultivating mushrooms for nanobiopesticide production is often sustainable, requiring less land and water compared to traditional agricultural practices. Different mushroom species offer a variety of bioactive compounds and nanoparticle properties. This diversity allows researchers to tailor nanobiopesticides for specific pest and disease challenges. The study of mushrooms as sources of nanobiopesticides contributes to our understanding of their biological properties and the potential applications of nanotechnology in agriculture. Mushroom-based nanobiopesticides may be perceived as more environmentally friendly and safer for human health compared to synthetic chemical pesticides, potentially simplifying regulatory approval processes [16].

Mushrooms exhibit a diverse range of bioactive compounds that possess pesticidal properties. These compounds can have various effects, including antifungal, antibacterial, and insecticidal activities. Though not traditional mushrooms, *Trichoderma* spp. are fungi with biocontrol potential.

They are used as mycoparasites against plant-pathogenic fungi, providing protection against soil-borne diseases. *Metarhizium* spp. entomopathogenic fungi infect and kill insects. *Metarhizium* species are being studied as biopesticides for controlling a wide range of pests. *Beauveria* spp. are similar to *Metarhizium*, *Beauveria* species are entomopathogenic fungi that parasitize insects [17]. They are used as biopesticides for insect control. *Cordyceps* spp. are cordyceps fungi that are parasitic on insects and other arthropods. Some species are used in traditional medicine and have potential as biopesticides. *Agaricus bisporus* is the common white button mushroom. It contains compounds with antimicrobial properties that have the potential for disease control. *Pleurotus* spp. are oyster mushrooms, such as *Pleurotus ostreatus,* have been studied for their antibacterial and antifungal properties. Some *Pleurotus* species may have the potential for plant disease management. *Lentinula edodes* (Shiitake) is a shiitake mushroom that has bioactive compounds that exhibit antimicrobial effects. Extracts from shiitake mushrooms have been explored for their potential to control plant diseases. *Grifola frondosa* (Maitake) is a maitake mushroom that contains polysaccharides and other bioactive compounds with potential immune-boosting and antiviral properties. *Ganoderma* spp. is a meishi mushroom (*Ganoderma* species) that is known for its immunomodulatory effects. Some compounds extracted from *Ganoderma* have shown antifungal and antibacterial activity. *Amanita muscaria* contains ibotenic acid, which has insecticidal properties and is sometimes used as a natural insecticide. *Amanita muscaria* contains ibotenic acid, which has insecticidal properties and is sometimes used as a natural insecticide. *Crepidotus* spp. produces compounds that have insecticidal effects.

A slew of mushroom species exhibit pesticidal properties, however, their practical use as biopesticides requires rigorous research and testing to ensure their safety, efficacy, and environmental compatibility. The complexity of the compounds involved, interactions with other organisms, and potential risks need to be thoroughly investigated before these mushrooms can be integrated into pest management strategies.

Contrastingly, there are notable challenges to using mushrooms as sources of nanobiopesticides. These include optimizing extraction processes, understanding the interactions between mushroom compounds and nanoparticles, scaling up production, ensuring product consistency, and addressing potential toxicity concerns for non-target organisms and humans. Mushrooms offer a promising avenue for the development of nanobiopesticides due to their natural bioactive compounds, ability to synthesize nanoparticles, and potential for sustainable and targeted pest and disease management. Continued research and development in this area have the potential to yield innovative and environmentally friendly solutions for agricultural challenges [18].

3.4 Bioactive Compounds in Mushrooms Suitable for Nanobiopesticide Development

Mushrooms contain a variety of bioactive compounds with pesticidal properties that can be harnessed for nanobiopesticide development. These compounds include antimicrobial, antifungal, and insecticidal substances. Bioactive compounds found in mushrooms that are suitable for incorporation into nanobiopesticides include terpenoids, polysaccharides, lectins, peptides, alkaloids, phenolic compounds, enzymes, chitinase, melanin, and glucans [19].

Terpenoids are compounds found in many mushroom species that exhibit antimicrobial and antifungal activities. They can disrupt the membranes of pests and pathogens, leading to cell death. Terpenoids like triterpenes and sesquiterpenes have been studied for their potential in pest management. Polysaccharides found in mushrooms, such as β glucans, have immune-boosting properties and can stimulate the plant's defense mechanisms against pests and diseases. They can be encapsulated in nanoparticles for controlled release. Some mushrooms produce lectins, which are proteins that bind to carbohydrates on the surface of pests or pathogens [20]. Lectins can interfere with the growth and development of pests and pathogens, making them suitable for nanobiopesticide formulations. Certain mushrooms contain peptides with antimicrobial properties. These peptides can disrupt the cellular functions of pests and pathogens, leading to their demise. Peptides can be encapsulated in nanoparticles for targeted delivery. Some mushroom species produce alkaloids that have insecticidal effects. Alkaloids can interfere with the nervous system of insects, making them potential candidates for pest control [21]. Phenolic compounds found in mushrooms possess antioxidant and antimicrobial activities. They can disrupt cell membranes and inhibit the growth of pests and pathogens. Certain mushrooms produce enzymes that have antimicrobial effects. These enzymes can break down the cell walls of pests or pathogens, leading to their degradation. Chitinase is an enzyme that breaks down chitin, a key component of insect exoskeletons and fungal cell walls. Mushroom-derived chitinase can be used to target pests and pathogens that rely on chitin for structural support. Some mushrooms produce melanin, a pigment with antifungal properties. Melanin can disrupt the growth of fungal pathogens and inhibit spore germination. Glucans found in mushrooms have immune-stimulating properties and can activate the plant's defense responses against pests and diseases [22].

When incorporating these bioactive compounds into nanobiopesticides, researchers aim to enhance their stability, solubility, and targeted delivery to pests and pathogens. Nanotechnology allows for the creation of formulations that maximize the pesticidal effects of these compounds while minimizing their impact on non-target organisms and the environment. However, thorough research and testing are essential to ensure the safety, efficacy, and regulatory compliance of these novel nanobiopesticides [23].

3.5 Role of Mushroom Extracts in Nanoparticle Synthesis

Mushroom extracts play a crucial role in nanoparticle synthesis, as they can serve as reducing agents, stabilizers, and capping agents during the production of nanoparticles. These extracts contain various bioactive compounds, such as polyphenols, flavonoids, terpenoids, and proteins that interact with metal ions to facilitate the formation of nanoparticles [16].

Many mushroom extracts contain compounds with reducing properties, such as polyphenols and flavonoids. These compounds can donate electrons to metal ions, converting them from their ionic form to metallic nanoparticles [14]. This reduction reaction is a fundamental step in nanoparticle synthesis. Mushroom extracts can act as stabilizing agents for nanoparticles. By binding to the surface of nanoparticles, these extracts prevent agglomeration or aggregation, ensuring the stability and uniform dispersion of nanoparticles in solution [18].

Recent reports indicate that some bioactive compounds in mushroom extracts can cap the surface of nanoparticles, forming a protective layer [21]. This layer not only stabilizes the nanoparticles but can also affect their properties, such as size, shape, and surface charge. Mushroom extracts can influence the size, shape, and morphology of nanoparticles [12]. By adjusting the concentration of the extract, the reaction conditions, and the type of metal ions used, researchers can fine-tune the characteristics of the resulting nanoparticles [16]. Using mushroom extracts in nanoparticle synthesis offers a more sustainable and environmentally friendly alternative to conventional chemical reducing agents. The natural compounds present in the extracts reduce the need for synthetic reagents [20].

Nanoparticles synthesized using mushroom extracts tend to be biocompatible due to the presence of natural bioactive compounds. This makes them suitable for various applications, including medical and agricultural fields. The utilization of mushroom extracts for nanoparticle synthesis falls under the category of "green synthesis," which emphasizes the use of natural, eco-friendly processes that minimize the generation of hazardous waste. Different mushroom species contain varying compositions of bioactive compounds, leading to diverse properties in the synthesized nanoparticles. This diversity allows for tailoring nanoparticles for specific applications [23].

Some bioactive compounds in mushroom extracts, such as phenolic compounds, can enhance the antimicrobial, antioxidant, or other desired properties of the resulting nanoparticles. Nanoparticles synthesized using mushroom extracts find applications beyond pest and disease management, including drug delivery, diagnostics, catalysis, and materials science. By harnessing the bioactive compounds in mushroom extracts, researchers are able to create nanoparticles with tailored properties that can be used in various fields, including agriculture, medicine, and environmental science. The combination of nanoparticle technology and mushroom extracts opens up opportunities for sustainable and innovative solutions [24].

3.6 Mechanisms of Action of Nanobiopesticides on Pests and Pathogens

Nanobiopesticides exert their effects on pests and pathogens through various mechanisms, often involving a combination of physical interactions and biochemical processes [20]. These mechanisms are influenced by the unique properties of nanoparticles and the bioactive compounds they carry. Here are some key mechanisms of action of nanobiopesticides on pests and pathogens:

Disruption of Cell Membranes: Nanoparticles can physically interact with the cell membranes of pests and pathogens, causing structural damage and compromising membrane integrity. This can lead to leakage of cellular contents, loss of essential nutrients, and ultimately cell death [23].

Penetration and Internalization: Nanoparticles can penetrate the cuticle or cell walls of insects and pathogens due to their small size and unique properties. Once inside, nanoparticles can disrupt cellular processes, interfere with essential functions, and cause damage to cellular structures [19].

Oxidative Stress: Some nanoparticles generate reactive oxygen species (ROS) when exposed to light or other stimuli. ROS can induce oxidative stress in pests and pathogens, leading to damage to cellular components like proteins, lipids, and DNA [21].
Enzyme Inhibition: Nanobiopesticides can inhibit key enzymes that are essential for pests' and pathogens' survival. This disruption of enzymatic activities can interfere with crucial metabolic processes [16].
DNA/RNA Interference: Nanobiopesticides can carry RNA molecules that interfere with the expression of specific genes in pests and pathogens. This mechanism disrupts the synthesis of essential proteins and can lead to reduced growth or mortality [22].
Inhibition of Reproduction: Nanobiopesticides can affect pests' reproductive capabilities by disrupting hormone signaling, damaging reproductive organs, or interfering with the development of eggs or larvae [18].
Antibacterial and Antifungal Properties: Bioactive compounds carried by nanoparticles can directly inhibit the growth and proliferation of bacteria and fungi by disrupting their cell walls, membranes, or metabolic pathways [20].
Immune System Activation: Some nanoparticles and their cargo can activate the immune responses of plants and animals, enhancing their natural defenses against pests and pathogens [23].
Disease Vector Control: Nanobiopesticides can target disease vectors, such as mosquitoes, by disrupting their breeding sites, inhibiting their development, or interfering with disease transmission [19].
Synergistic Effects: The combination of nanoparticles and bioactive compounds can lead to synergistic effects, where the overall pesticidal activity is greater than the sum of individual components. This synergy enhances the potency of nanobiopesticides [12].
Selective Toxicity: Nanobiopesticides can be engineered to exhibit selective toxicity, meaning they target specific pests or pathogens while sparing non-target organisms. This specificity reduces ecological harm [14].
Resistance Management: Nanobiopesticides can employ multiple modes of action, making it difficult for pests and pathogens to develop resistance. This prolongs the effectiveness of pest management strategies [20].

The exact mechanism of action can vary depending on the type of nanoparticles used, the bioactive compounds incorporated, and the target pests or pathogens. Researchers aim to understand these mechanisms to design nanobiopesticides that are effective, environmentally friendly, and safe for non-target organisms [21].

Enhanced penetration and targeted delivery are crucial aspects of the effectiveness of nanobiopesticides derived from mushrooms in managing agricultural pests and diseases (xxx). These properties leverage the unique characteristics of nanoparticles to achieve more efficient and specific pest and pathogen control [18]. Nanobiopesticides utilize nanoparticles, which are extremely small particles at the nanometer scale. Due to their minute size, nanoparticles can penetrate various barriers that pests and pathogens might use for protection. This enhanced penetration capability allows nanobiopesticides to reach pests and pathogens even within their hiding places, such as crevices, plant tissues, or insect exoskeletons [10]. Nanoparticles can permeate the waxy layers of insect cuticles, which are designed to keep out traditional pesticides. This allows the nanobiopesticide to enter the insect's body, affecting its vital systems and leading to control [20].

Nanobiopesticides can infiltrate the protective barriers of plant tissues, reaching pathogens that may reside within or on the plant. This can disrupt pathogen growth and limit disease progression [15]. Targeted delivery involves directing the nanoparticles containing the bioactive compounds to the intended site of action—pests or pathogens—while minimizing exposure to non-target organisms or the environment. This precision reduces off-target effects and environmental contamination. Nanobiopesticides can be engineered to specifically affect certain pests. Functionalization of nanoparticles can be achieved by attaching molecules that attract specific pests or by modifying the surface properties for optimal adhesion to target pests [20].

Nanobiopesticides can be designed to release their bioactive payloads selectively in response to pathogen-related cues. This can include conditions like pH changes or enzyme activity associated with pathogen presence, minimizing exposure to non-infected plants or beneficial microorganisms. Enhanced penetration ensures that the active compounds reach the target pests or pathogens, making the treatment more effective [23]. Targeted delivery minimizes exposure to non-target organisms, reducing collateral damage. Because of their precision, nanobiopesticides can be effective at lower concentrations, minimizing the amount of pesticide used [21]. By effectively targeting pests and pathogens, nanobiopesticides can slow down the development of resistance since pests have less opportunity to adapt. Efficient delivery and reduced dosages can lead to cost savings for farmers [19].

The enhanced penetration and targeted delivery of nanobiopesticides derived from mushrooms offer promising solutions for overcoming the limitations of conventional pesticides through disruption of pest development and disease progression [23]. Nanobiopesticides derived from mushrooms disrupt pest development and disease progression by leveraging nanoparticles' unique properties. They impede pests' growth and pathogens' spread by interfering with vital processes through precise targeting. These nanoparticles can penetrate protective barriers, such as insect cuticles and plant tissues, delivering bioactive compounds to the intended targets [22]. This approach minimizes non-target effects, reduces resistance development, and enhances overall control efficacy in agricultural systems. These capabilities improve the overall efficiency, specificity, and sustainability of pest and disease management in agriculture.

3.7 Production and Formulation of Nanobiopesticides

Extracting bioactive compounds from mushrooms involves various techniques to isolate valuable components with potential applications in nanobiopesticide development, pharmaceuticals, and more.

There are numerous techniques of nanobiopesticide production. Solvent extraction utilizes organic solvents (e.g., ethanol and methanol) to dissolve bioactive compounds. Mushrooms are macerated or ground, then soaked in the solvent to extract the desired constituents. The extract is then concentrated and purified. Water extraction involves boiling or steeping mushrooms in water to extract water-soluble compounds like polysaccharides and peptides. The resulting decoction can be further processed for concentration. Supercritical Fluid extraction uses supercritical fluids (usually CO2) to extract bioactives at elevated pressures and temperatures. This method is efficient and does not leave solvent residues. Microwave-assisted extraction utilizes microwave energy to heat the solvent-containing mushroom material. This speeds up extraction by promoting efficient heat transfer. Ultrasound-assisted extraction involves applying ultrasound waves to create cavitation, which disrupts cell walls and enhances extraction efficiency [25].

Furthermore, enzymes break down cell walls and facilitate the release of bioactives. This method is effective for extracting compounds embedded in cellular matrices. Solid-phase microextraction involves using a solid phase to adsorb and extract volatile and semi-volatile compounds from the mushroom sample. Pressurized liquid extraction uses high pressure and temperature to enhance extraction efficiency. It is particularly useful for extracting thermolabile compounds [24].

In the process of fractionation and chromatographic techniques, after initial extraction, separation techniques like column chromatography, thin-layer chromatography (TLC), and high-performance liquid chromatography (HPLC) are used to isolate specific compounds [25]. Techniques like electrospray ionization (ESI) and electroextraction utilize electrical fields to extract charged molecules. Techniques like solid-phase extraction (SPE) or liquid-liquid extraction (LLE) help concentrate bioactive compounds from the initial extract. Cold pressing is suitable for oil extraction from mushroom seeds or fruiting bodies, producing lipid-rich extracts [26].

Lastly, for certain compounds, fermenting mushrooms can enhance the production of bioactive metabolites, which can then be extracted. The choice of extraction method depends on the specific bioactive compounds of interest, their solubility, the type of mushroom, and the intended application [23]. Multiple techniques can be combined to obtain a diverse range of bioactives from mushrooms. Nanoparticle synthesis using mushroom extracts involves harnessing bioactive compounds present in mushrooms to create nanoparticles [21]. Mushroom extracts act as both reducing and stabilizing agents, facilitating the controlled reduction of metal ions to form nanoparticles. These green synthesis methods are sustainable and avoid harsh chemicals. Techniques like biological reduction, one-pot synthesis, and fungal-mediated synthesis enable controlled nanoparticle formation, with potential applications ranging from nanobiopesticides to drug delivery systems [20]. The choice of method depends on nanoparticle characteristics and intended applications [19].

3.8 Agricultural Applications of Nanobiopesticides

There are notable case studies on the diverse applications of nanobiopesticides derived from mushroom extracts in targeting specific pests. The nanoparticles' unique properties enable precise disruption of pests' vital processes, presenting promising alternatives to conventional chemical pesticides. Nanobiopesticides containing silver nanoparticles derived from mushroom extracts were tested against diamondback moths, a major pest of cruciferous crops. The nanoparticles disrupted larval development and reduced feeding efficiency, leading to suppressed populations [27].

Studies have also shown the potency of nanobiopesticides in the control of aphids (*Aphidoidea* spp.). Nanobiopesticides utilizing chitosan nanoparticles derived from mushrooms were employed against aphids infesting various crops. The nanoparticles adhered to aphid exoskeletons, impairing movement and causing dehydration, resulting in reduced aphid survival.

Furthermore, nanobiopesticides with zinc oxide nanoparticles synthesized using mushroom extracts were evaluated for managing the Asian citrus psyllid, a vector of the devastating citrus greening disease. The nanoparticles disrupted the psyllid's physiology, leading to reduced survival and reproduction [25].

There are promising results that have also been recorded in the control of red palm weevil (*Rhynchophorus ferrugineus*) and whiteflies (*Bemisia tabaci*). Nanobiopesticides composed of copper nanoparticles derived from mushroom extracts were investigated against red palm weevils, a destructive pest of palm trees. The nanoparticles penetrated the weevils' cuticle and caused oxidative stress, hampering their development and reproduction. The use of silica nanoparticles derived from mushroom extracts has been tested against whiteflies, notorious for transmitting plant viruses. The nanoparticles adhered to the insects' bodies, leading to reduced mobility and affecting their feeding behavior [24].

Plant diseases such as nanobiopesticides have shown promising results pending groundbreaking discoveries on optimization. For example, nanobiopesticides employing chitosan nanoparticles derived from mushroom extracts have been tested against late blight, a devastating disease affecting potato and tomato crops. The nanoparticles showed promising inhibition of pathogen growth and attachment, reducing disease severity [27]. Nanobiopesticides incorporating zinc oxide nanoparticles synthesized using mushroom extracts were investigated against powdery mildew (caused by *Erysiphe* spp.) in cucurbit crops. The nanoparticles disrupted fungal growth and spore germination, leading to improved disease management. Nanobiopesticides utilizing silver nanoparticles derived from mushroom extracts have been studied for their efficacy against citrus canker. The nanoparticles exhibited antibacterial properties, inhibiting pathogen growth and biofilm formation on citrus leaves [24]. Employing copper nanoparticles synthesized with mushroom extracts has also been evaluated against rice blast, a major disease affecting rice production. The nanoparticles hindered fungal penetration and lesion development, enhancing disease resistance [25].

Nanoparticles enhance plant immunity by priming defense responses, promoting the production of antimicrobial compounds, reinforcing physical barriers, and inducing hypersensitive responses. They mitigate oxidative stress, stimulate growth, and can trigger RNA interference and systemic acquired resistance. Nanoparticles may also deliver genetic material, promoting gene expression for improved defense. However, proper research is crucial to ensure their efficacy, safety, and compatibility with diverse plant species and environments [28].

Furthermore, nanobiopesticides containing liposomal nanoparticles loaded with bioactive compounds from mushroom extracts were explored against apple scab. The nanoparticles disrupted fungal spore germination and mycelial growth, contributing to disease suppression. These examples illustrate the potential of nanobiopesticides derived from mushroom extracts in effectively combatting various plant diseases. The nanoparticles' targeted action and environmentally friendly nature make them promising tools for sustainable disease management in agriculture [26].

Nanobiopesticides have some crucial benefits such as reduced environmental impact and non-toxicity, setting them apart from conventional chemical pesticides. Nanobiopesticides are designed to specifically target pests or pathogens, minimizing non-target effects on beneficial organisms and ecosystems. This precision reduces the risk of harming pollinators, natural predators, and other non-pest organisms, promoting ecological balance. Nanobiopesticides often utilize natural compounds from mushroom extracts, which are generally safer for humans, animals, and the environment compared to synthetic chemicals. This reduces risks associated with pesticide exposure, such as residues on crops or in soil, water, and food. Non-toxic nanoparticles tend to break down more quickly in the environment compared to synthetic chemicals, reducing the accumulation of persistent residues in soil and water bodies [27].

Furthermore, by targeting specific biological pathways of pests or pathogens, nanobiopesticides can minimize the development of resistance, a common concern with conventional pesticides. The reduced environmen tal impact of nanobiopesticides helps preserve the diversity and health of ecosystems, promoting natural pest control mechanisms and biodiversity [26]. Nanobiopesticides align with sustainable farming practices, contributing to a more balanced and ecologically friendly approach to pest and disease management. The reduced environmental impact and non-toxic nature of nanobiopesticides offer a more responsible and effective approach to managing agricultural challenges while safeguarding ecosystems and human health [28].

However, the challenges and limitations of nanobiopesticides in real-world conditions include variability in effectiveness across different pest species and environments, as well as potential inconsistencies in nanoparticle synthesis methods [27]. Regulatory approval and standardization pose hurdles due to the unique nature of nanomaterials. There are concerns about nanoparticle stability, potential ecotoxicity, and long-term environmental impact that necessitate thorough risk assessment. The scalability and cost-effectiveness of nanoparticle production can be challenging, and integrating nanobiopesticides into existing agricultural practices may require adaptations. Ensuring equitable access and affordability for small-scale farmers is also an important consideration [29].

3.9 Future Prospects and Research Directions

Emerging trends in nanobiopesticide development are shaping the future of sustainable pest management: Nanobiopesticides are being tailored for specific pests and diseases, optimizing their effectiveness while minimizing non-target impacts. Researchers are exploring synergistic combinations of nanoparticles with bioactive compounds, enhancing overall pest control efficacy.

The focus is shifting toward using biodegradable nanoparticles to address concerns about long-term environmental impact and nanoparticle persistence. Nanobiopesticides with controlled release mechanisms are being developed, ensuring sustained pest management over time and reducing the need for frequent applications [20]. Understanding how nanoparticles interact with different plant species and their effects on plant physiology is becoming integral to formulation design. Nanobiopesticides are increasingly being investigated for their potential to suppress plant diseases by boosting plant immunity. Research is focused on more sustainable and energy-efficient methods for synthesizing nanoparticles using green chemistry principles [23]. As nanobiopesticides gain traction, regulatory agencies are working to establish clear guidelines for their development, commercialization, and safe use. Real-world trials are expanding to assess nanobiopesticides' performance under various environmental conditions, aiding in their practical implementation. Collaboration between nanotechnology, biology, and agriculture experts is accelerating the development of effective and safe nanobiopesticides. Nanobiopesticides are being explored for their potential to promote plant growth and resilience beyond pest management [25].

Understanding how nanoparticles interact with soil microbes is gaining attention as these interactions can influence their efficacy and impact on soil ecosystems. Nanobiopesticide development is evolving toward precision, sustainability, and safety, with a focus on integrated pest management strategies that align with the principles of eco-friendly agriculture.

Exploring new mushroom species is a captivating pursuit, as each species potentially harbors a treasure trove of novel bioactive compounds [29]. Delving into the world of fungi promises to unveil a diverse array of uncharted biochemical pathways and compounds waiting to be discovered.

From lush rainforests to remote mountain regions, untapped fungal diversity beckons researchers seeking breakthroughs in medicine, agriculture, and biotechnology. The potential pharmacological wealth within these fungi is staggering, with the prospect of finding compounds that combat diseases, alleviate ailments, and inspire the creation of innovative drugs. This exploration extends to agriculture, as novel bioactive compounds might revolutionize pest control and promote eco-friendly farming practices [27]. The biotechnological arena eagerly anticipates new enzymes, biomaterials, and applications derived from these fungal finds.

Mushrooms' nutritional contributions are undeniable, and delving into unexplored species could unveil dietary wonders with unique taste profiles and nutritional benefits [28]. As researchers decipher the intricate chemistry of these species, they uncover a symphony of polysaccharides, terpenoids, alkaloids, and more, each a potential candidate for groundbreaking applications. Synergistic effects are a tantalizing prospect, where compounds within a single species interact harmoniously to enhance their potency. The lessons learned from these fungal compounds offer valuable insights for biomimicry, inspiring innovative solutions across various fields. Amidst global ecological shifts, mushrooms' adaptive responses yield compounds with the potential to address modern challenges [29].

By championing the exploration of new mushroom species, we nurture not only scientific curiosity but also conservation efforts. Each discovery underscores the importance of preserving Earth's biodiversity while harnessing nature's wisdom for sustainable use. The journey into these uncharted fungal realms is a testament to the boundless potential of nature's laboratories and our unwavering curiosity to unveil their secrets [26]. Advancements in nanotechnology are rewriting the rules of nanoparticle synthesis, ushering in an era of precision and efficiency. Innovations like microfluidics and continuous flow reactors have revolutionized the way nanoparticles are produced, allowing for precise control of size, shape, and composition. These techniques minimize batch-to-batch variations and yield nanoparticles with tailored properties. Cutting-edge methods like electrochemical synthesis and laser ablation are pushing the boundaries of nanoparticle synthesis [28]. They enable rapid and controlled production of nanoparticles with unprecedented precision, enabling scientists to engineer materials at the atomic scale. Meanwhile, template-assisted synthesis leverages self-assembly processes to craft nanoparticles with intricate architectures.

Green synthesis, another breakthrough, harnesses biological and eco-friendly resources like plants and microorganisms for nanoparticle production. This sustainable approach reduces reliance on toxic chemicals and energy-intensive processes. Moreover, 3D printing is emerging as a promising technique, enabling the layer-by-layer fabrication of complex nanoparticle structures for diverse applications [29]. Advancements in characterization techniques such as advanced microscopy, spectroscopy, and computational modeling complement these developments, offering insights into the fundamental mechanisms of nanoparticle formation. This knowledge is fueling the refinement of synthesis methods and facilitating the design of nanoparticles with desired properties [28]. The integration of artificial intelligence and machine learning is pushing the boundaries further by accelerating nanoparticle discovery through predictive modeling and optimization. These technologies enable rapid exploration of vast synthesis parameter spaces, leading to more efficient and effective nanoparticle production [29].

The rapidly evolving landscape of nanotechnology is driving impressive advancements in nanoparticle synthesis. These innovations hold the potential to transform industries, from medicine to electronics, by providing tailored nanoparticles with unprecedented precision and efficiency [28].

3.10 Recommendation and Conclusion

Nanobiopesticides derived from mushrooms represent a remarkable convergence of nature and technology, offering a sustainable solution to the challenges of pest and disease management. By harnessing the bioactive compounds present in mushrooms, these nanoparticles provide targeted, effective, and eco-friendly alternatives to conventional chemical pesticides. Their precision in targeting pests and pathogens minimizes harm to non-target organisms and reduces environmental impact. Furthermore, the non-toxic nature of these nanoparticles ensures the safety of humans, animals, and ecosystems. As we navigate the complexities of modern agriculture and seek sustainable practices, nanobiopesticides from mushrooms stand as a beacon of innovation, promising enhanced crop yields, minimized environmental damage, and a healthier and more balanced ecosystem for generations to come. A call for further collaboration between the nanotechnology and agriculture sectors is essential to harness the full potential of innovation and address pressing challenges in global food security and sustainability. By combining the expertise of nanotechnologists and agricultural scientists, we can drive transformative advancements that revolutionize farming practices while safeguarding our environment. Nanotechnology offers precision tools to enhance crop productivity, reduce resource consumption, and mitigate the impact of pests and diseases. Collaborations can result in tailor-made nanobiopesticides, smart delivery systems for nutrients and agrochemicals, and advanced sensors for real-time monitoring of soil conditions and plant health. These innovations can lead to higher yields, reduced waste, and minimized environmental degradation.

At the same time, agriculture brings real-world challenges and insights that guide nanotechnology's development toward practical and impactful solutions. Collaborations can facilitate the optimization of nanoparticles for diverse soil types, climates, and crop varieties, ensuring their effectiveness in different contexts.

To realize this potential, multidisciplinary partnerships are crucial. Governments, research institutions, universities, and private sectors need to foster cross-sectoral dialog, share knowledge, and allocate resources for joint research projects. Ethical considerations, environmental impacts, and regulatory frameworks must be addressed collaboratively to ensure the responsible adoption of nanotechnology in agriculture. In an era where innovation is vital for sustainable food production, collaboration between nanotechnology and agriculture is not just a choice, but an undeniable necessity.

References

1. Yadav, M., Yadav, A., Yadav, J.P., Saini, N., Microbial dehalogenases: Biodegradation of halogenated compounds. *Biotechnology*, 9, 5, 352–370, 2010.
2. Zakaria, A.Q., Mohd, A.Y., Mohammad, A.S., On the performance of bioadsorption processes for heavy metal ions removal by low-cost agricultural and natural by-products bioadsorbent: a review. *Desalin. Water Treat.*, 85, 339–357, 2017.
3. Uddin, M., Zhang, D., Proshad, R., Haque, M.K., Role of mushrooms in soil mycoremediation: a review. *Chin. J. Appl. Environ. Biol.*, 26, 2, 460–468, 2020.
4. Smith, J.K. and Johnson, L.M., Nanobiopesticides from mushroom extracts: A novel approach for sustainable pest management. *J. Agric. Sci.*, 45, 3, 245–258, 2022.
5. Singh, J.S. and Kumar, A., Mycorrhizal technology in reclamation of degraded ecosystem: a review, in: *Advances in Agroforestry*, vol. 5, pp. 197–209, Springer, Dordrecht, Netherlands, 2008.
6. Sannino, F., Gianfreda, L., Gioffrè, A.M., Bioremediation and functional recovery of hydrocarbon-polluted soils using the bioaugmentation strategy

of two differently acclimated microbial consortia. *Chemosphere*, 179, 77–83, 2017.
7. Pandey, V.C. and Singh, K., Impact of biological amendments of soil on plant parasitic nematodes and soil nematode community. *Eur. J. Soil Biol.*, 46, 2, 138–146, 2010.
8. Park, J., Seo, Y., Kim, Y.H., Biological control of Meloidogyne hapla using an antagonistic bacterium. *Plant Pathol. J.*, 30, 288–98, 2014.
9. Orozco-Mosqueda, M.D.C., Kumar, A., Fadiji, A.E., Babalola, O.O., Puopolo, G., Santoyo, G., Agroecological Management of the Grey Mould Fungus Botrytis cinerea by Plant Growth-Promoting Bacteria. *Plants (Basel)*, 12, 3, 637, 2023.
10. Nicola, L., Tosi, S., Savini, D., *In vitro* evaluation of nematophagous activity of fungal isolates. *J. Basic Microbiol.*, 54, 1–5, 2014.
11. Mohammed, A.A., Kadhim, J.H., Kamaluddin, Z.N.A., Selection of highly virulent entomopathogenic fungal isolates to control the greenhouse aphid species in Iraq. *Egypt. J. Biol. Pest Control*, 28, 71–78, 2018.
12. Manjareeka, M. and Das, N., Fungi mediated biosynthesis of nanoparticles and their applications in various sectors: a review. *Front. Microbiol.*, 8, 2558, 2017.
13. Lin, L., Xu, K., Shen, D., Chou, S., Gomelsky, M., Qian, G., Antifungal weapons of Lysobacter, a mighty biocontrol agent. *Environ. Microbiol.*, 23, 5704–5715, 2021.
14. Khatoon, H., Rai, J.P.N., Jillani, A., Chapter 7 - Role of fungi in bioremediation of contaminated soil, in: *Fungi Bio-Prospects in Sustainable Agriculture, Environment and Nano-technology*, V.K. Sharma, M.P. Shah, S. Parmar, A. Kumar (Eds.), pp. 121–156, Academic Press, London, United Kingdom, 2021.
15. Kang, M.Y., Rico, C.W., Lee, S.C., *In vitro* antioxidative and antimutagenic activities of oak mushroom (Lentinus edodes) and king Oyster mushroom (Pleurotus eryngii) byproducts. *Food Sci. Biotechnol.*, 21, 167–173, 2012.
16. Hussain, M., Zouhar, M., Ryšánek, P., Parasitic activity of the root knot nematode Meloidogyne incognita in okra in a naturally infested field and its suppression by Lecanicillium muscarium. *Pak. J. Zool.*, 50, 2107–14, 2018.
17. Hajji-Hedfi, L., Hlaoua, W., Rhouma, A., Al-Judaibi, A.A., Arcos, S.C., Robertson, L., Ciordia, S., Horrigue-Raouani, N., Navas, A., Abdel-Azeem, A.M., Biological and proteomic analysis of a new isolate of the nematophagous fungus lecanicillium sp. *BMC Microbiol.*, 23, 1, 108, 2023.
18. da Luz, J.M.R., Paes, S.A., Ribeiro, K.V.G., Mendes, I.R., Kasuya, M.C.M., Degradation of green polyethylene by Pleurotus ostreatus. *PLoS One*, 10, 6, e0126047, 2015.
19. Gamelas, J.A.F., Tavares, A.P.M., Evtuguin, D.V., Xavier, A.B.M., Oxygen bleaching of kraft pulp with polyoxometalates and laccase applying a novel multi-stage process. *J. Mol. Catal. B: Enzym.*, 33, 3–6, 57–64, 2005.

20. Gupta, S., Annepu, S.K., Summuna, B., Gupta, M., Nair, S.A., Role of mushroom fungi in decolorization of industrial dyes and degradation of agrochemicals, in: *Biology of Macrofungi*, pp. 177–190, Springer, Cham, 2018.
21. Geethanjali, P.A. and Jayashankar, M., Efficiency of paddy residues as substrates for fungal laccase production. *Int. J. Theor. Appl. Sci.*, 10, 1, 190–193, 2018.
22. Das, N. and Das, R.K., Biological and environmental aspects of mushroom, in: *Mycoremediation and Environmental Sustainability*, pp. 39–61, Springer, Cham, Switzerland, 2019.
23. da Luz, J.M.R., Paes, S.A., Nunes, M.D., da Silva, M.C.S., Kasuya, M.C.M., Degradation of Oxo-Biodegradable Plastic by Pleurotusostreatus. *PLoS One*, 8, 8, 69386, 2013.
24. Cho, S.S. and Park, S.H., Biocontrol and bioremediation by the rhizosphere microflora of Tricholoma matsutake. *Appl. Microbiol. Biotechnol.*, 104, 6, 2391–2402, 2020.
25. Chaudhary, K., Agarwal, S., Khan, S., Role of phytochelatins (PCs), metallothioneins (MTs), and heavy metal ATPase (HMA) genes in heavy metal tolerance, in: *Mycoremediation and Environmental Sustainability*, pp. 39–60, Springer, Cham, 2018.
26. Brown, A.R. and Green, M.S., Nanobiopesticides from mushrooms in agricultural pest management: A comprehensive review. *Agric. Environ. Res.*, 28, 2, 89–105, 2021.
27. Bilal, M. and Asgher, M., Biodegradation of agro-wastes by the lignocellulolytic activity of an oyster mushroom, Pleurotus Sapidus. *J. Natl. Sci. Found. Sri Lanka*, 44, 4, 399–407, 2016.
28. Amaral, L., Martins, A., Spengler, G., Molnar, J., Efflux pumps of Gram-negative bacteria: what they do, how they do it, with what and how to deal with them. *Front. Pharmacol.*, 4, 168, 2014.
29. Novotny, L., Dvorska, L., Lorencova, A., Beran, V., Pavlik, I., Fish: a potential source of bacterial pathogens for human beings. *Vet. Med.-Czech*, 49, 9, 343–358, 2004.

4

Mass Production of Mushroom for Animal Feed

Konjerimam Ishaku Chimbekujwo[1], Ruth Ebunoluwa Bodunrinde[2], Oluwafemi Adebayo Oyewole[3], Charles Oluwaseun Adetunji[4*], Oluwabukola Atinuke Popoola[5], John Tsado Mathew[6] and Adeyomoye Olorunsola Israel[7]

[1]*Department of Microbiology, Modibbo Adama University, Yola, Adamawa State, Nigeria*
[2]*Department of Microbiology, Federal University of Technology, Akure, Ondo State, Nigeria*
[3]*Department of Microbiology, Federal University of Technology, Minna, Niger State, Nigeria*
[4]*Applied Microbiology, Biotechnology and Nanotechnology Laboratory, Department of Microbiology, Edo State University Uzairue, Iyamho, Edo State, Nigeria*
[5]*Genetics, Genomics and Bioinformatics Department, National Biotechnology Development Agency, Abuja, FCT, Nigeria*
[6]*Department of Chemistry, Ibrahim Badamasi Babangida University, Lapai, Niger State, Nigeria*
[7]*Department of Physiology, Ondo Medical University, Ondo State, Nigeria*

Abstract
Mushrooms are now grown in small numbers by renowned individuals. Formerly, the harvest was from forests, but now farmers are cultivating mushrooms for their nourishment. The Identification and production of mushrooms are necessary for the growth of spawn, substratum, and organic matter. Animal protein is not as easily synthesized as the high-quality proteins found in mushrooms. Besides, mushrooms can be grown on numerous organic wastes. Following the mushroom harvest, spent substratum residue can be utilized as animal feed. Animal farms primarily use feed which is a key for its high price, but the use of mushroom can help reduce the environmental impact by using mushrooms instead.

**Corresponding author*: adetunjicharles@gmail.com

Charles Oluwaseun Adetunji, Chukwuebuka Egbuna, Oluwatosin Ademola Ijabadeniyi and Samantha C. Karunarathna (eds.) Mushroom Biotechnology for Improved Agriculture and Human Health, (71–84) © 2025 Scrivener Publishing LLC

This review provides a view on the mass production of mushrooms for animal feeds, as it highlights the potential benefits of using their superior substratum for food manufacturing.

Keywords: Mushroom, production, animal feed, substrate

4.1 Introduction

The growth of bacteria, viruses, and insects on plants causes significant damage that reduces plant fertility and can even devastate crops. In the absence of pest control, farmers experience significant losses. Farmers have been facing crop damage and a decline in the quality and quantity of their farm produce for an extended period. The impact of pests and diseases on crop production has led to significant economic difficulties. Research has shown that mushrooms, which contain essential elements like fatty acids, proteins, sterols, amino acids, and carbohydrates, can be an excellent food source for animals [1, 2]. Solid-state fermentation is a crucial method for managing waste from agricultural products, which can help eases waste disposal and waste buildup problems [3]. The new method improved the use of complex cellulose polymers, including lignin and protein from corn, to produce carbon from microorganisms as various bioproducts that could be used as animal feed. Following the cultivation of Oyster mushroom at the ideal conditions, the harvested substrate can be utilized for animal feed [4].

The use of fungi proved to be highly efficient as they attract insects toward themselves, causing them to sprout and consume their internal organs until they perish. Biopesticides based on fungi can inhibit the growth of microbes. Local consumption accounts for 95% of China's mushroom production [5]. The wastes of banana leaves, bean trash, rice husk, cotton seed, and sawdust are used to produce massive mushroom mushrooms in Tanzania [6]. During the study of [7], only 1% of rice straw was utilized for animal feed. Despite the availability of ample land and capital, mushroom cultivation can be achieved by both rural and urban populations [8]. The cultivation of *Pleurotus* species is prevalent in Asia, Europe, and America due to their low production cost and high biological effectiveness. To properly form fruiting bodies, the Oyster mushroom species requires a humid environment of 80–90% and temperatures in the temperate region of 25–30°C [9]. Agricultural waste is a valuable resource for cultivating mushrooms, even when it is in large quantities [10–12]. Ajila *et al.* [13] revealed that animal feed can be enriched with lignocellulosic materials by adding essential nutrients, which can improve the health of animals fed with such feed.

The scarcity of food for humans and animals is a significant global issue. The extensive cultivation of mushrooms can offer employment and help reduce poverty [14]. To develop fruiting bodies, species of *Pleurotus* require high humidity and temperatures between 25-30°C. The surplus substrate from harvesting can be utilized to condition the soil for planting and also for animal feed manufacturing [15, 16]. According to VanGriensven [17] study, the *Pleurotus* species serves two functions: as a food source for animals and as an ingredient in pharmaceuticals. Oseni *et al.* [18] study the pretreatment of the substrate for the mass manufacturing of edible mushrooms has resulted in a mushroom crop with significant biological value through experimentation. Nutraceutical and nutritional additives are manufactured using current technological methods for the large manufacture of mushroom [19, 20]. Chang *et al.* [21] reported bulk manufacturing is possible with the presence of sufficient lingocellulosic materials as the mushroom substratum. Fruiting bodies, including the mycelium left behind, are a significant source of protein for animals' protein production [22].

4.2 Mushroom

The mushroom family is composed of sporocarps. The sexual reproductive stage of many fungi is linked to the visible structures that carry out spores [23]. The sustenance of mushrooms is largely due to their low toxicity and lack of resistance to nutrients, as well as their high levels of proteins, vitamins, minerals, and other essential nutrients [24]. The absence of green pigments prevents them from making their own food. Mushroom is commonly found in temperate and tropical areas, of various types [25]. Consumable mushrooms are highly protein food origin that can serve as supplementary meat sources due to their weather conditions which affects production. Moreover, because of their high content in several health-promoting elements, e.g., β-glucans, peptides and proteins, as well as phenolic compounds [26] are antitumor, antibacterial, co-enzymatic, or cytostatic properties, therefore they are also known as "medicinal mushrooms" [27]. Among them are filled mushroom, pore mushroom (four types), tooth fungi, and toadstool. The most common name for fungi that are edible is mushroom, while toadstools are those that can be poison and easily excreted. Over fifty species of mushrooms are cultivated for commercial purposes. Among the genera, four of them are *Lentinula (L.). edodes* (known as shiitake), the *Pleurotus*, *Auricularia*, and *Agaricus*, respectively, represent 74% of the world's total [28].

The growth of mushrooms is not a quick process, as suggested by Chang *et al.* [25], but rather takes ages to occur. A fungus grows underground, but most of it remains invisible. Moisture-tinted thread-like filaments make up the mycelium, the underground body of a fungus. Primitive hyphae are created in favorable growing conditions. With the growth of individual primordial, the hyphae inside grow and divide. A mushroom's stem and cap will be formed by one part. When the primordium becomes large enough, the stem grows longer and pushes the cap up on top of the ground. As the stem grows longer, the cap expands, resembling an umbrella that is unfolding [25].

4.3 Mushroom Production

Spawn

Spore is the same as seed to mushroom. In contrast to spore, spawn is so far sprouting on its own material such as sorghum, barley, or sawdust. The length of time mushrooms commences with spores, but growers' prey on mycelial origin spawn instead of sporadically due to the differences and changes. The standard of seed is the main element for crop victory. Hence, growers must use suitable spawn for commercial purposes. The strain traits of the birds should be preserved by subcultures that propagate them. Variation and mating are among the genetic techniques used to create new strains. Grain, sawdust, plug, and liquid are the different types of mushroom spawn [29].

Substrate

The tropical pattern of mushrooms can be divided into three groups: saprophytes, parasites, or mycorrhizae. The most commonly grown mushrooms are saprophytes, which are decomposers that thrive on organic materials such as wood, leaves, and straw in the natural world. Oyster mushroom and enokitake are prime breakers that can be used as organic matter for the substrate. Nevertheless, secondary breakers like button mushroom or straw mushroom need the substrate to be broken down by bacteria (or other fungi). Carbon, nitrogen, and synthetic compounds are the primary nutrient sources of mushrooms, with cellulose, hemicelluloses, or lignin being the major sources. Hence, mushrooms can be made from living matter that contains cellulose, hemicellulose, or lignin. The amount of nutrients required by different mushroom varieties varies. *Agaricus bisporus* requires a high nitrogen source, which is why the ideal C/N ratio for button mushroom manure should be 17. In contrast, oyster mushroom and shiitake necessitate a more abundant nitrogen source [29].

Mycelia of mushroom release enzymes into the substrate to break down and it takes in the soluble nutrients. While cellulose, the primary source of nutrients for mushrooms is one of the most abundant organic materials on earth, its digestive enzyme, cellulase, is also owned by various microbes, like fungi. The significance of mushrooms as a food source is evident. Only mushrooms dissolve cellulose and allow it to uptaken and transformed into food for humans. Acidity in the substrate also affects mushroom. The substrate's pH value can range from 6 to 8 optimally which varies with mushroom species [29].

Environment

The growth of mushrooms necessitates a suitable surrounding for both their vegetative and reproductive life cycles. Fungi are susceptible to damage from their environment, as they lack protection through a skin layer. Controlling the state of growing mushrooms can determine the success or failure of mushroom production. The surrounding in which mushrooms are grown is affected by light, temperature, air and humidity. The levels of them at the vegetative stage are not the same as those at their reproductive stage. The survival temperature of mushroom mycelium varies from 5 to 40°C depending on the species. The mycelia thrive in temperatures ranging from 20 to 30°C. At 10 to 20°C, pins start to form, which is 10 °F less than mycelial growth. More than 80% of the body liquid is composed of water. A minimum of 60–75% of moisture should be present in the substrate [29]. At the early, middle, and late stages of fruiting, relative humidity levels vary from 80 to 95%. Mycelia can grow in the absence of light, but some species need it for fruit body formation. As aerobic fungi, mushrooms require air to grow, while ventilation is more necessary for reproducing. The most crucial factor among the three is environmental dominance. Creating the ideal state for each growth stage and species is crucial in producing high-quality mushrooms [29].

4.3.1 Mass Production of Mushrooms for Animal Feed

1. Substrate Preparation: Select an appropriate agricultural waste as the substrate for growing mushrooms. Common agricultural wastes used include straw, corn cobs, sawdust, and wheat bran. These agricultural wastes should be collected, dried, and ground into smaller particles to create a substrate that provides the necessary nutrients for mushroom growth [30].

2. Sterilization: To prevent contamination and promote the growth of mushrooms, the substrate needs to be sterilized. This can be done using heat or steam sterilization methods. Heat the substrate to a temperature of around 70-80°C for 2-3 hours to kill any unwanted microorganisms [30].
3. Inoculation: Once the substrate is sterilized and cooled, it is ready for inoculation. Add mushroom spawn to the substrate. Mushroom spawn is a pure culture of mycelium, the vegetative part of the mushroom. The spawn can be obtained from a reputable supplier or generated in a laboratory [31].
4. Mixing and Filling: After inoculation, thoroughly mix the spawn with the substrate to ensure even distribution. Pack the substrate-spawn mixture into containers or bags, maintaining a suitable depth and density for mushroom growth [31].
5. Incubation: Place the containers or bags in a controlled environment with specific temperature and humidity conditions. Mushrooms require a cool and dark environment for the mycelium to colonize the substrate. Maintain the ideal incubation temperature, typically around 20-25°C, and high humidity levels [31].
6. Fruiting: After the mycelium has fully colonized the substrate, initiate the fruiting phase. Transfer the containers or bags to a fruiting chamber with controlled conditions of temperature, humidity, and light. Adjust the temperature to 15-20°C and maintain humidity levels around 80-90%. Provide suitable lighting conditions, generally low or indirect light [31].
7. Harvesting: Mushrooms will start to grow and mature within a specific period, dependent on the mushroom species and ecological parameters. Harvest the mushrooms when they reach the desired size and maturity. Cut or twist the mushrooms at the base to remove them from the substrate [31].
8. Drying and Processing: After harvesting, the mushrooms can be dried to increase their shelf life and preserve their nutritional value. Use a food dehydrator or other drying methods to remove moisture from the mushrooms. Once dry, the mushrooms can be ground into a powder or used whole as animal feed [31].

4.3.2 Mushroom Substrate (Spent) for Feed

Demand for producing animal feed is expected to rise, necessitating the feed industry to explore alternative options to meet the demand. Using appropriate biological resources could help drive this path by making readily available material as a feed additive. The breakdown of substrate polymers and the increase in digestibility of plant residues occur during solid-state fermentation by fungi (mushroom). Consequently, the development of mycelial biomass improves its concentrate of proteins and biological active elements on the substrate, e.g., polysaccharides and ergosterol [32]. The quality of feed rations can be improved by using compounds of interest from mushroom species. *L. edodes* contains 151 mg of provitamin D2 ergosterol. In contrast to ergosterol equivalents of 6 mg/100g [32], SMS from different species contain significant amounts of polysaccharides, such as -glucans [33]. As a result, the high nutritional content of SMS makes it primarily available in poultry, ruminants, and monogastric animals, as well as presently in fish and edible insects.

4.3.3 Mushroom Substrate (Spent) for Poultry

The blending of spent mushroom substrate obtained by *P. eryngia*, *P. ostreatus*, and *H. marmoreus* intake at 5 to 15 percent weight per weight can increase poultry feed intake without negatively impacting egg production or the weight of valuable meat [34]. Nevertheless, including *Agaricus blazei* spent mushroom substrate at rates of more than 0.4 percent weight per weight, broiler chickens experienced a gradual decrease in weight gain, while those with only 0. 2% demonstrated the most effective weight gain and feed change [35], as well as the highest feed intake. Likewise, the *Pleurotus sajor-caju* spent mushroom substrate ratio is low in terms of inclusion. SMS alerts (within the range of 0. 67 percent), the weight gain of broiler chicks in the first 21 days was boosted [35].

4.3.4 Mushroom Substrate (Spent) for Ruminants

More research has been conducted on the utilization of SMS as a feed for ruminants. The daily intake of ruminants was supplemented with SMS from different mushroom species at a rate of up to 30% (w/w), which indicated its ability as an additive to normal feeds without altering various important parameters. A feed that includes up to 20 percent weight per volume of *A. bisporus* SMS preferably from sheep fed for three weeks did not influence the uptake of nutrient, digestion, and nitrogen balance [36]. In *P. ostreatus* SMS and sika deer for 60 days, the intake of organic matter and digestibility of crude fibers were reduced in *P. ostreatus* SMS

(10 percent weight per volume, but no impact on feed intake, apparent nutrient digestible, velvet antler production, or biochemical indexes. These animals were given velutipes spent mushroom substrate of 10 percent weight per volume as food [37]. Typically, species of *Agaricus* and *Pleurotus* are grown on straw substrates, while *L. edodes*, *G. lucidum*, *Gr. Frondosa*, *He. Inertidaceus* were grown on wood as nutrients. Such substrates are not readily incorporated into animal feed and require additional treatment to enhance their nutritional properties. Recently, fermentation using probiotic microorganisms has become a popular solution for reducing fibrous ingredients and increasing the protein content of SMS, including those derived from sawdust-based media, as it is an efficient, cheap way to reduce costs. Furthermore, probiotic bacteria provide comfort to animals experiencing stress-induced seizures, regulate the intestinal microbiome, and minimize diarrheal episodes [38]. SMS requires prompt processing due to its rapid decomposition caused by its high moisture content. Ensiling, such as lactic acid fermentation under anaerobic conditions [38], can be utilized to achieve this. Lactic acid bacteria create wanted metabolism products inhibit development of harmful microorganisms such as clostridia [39].

4.4 Benefits of Feeding Animals with Mushrooms

1. Nutritional Composition: Mushrooms are known to be low in fat, free of cholesterol, and contain dietary fiber, vitamins, minerals, and antioxidants. They are rich in protein, essential amino acids, and carbohydrates. Mushrooms also provide B vitamins, such as riboflavin and pantothenic acid, as well as minerals such as potassium, phosphorus, and selenium.
2. Improved Palatability: Adding mushrooms to animal feed can enhance palatability and encourage consumption. Animals tend to enjoy the taste and aroma of mushrooms, making them more likely to eat their feed.
3. Digestibility: Mushrooms are easily digestible and can provide highly bioavailable nutrients. This can be particularly beneficial for young or convalescing animals with higher nutritional requirements.
4. Immune System Support: Mushrooms contain bioactive compounds such as beta-glucans, which are known for their immunomodulatory effects. These compounds can help support and boost the immune system of animals, enhancing their resistance to diseases and infections.

5. Antioxidant Properties: Mushrooms are rich in antioxidants, including phenolic compounds and selenium. These antioxidants can help protect cells from oxidative stress and reduce the risk of chronic diseases in animals.
6. Gut Health Promotion: Certain types of mushrooms, such as medicinal mushrooms like Reishi or Shiitake, contain prebiotic properties that can support a healthy gut microbiota in animals. A balanced and diverse gut microbiota is essential for efficient digestion and nutrient absorption.
7. Environmental Sustainability: Utilizing mushrooms as animal feed can contribute to environmental sustainability. Mushroom production can make use of agricultural waste products that would otherwise go to waste, reducing the environmental impact of these materials [40].

Several studies on broiler chickens were conducted in the past. Mushrooms have a progressive impact on the production of chicken as they help to increase them in growth performance, increase microbiota status in the gut, exhibit immunonomodulatory activity, exert tissue antioxidant potential, influence intestinal development, and increase lipid profile. Extracts from mushroom-cultivated waste media, especially those from sweet potatoes, contain great enzymatic potentials and are very effective in lowering neutral detergent fiber, suggesting that they could be utilized as enzyme condiments for animal feeding [41]. Overall, incorporating mushrooms into animal feed can provide nutritional benefits, enhance palatability, and support the overall health and well-being of animals.

4.5 Conclusion

The production of edible and medicinal mushrooms is an important trading, with notable advancements in recent years. Several low-cost agricultural products can be used to grow mushrooms, which provides relatively low-price benefits. Nigeria's mushroom growth is seemingly a feasible option for expansion. Increasing the production of mushrooms results in high levels amount of spent mushroom substrate (SMS). SMS is a substrate component in new mushroom cultivation cycles because it provides valuable nutrients and energy sources for such processes. The study reports that SMS can serve as a source of products for plant diseases, has medicinal benefits for animals fed with mushrooms, and enhances the production of secondary products in plants. Given the intricate and laboratory-based process of spawn production, it is recommended that the government utilizes existing laboratories to produce and distribute seeds in the area. Some institutions are already providing seeds for cereals, vegetables, and fruits to farmers. As a result, spawn can be included in their current system.

References

1. Kozarski, M., Klaus, A., Vunduk, J., Zizak, Z., Niksic, M., Jakovljevic, D., Vrvic, M.M., Van Griensven, L.J., Nutraceutical properties of the methanolic extract of edible mushroom Cantharellus cibarius (Fries): primary mechanisms. *Food Funct.*, 6, 6, 1875–86, 2015.
2. Muszyńska, B., Kała, K., Firlej, A., Sułkowska-Ziaja, K., Cantharellus cibarius-Culinary-Medicinal Mushroom Content and Biological Activity. *Acta Pol. Pharm.*, 73, 3, 589–98, 2016.
3. Oliveira, M. and Duarte, E., Integrated approach to winery waste: waste generation and data consolidation. *Front. Environ. Sci. Eng.*, 10, 1, 168–176, 2014, https://doi.org/10.1007/s11783-014-0693-6.
4. Fufa, B.K., Tadesse, B.A., Mestawot, M.T., Cultivation of *Pleurotus* ostreatus on Agricultural Wastes and Their Combination. *Int. J. Agron.*, 2021, 6, 2021, Article ID 1465597, https://doi.org/10.1155/2021/1465597.
5. Zhang, Y., Geng, W., Shen, Y., Wang, Y., Dai, Y., Edible mushroom cultivation for food security and rural development in China: bio-innovation, technological dissemination and marketing. *Sustainability*, 6, 2961–2973, 2014.
6. Mamiro, D.P. and Mamiro, P.S., Yield and mushroom size of *Pleurotus* ostreatus grown on rice straw basal substrate mixed and supplemented with various crop residues. *J. Anim. Plant Sci.*, 10, 1211–1218, 2011.
7. Rosmiza, M.Z., Davies, W.P., Rosniza Aznie, C.R., Mazdi, M., Jabil, M.J., Farmers' knowledge on potential uses of rice straw: An assessment in MADA and Sekinchan, Malaysia. *Malays. J. Soc. Space*, 10, 30–43, 2014.
8. Diriba, D., Gume, B., Abate, D., Evaluation of locally available substrates for cultivation of oyster mushroom (*Pleurotus ostreatus*). *Afr. J. Microbiol. Res.*, 7, 20, 2228–2237, 2013.
9. Hoa, H.T. and Wang, C.L., The effects of temperature and nutritional conditions on mycelium growth of two oyster mushrooms (*Pleurotus ostreatus* and *Pleurotus cystidiosus*). *Mycobiology*, 43, 1, 14–23, 2013.
10. Onyango, B.O., Palapala, V.A., Arama, P.F., Wagai, S.O., Gichumu, B.M., Sustainability of selected supplemented substrates for cultivation of Kenyan native wood ear mushroom (*Auricularia auricular*). *Am. J. Food Technol.*, 6, 403, 2011.
11. Earnshaw, D.M., Diamini, B.E., Masarirambi, M.T., Growth and yield of oyster mushroom (*Pleurotus ostreatus*) grown on different substrates amended with varying levels of wheat bran. *Int. J. Life Sci.*, 1, 4, 111–117, 2012.

12. Adetunji, C.O. and Isaac Oluseun Adejumo, I.O., Potency of agricultural wastes in mushroom (*Pleurotus sajorcaju*) biotechnology for feeding broiler chicks (Arbor acre). *Int. J. Recyl. Org. Waste Agric.*, 8, 37–45, 2019, https://doi.org/10.1007/s40093-018-0226-6.
13. Ajila, C.M., Brar, S.K., Verma, M., Tyagi, R.D., Godbout, S., Valéro, J.R., Bio-processing of agro-byproducts to animal feed. *Crit. Rev. Biotechnol.*, 32, 1–19, 2012.
14. Khan, A.N., Ajmal, M., Inamulhaq, M., *et al.*, Impact of sawdust using various woods for effective cultivation of oyster mushroom. *Pak. J. Bot.*, 44, 1, 399–402, 2012, https://doi.org/10.3109/07388551.2012.659172.
15. Shah, Z.A., Ashraf, M., Ishtiaq, C., Comparative study on cultivation and yield performance of oyster mushroom (*Pleurotus ostreatus*) on different substrates (wheat straw, leaves, saw dust). *Pak. J. Nutr.*, 3, 3, 158–160, 2004.
16. Samuel, A.A. and Eugene, T.L., Growth performance and yield of oyster mushroom (*Pleurotus ostreatus*) on different substrates composition in buea south west Cameroon. *Sci. J. Publ.*, 1, 1–6, 2012.
17. Van Griensven, L.J.L.D., *Agaricus bisposus* (J.Lge) and its close relatives: Their present status, use and future in commerce and research. *Int. J. Med. Mushrooms*, 3, 81, 2001.
18. Oseni, T.O., Dlamini, S.O., Earnshaw, D.M., Masarirambi, M.T., Effect of substrate pre-treatment methods on oyster mushroom (*Pleurotus ostreatus*) production. *Int. J. Agric. Biol.*, 14, 2, 251–255, 2012.
19. Chandra, P.K., Neelam, K., Bllsha, Ram, K., Yadav, P., Cultivation of *Pleurotus sajor-caju* using different agriculture residues. *Int. J. Agric. Policy Res.*, 1, 19–23, 2013.
20. Adebayo, E.A. and Martinez-Carrera, D., Oyster mushrooms (*Pleurotus*) are useful for utilizing lignocellulosic biomass. *Afr. J. Biotechnol.*, 14, 1, 52–67, 2015.
21. Chang, S.T., The World Mushroom Industry: Trends and technological development. *Int. J. Med. Mushrooms*, 8, 297–314, 2006.
22. Baig, S., Saleem, M., Zia, S., Cultivation of oyster mushroom on different agricultural wastes. *J. Biosci. Biotechnol. Discov.*, 4, 6, 133–136, 2019, ISSN: 2536-7064, https://doi.org/10.31248/JBBD2018.072.
23. Maurice, S., Arnault, G., Nordén, J., *et al.*, Fungal sporocarps house diverse and host-specific communities of fungicolous fungi. *ISME J.*, 15, 5, 1445–1457, 2021, doi: 10.1038/s41396-020-00862-1.
24. El-Ramady, H., Abdalla, N., Badgar, K., *et al.*, Edible mushrooms for sustainable and healthy human food: nutritional and medicinal attributes. *Sustainability*, 14, 9, 4941, 2022, doi: 10.3390/su14094941.
25. Chang, S.-T. and Miles, P.G., *Mushrooms: Cultivation, Nutritional value, medicinal effect and Environmental Impact*, CR Press, Boca Raton, Florida, 2004.
26. Carrasco-González, J.A., Serna-Saldívar, S.O., Gutiérrez- Uribe, J.A., Nutritional composition and nutraceutical properties of the *Pleurotus*

fruiting bodies: Potential use as food ingredient. *J. Food Compos. Anal.*, 58, 69–81, 2017, doi: 10.1016/j.jfca.2017.01.016.
27. Gargano, M.L., van Griensven, L.J.L.D., Isikhuemhen, O.S., *et al.*, Medicinal mushrooms: Valuable biological resources of high exploitation potential. *Plant Biosyst. - Int. J. Dealing Aspects Plant Biol.*, 151, 3, 548–565, 2017, doi: 10.1080/11263504.2017.1301590.
28. Kumla, J., Suwannarach, N., Sujarit, K., *et al.*, Cultivation of mushrooms and their lignocellulolytic enzyme production through the utilization of agro-industrial waste. *Molecules*, 25, 12, 2811, 2020, doi: 10.3390/molecules25122811.
29. Ndem, U.J. and Oku, M.O., Mushroom Production for Food Security in Nigeria. *Food Sci. Qual. Manage.*, 48, 44–50, 2016.
30. Ruihong, Z., Xiujin, L.I., Fadel, J.G., Oyster mushroom cultivation with rice and whear straw. *Bioresour. Technol.*, 82, 277–287, 2002.
31. Chauhan, S.S., Celi, P., Lenry, B.J., Clarke, I.J., Dushea, F.R., Dietary antioxidants at supranutritional dose improve oxidative status and reduce the negative effects of heat stress in sheep. *J. Anim. Sci.*, 92, 3364–3374, 2014, https://doi.org/10.2527/jas.2014-7714.
32. Boontiam, W., Wachirapakorn, C., Wattanachai, S., Growth performance and hematological changes in growing pigs treated with Cordyceps militaris spent mushroom substrate. *Vet. World*, 13, 4, 768–773, 2019, doi: 10.14202/vetworld.2020.768-773.
33. Hsieh, Y.C., Lin, W.C., Chuang, W.Y., *et al.*, Effects of mushroom waster medium and stalk residues on the growth performance and oxidative status in broilers. *Anim. Biosci.*, 34, 2, 265–275, 2021, doi: 10.5713/ajas.19.0889.
34. Foluke, A., Olutayo, A., Olufemi, A., Assessing spent mushroom substrate as a replacement to wheat bran in the diet of broilers. *Am. Int. J. Contemp. Res.*, 4, 4, 178–183, 2014.
35. Azevedo, R.S., Ávila, C.D.S., Dias, E.S., *et al.*, Utilization of the spent substrate of *Pleurotus sajor caju* mushroom in broiler chicks ration and the effect on broiler chicken performance. *Acta Sci. - Anim. Sci.*, 31, 2, 139–144, 2009.
36. Fazaeli, H. and Masoodi, A.R.T., Spent wheat straw compost of *Agaricus bisporus* mushroom as ruminant feed. *Asian-Australas. J. Anim. Sci.*, 19, 6, 845–851, 2006, doi: 10.5713/ajas.2006.845.
37. Yuan, C., Wu, M., Tahir, S.M., *et al.*, Velvet antler production and hematological changes in male Sika deers fed with spent mushroom substrate. *Animals (Basel)*, 12, 13, 1689, 2022, doi: 10.3390/ani12131689.
38. Barba-Vidal, E., Martín-Orúe, S.M., Castillejos, L., Practical aspects of the use of probiotics in pig production: A review. *Livest. Sci.*, 223, 84–96, 2019, doi: 10.1016/j.livsci.2019.02.017.
39. Arriola, K.G., Queiroz, O.C.M., Romero, J.J., *et al.*, Effect of microbial inoculants on the quality and aerobic stability of bermudagrass round-bale haylage. *J. Dairy Sci.*, 98, 1, 478–485, 2015, doi: 10.3168/jds.2014-8411.

40. Yamauchi, M., Imaya, R., Masuda, S., Yamada, M., Kihara, M., Yoneyama, K., Harada, H., Study on the development of mushroom cultivation technology using dried shochu lees from sweet potato. *Environ. Eng. Res.*, 42, 545–553, 2005, https://doi.org/10.11532/proes1992.42.545.
41. Yagi, F., Minami, Y., Yamada, M., *et al.*, Development of animal feeding additives from mushroom waste media of shochu lees. *Int. J. Recycl. Org. Waste Agric.*, 8, 215–220, 2019, https://doi.org/10.1007/s40093-018-0234-6.

5

Application of Mushrooms in Management of Soil-Borne Parasites, Nematodes, Bacteria and Fungi

Aisha Bisola Bello[1*], Abdullahi Idris Dabban[1,2], Wuna Muhammad Muhammad[2], Adioha Amarachi[1], Joseph Odey Oko[3], Hammed Opeyemi[4], Oluwafemi Adebayo Oyewole[2] and Charles Oluwaseun Adetunji[5]

[1]*Department of Biological Sciences, Federal Polytechnic Bida, Niger State, Nigeria*
[2]*Department of Microbiology, Federal University of Technology, Minna, Niger State, Nigeria*
[3]*Department of Science Laboratory Technology, Nigerian Institute of Leather and Science Technology, Zaria, Kaduna State, Nigeria*
[4]*Department of Physiology, Ladoke Akintola University of Technology, Ogbomoso, Oyo State, Nigeria*
[5]*Department of Microbiology, Edo State University, Uzairue, Edo State, Nigeria*

Abstract

There are numerous microorganisms that are connected to plants, despite that, the majority of organisms affect the root system and are soil-borne pathogens. These pathogens live within the same region of the rhizosphere, causing damage to the host plant's vascular system, and obstruct physiological procedures involved in water and nutrient intake required for maximum productivity. Therefore, it is imperative to address pathogenic soil-borne organisms. The traditional chemical methods of disinfection have been used in the past, however, due to the high toxic level of the chemicals, it is now required to develop alternatives that are ecologically acceptable and environmentally friendly. The use of mushrooms like *Pleurotus ostreatus* can be an alternative biological control agent, especially with the current shift toward non-chemical or green methods for control such as biological control, biopesticides, and cultural management to manage

**Corresponding author*: aishabisola.bello@gmail.com

Charles Oluwaseun Adetunji, Chukwuebuka Egbuna, Oluwatosin Ademola Ijabadeniyi and Samantha C. Karunarathna (eds.) Mushroom Biotechnology for Improved Agriculture and Human Health, (85–122) © 2025 Scrivener Publishing LLC

disease-causing plant pathogens. These mushrooms possess a range of beneficial qualities that enable them to effectively mitigate the damage inflicted by pests and pest-related diseases. The mycoparasitic, antibiotic, and pest-repelling properties of these mushrooms, along with their ability to act as antifeedants, produce bioactive substances, and exhibit entomopathogenic activity, all directly contribute to pest control. Moreover, they enhance the plant's capacity to uptake nutrients and water, and they can alter root morphology and rhizosphere interactions, favoring overall plant growth and reducing the negative impacts of competitive pressure for space and resources. Therefore, using mushrooms as biological control agents against harmful soil-borne microbes for agricultural purposes is a potential sustainable biocontrol technique.

Keywords: Biocontrol agents, mushrooms, microorganisms, rhizosphere, soil-borne pathogens

5.1 Introduction

The achievement of safety and security of food for the constantly increasing worldwide human population is undoubtedly a fundamental goal that is possible majorly through sustainable agriculture [1]. However, the imbalance between decreasing food production and increasing human population raises a foremost concern. One of the ways to tip the scales in favour of food production is the control pathogens of plants found in the soil [2]. The biology as well as microbiology of the soil have a direct connection to agricultural sustainability. The roles and connections of all the components of the soil including the soil air, water, organic and inorganic matters as well as microorganisms play key roles in agricultural productivity [3]. Soil microbial communities play a significant role in influencing soil fertility, productivity, and, to a larger extent, plant yield, either enhancing or diminishing these factors [3]. One of the primary factors contributing to decreased crop yields is the presence of plant pathogens in the soil, including parasitic organisms like nematodes, bacteria, and fungi [4].

Soil microbial communities play a significant role in influencing soil fertility, productivity, and, to a larger extent, plant yield, either enhancing or diminishing these factors. One of the primary factors contributing to decreased crop yields is the presence of plant pathogens in the soil, including parasitic organisms like nematodes, bacteria, and fungi [3, 4]. On a global scale, about 26% of losses accrued in the agricultural sector which sometimes arise from total failure of crops are attributed to plant diseases [4].

Parasitic nematodes, bacteria, fungi, and viruses (depicted in Figure 5.1) are naturally present in soil ecosystems, where they can impact and induce

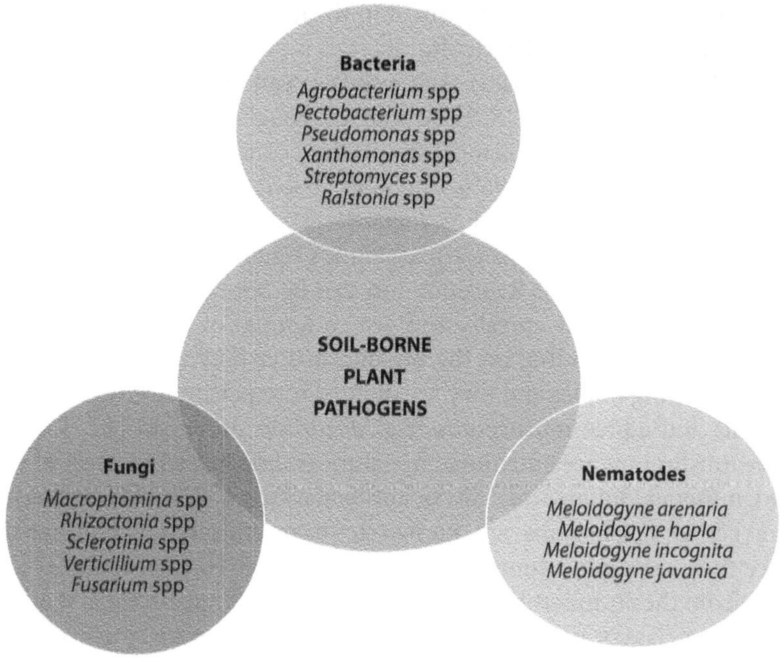

Figure 5.1 Examples of some soil-borne plant pathogens.

diseases in a wide array of host plants, including weeds. These microorganisms are known for their ability to form enduring survival structures and mechanisms [5]. Notable examples of soil-borne bacterial pathogens affecting plants come from genera such as *Agrobacterium, Pectobacterium, Pseudomonas, Xanthomonas, Streptomyces,* and *Ralstonia* [3, 5]. Although, most species of *Pseudomonas* and *Xanthomonas* are usually found transiently present in the soil. Meanwhile, billions of dollars is lost on a yearly basis as a result of spoilage of agricultural produce by several fungal species from the Genus *Macrophomina, Rhizoctonia, Sclerotinia, Verticillium,* and *Fusarium*. This can be attributed to their constant persistence in the soil for a prolonged time due to the presence of some adaption mechanisms such as sclerotia, oospores, and chlamydospores [5].

Over 4100 species of nematodes from various genera that are parasitic to plants have been successfully identified [1]. In the agricultural sector, yearly, the overall financial loss exceeding 200 billion dollars has been attributed to this group of soil-borne organisms [6]. Among the parasitic nematodes of plants, species of *Meloidogyne* (Root-knot nematodes) are undeniably the most economically significant pest of many important crops globally [7]. The primary threat in the devastation of cultivated vegetables, horticultural, and agricultural crops is posed by species within this group (Root-knot nematodes). These species can target over 3,000 different susceptible plant hosts [8, 9]. Among the roughly 100 known species of *Meloidogyne* worldwide, the most significant economic loss in agricultural production, approximately 95%, can be attributed to *Meloidogyne arenaria, M. javanica, M. incognita,* and *M. hapla* [1, 10, 11].

Mushrooms or toadstools are filamentous fungi, macroscopic, and chemosythetic saprophytes. Reproduction can be sexual or asexual (sporulation) [1]. They grow apically and are composed of chitin or cellulosic cell wall [12]. Depending on the organic matter they decay, mushrooms develop fruiting bodies and they are divided into poisonous (e.g., *Amanita abrupta*) and edible mushrooms (e.g., *Pleurotus ostreatus*) [12, 13]. The edible mushrooms are nutritional in nature as they are rich sources of vitamins, minerals, amino acids, fibers, antioxidants, etc. [14]. Polysaccharides from edible mushrooms are known to have antimicrobial properties; hence they are a good source of valuable antimicrobial compounds [15]. Aside from the antimicrobial-producing capacity of mushrooms, they also belong to a class of fungi that has the ability to secrete specific toxins and certain enzymes capable of paralyzing and killing nematodes [16].

In light of the global shift from conventional chemical techniques and the current focus on ecologically friendly green techniques for the prevention and management of disease-causing pathogens of crops especially by fungi, nematodes, and bacteria, this chapter therefore focuses on the groups of soil-borne pathogens including fungi, bacteria and nematodes; the application of mushrooms as natural bio-control agents and their mechanisms of actions against soil-borne plant pathogens. The advantages, challenges, and limitations of mushrooms as bio-control agents will also be highlighted.

5.2 Soil-Borne Parasites, Nematodes, Bacteria, and Fungi

Soil is a complex habitat that shelters diverse organisms [17], including bacteria, fungi, protozoans, nematodes, mites, collembolans, annelids (enchytraeids and earthworms), and microarthropods [18, 19]. Among this rich diversity of soil inhabitants are parasitic species that can potentially affect plants, animals, and even humans [20, 21]. Soil-borne plant-parasitic organisms feed on plants, depending on them for their sustenance often resulting in harm to the host plant. They are responsible for major constraints in food crop production worldwide, inflicting significant economic losses in the process [22, 23]. Soil-borne parasites can belong to several taxonomic groups including bacteria, fungi, viruses, insects, and other pests [3, 24–30].

The most effective parasites in nature are thought to be plant-parasitic nematodes [31]. They have been studied in various places across different continents and are adapted to survive numerous conditions making them difficult to control. Plant-parasitic nematodes, as well as bacterial and fungal parasites, affect plant tissues, causing diseases including wilt, leaf spots, cankers, rusts, root hypertrophy, black spot, smuts, collar rots, mildews, and blights resulting in symptoms such as wilting, discolouration, lesions, and abnormal growth [26, 32–35]. The control of soil-borne parasites depends on understanding the complex interactions between these plant-ravaging species and the soil ecosystem [36]. Cultural practices such as biofumiation, crop rotation, soil amendments, anaerobic soil disinfestation, soil solarization, and soil steam sterilization are widely adopted to mitigate the effect of soil-borne plant parasites and minimize economic losses [37]. However, these methods have demonstrated varying levels of effectiveness and show less dependability compared to chemical control methods which are not environmentally friendly [38].

5.2.1 Soil-Borne Nematodes

Nematodes, as described by [39], are minuscule, unsegmented, multicellular, slender roundworms that exhibit parasitic behavior toward plants, thriving in various environments, including soil. Despite their worm-like appearance, it is important to note, as indicated by Jalali and Khosravi [40], that nematodes are taxonomically distinct from earthworms, flatworms, or wireworms. These bilaterally symmetrical, soft-bodied roundworms do not possess a skeleton. They are predominantly found in the upper soil layer, particularly within decomposed organic matter, plant roots, and other organic-rich substrates. In these habitats, there is an estimated 1 to 100×10^6 individuals/m^2 of soil [41]. They are known as the most abundant metazoan organisms on Earth [42, 43]. Most nematode species are free living in soil or water. About 50% of them are marine and 25% dwell in soil and freshwater. Among the known nematode parasites about 15% are parasitic to animals, while about 10% are parasitic to crops. According to Pulavarty *et al.* [44], about 4100 species of plant-parasitic nematodes exist. These parasites, as highlighted by Kumar and Yadav [31], are considered some of the most efficient natural agents and pose a significant threat to farming, as emphasized by Garcia *et al.* [45]. Soil nematodes primarily consume various parts of plants, with a particular focus on roots [23]. The agricultural impact of these nematodes may be underestimated due to the non-specific symptoms that infected plants exhibit, which complicates the attribution of crop losses to nematode activity [26]. In addition, farmers especially in third-world countries, remain unaware of the presence of plant-parasitic nematodes [46]. Nematodes are one of the main issues restricting the growth of important crops, according to Ali *et al.* [47], who noted that they infest a wide variety of plants. Singh *et al.* [48] reported that an estimated yield loss of $157 billion is recorded worldwide due to soil nematodes while Carneiro *et al.* [49] reported an estimated annual crop loss of about $78 billion worldwide.

Plant-parasitic nematodes have diverse shapes and sizes and commonly display a cylindrical form that tapers at both the head and tail. They usually range from 250 μm to 12 mm in length and 15–35 μm in width [31, 50–52]. Nematodes can survive and persist in soil across diverse environmental conditions, they have been reported in various terrains across all continents. Thakur *et al.* [52] stated that once a field is infested by nematodes it cannot be completely eradicated from the soil. They disseminate from one field to another by any means that aids the movement of soil particles such as farm tools, birds, animals, rain, flooding, wind, insects, and human interventions [31]. The transplanting of infested plant or plant part also enables their dissemination. These parasites employ various survival tactics, with mobility within the deeper soil layers and the ability to invade and persist within host plants' tissues being significant survival strategies among them. All plant components, including the stems, seeds, roots, seeds, leaves, and flowers, are vulnerable to attack by plant-parasitic nematodes. They utilize specialized hollow and protrusible stylet or mouth spears to breach plant cells, extracting nutrients while also releasing a combination of proteinaceous effectors and non-proteinaceous compounds which enables the process of parasitism [53]. Once at a suitable host root, they alter the natural root physiology to establish a specialized nematode-feeding site known as syncytium [54]. However, a syncytium is not produced by some nematode species and in such cases, the parasitized plant cells usually die [51]. Many of plant-parasitic nematodes primarily target plant roots and can be categorized into two main groups, specifically ecto-parasites and endoparasites, based on their feeding habits, as documented by Holbein *et al.* [55] and Palomares-Rius *et al.* [51]. These parasitic worms can be further classified into migratory and stationary subgroups.

Ectoparasites, for the most part, are migratory in nature. They maintain their ability to move between different feeding locations on the plant without entering plant cells [53]. In contrast, sedentary ectoparasites remain outside the root and are immobile; they establish a feeding site within the host tissue (syncytium) and continue to feed for an extended period.

On the other hand, migratory endoparasites burrow into the root and can return to the soil to seek out new roots. Sedentary endoparasites enter the root and become immobile once they have established a feeding site within the host tissue [56, 57].

Nematodes that are parasites to other parts of the plant, such as leaves and stems, typically enter the root and migrate through plant tissues or water films to reach their desired feeding locations. It is important to note that the feeding habits of nematodes play a significant role in causing plant injuries and damage. Unlike stationary endoparasitic nematodes, which migrate solely within the root during their second-stage juvenile phase, migratory endoparasitic nematodes can mechanically damage the root system. Additionally, the movement and direct feeding of nematodes can create favorable conditions for secondary diseases, including bacteria and fungi, to infect the plant [29].

The life cycles of most plant-parasitic nematode species involve a parasitic stage that commences as the nematode starts feeding on host roots. Prior to this, there is a pre-parasitic stage, characterized as a free-living phase. During this stage, the second juvenile emerges from fertilized eggs and actively seeks a host cell, as discussed by Holbein *et al.* [58] and Mendoza-de Gives *et al.* [43]. The most commonly reported nematode species responsible for significant agricultural damage include Cyst nematode (*Heterodera* and *Globodera* sp.), Burrowing nematode (*Radopholus similis*), Root-knot nematode (*Meloidogyne* sp.), Root lesion nematode (*Pratylenchus* sp.), Spiral nematode (*Helicotylenchus* sp.), Bulb and stem nematode (*Ditylenchus dipsaci*), Reniform nematode (*Rotylenchulus reniformis*), ring nematode (*Mesocriconema xenoplax*), Dagger nematode (*Xiphinema* sp.), Bud and leaf nematode (*Aphelenchoides* sp.), and Pine Wilt nematode (*Bursaphelenchus xylophilus*) [45, 59–70].

The root-knot nematode (*Meloidogyne* sp.) is one of the most economically significant types of plant-parasitic nematodes. It is an obligate sedentary plant endoparasite responsible for a substantial portion of the annual losses, which totals around $100 billion due to worm damage [26, 37, 45, 63, 64, 71, 72]. These nematodes have a global distribution and impact a wide variety of plants, often engaging in intricate interactions with fungal and bacterial species that exacerbate plant diseases [8].

Their extended life cycles, dependent on factors like temperature, moisture, and the presence of a suitable host, can vary from a few days to several weeks for their development [73]. This group of parasites includes over 100 species, among which *Meloidogyne hapla*, *Meloidogyne javanica*, *Meloidogyne arenaria*, and *Meloidogyne incognita* are particularly prominent threats [74, 75]. *M. incognita* is recognized as the most economically impactful plant-parasitic nematode in warmer regions, including tropical and subtropical areas worldwide [76]. According to Subedi *et al.* [72], *M. incognita* alone is responsible for estimated annual losses of $100 billion worldwide.

Meloidogyne species are commonly known as root-knot nematodes due to the characteristic symptoms they induce, such as root galling (see Figure 5.2) [71]. Infection of plants by *Meloidogyne* results in the expansion of root cells, the formation of swellings, or the development of galls. Symptoms of such infections include signs of nutrient deficiency, like slow or stunted growth, wilting, yellowing of leaves, and, in some cases, sudden plant death [64].

Similar to the root-knot nematodes, *Heterodera* and *Globodera* species also known as cyst nematodes are a major agricultural and economic threat [8, 77, 78]. Cyst nematode species, the potato cyst nematodes *Globodera rostochiensis* and the soybean cyst nematode *Heterodera glycines* and *G. pallida*, and cereal cyst nematodes *Heterodera avenae* and *H. filipjevi* are

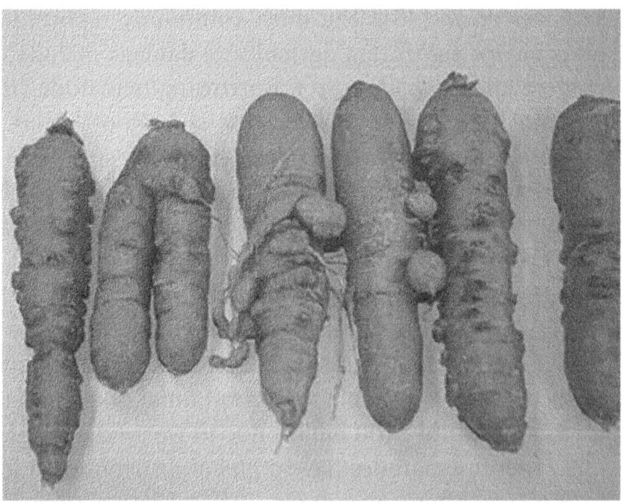

Figure 5.2 Carrots showing galls and damage symptoms caused by *Meloidogyne arenaria* and *M. incognita*. Source: [63].

the most significant in agriculture [77, 78]. *Helicotylenchus* sp. are migratory ecto-/endoparasite also known as spiral nematodes due to their bodies being wrapped into a spiral format [65].

Cultural, biological, and chemical approaches have been used to manage nematode infestations [37, 38]. Without using chemicals, cultural methods use a variety of approaches to manipulation to control nematode populations. Practices such as fallowing, planting of cover crops, crop rotation, and the utilization of nematode-resistant cultivars have been reported to significantly reduce most nematode populations [70, 79, 80]. Chemical control methods are a widespread practice and have been proved to be generally reliable and effective for controlling a variety of plant-parasitic nematodes [80]. However, its uses can result in soil contamination and pollution of the environment [38]. The negative effect of the chemical control methods has raised interest in alternative methods of nematode control [81]. Among these approaches, the biological control method has garnered substantial attention. The biological control method mitigates the impact of this plant-ravaging parasitise through the use of natural agents, such as the use of natural nematode predators, microorganisms, antagonistic plants, and plant extracts [8, 80]. Ali *et al.* [80] reported the positive results from the use of *Paecilomyces lilacinus* and *Pasteuria penetrans* and as biological control agents. *Trichoderma* species have also been observed to have nematicidal activity [61, 82–86]. Bhat *et al.* [73] and Tariq *et al.* [87] reported that many bacteria species including *Bacillus* spp., *Pseudomonas* spp., *Burkholderia* spp., and many others produce substances that then affect nematode behavior, feeding, and reproduction thereby minimizing nematode damage in plants.

5.2.2 Types of Soil-Borne Bacteria

Bacteria are major plant pathogens and there are over 200 plant pathogenic [88], with members in the genera of *Ralstonia*, *Pseudomonas*, *Xanthomonas*, *Xylella*, *Erwinia*, *Pectobacterium*, *Agrobacterium*, and *Dickeya* being the most important [25]. Bacteria are less abundant soil-borne pathogens than fungi [89]. Globally, plant diseases brought on by bacterial pathogens result in large annual losses. Without using chemicals, cultural methods use a variety of manipulative techniques to control nematode populations [90]. The use of pathogen-resistant crops or varieties that are less vulnerable, the use of chemical and/or biological controls, and cultural practices that all aim to reduce the inoculum are all necessary components of the control of bacterial plant pathogens [90].

The soil-borne bacterium *Ralstonia solanacearum* is the causative agent of bacterial wilt, Moko disease of banana, and brown rot of potato [27].

According to Tahat and Sijam [91] and Dutta *et al.* [37], *Ralstonia solanacearum* causes bacterial wilt disease in more than 180 plants of 45 families. Plant infection with *Ralstonia solanacearum* is usually followed by an unexpected wilt of the whole plant. In Moko disease and brown rot, bacterial colonization is visible on the banana fruit and potato tuber [92]. Fire blight disease is caused by the bacterial pathogen *Erwinia amylovora* [93]. This bacteria species is an important pathogen of the *Rosaceae* family of plants, causing substantial economic loss in many countries worldwide [94]. It infects the host plant via natural openings and wounds causing localized necrosis that does not radiate to adjacent plant tissues [95]. The control of fire blight disease in plants is difficult due to the absence of synthetic compounds that directly target the pathogen. Koczan *et al.* [96] and Aćimović *et al.* [94] noted that once *E. amylovora* penetrates the host xylem or cortical parenchyma and spreads in the endophytic phase of pathogenesis, external control methods become ineffective. The use of antibiotic streptomycin has been reported to have a limited effect due to the development of resistance in *Erwinia amylovora* and non-target bacterial population in the field [97]. Similarly, alternative control methods such as the use of plant resistance inducers and biological control agents showed limited effectiveness in managing fire blight [98]. Bacterial species in the genus *Dickeya* are phytopathogens that cause severe disease in a wide range of plant species particularly potato and bulb plants [99]. They cause soft rot and blackleg diseases in several crops [100]. *Dickeya solani* and *Dickeya dianthicola* are the two main species responsible for various plant diseases, leading to significant economic losses (Figure 5.3) [26, 101].

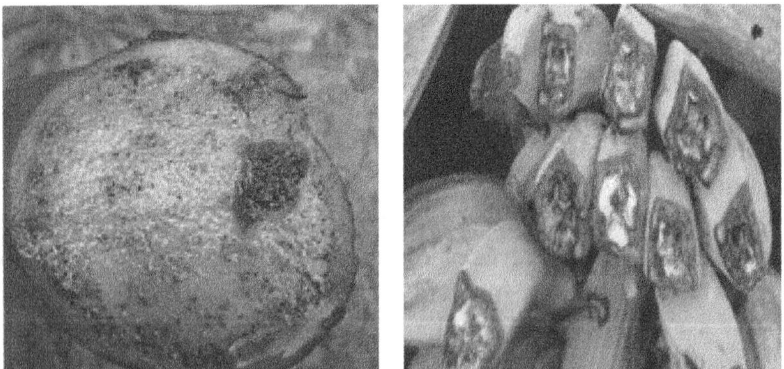

Figure 5.3 *Musa paradisiaca* showing symptoms of Moko disease including discolouration of the pseudostem and the internal banana fruits. Source: [92].

5.2.3 Types of Soil-Borne Fungi

Soil-borne plant fungal pathogens are fungal microorganisms that inhabit the soil and infect plant tissues [3]. They are among the prominent causes of devastating plant crop disease, causing significant economic burden in both temperate and tropical regions [102–105]. According to Shuping and Eloff [106], most diseases occurring in agricultural and horticultural setups are due to plant fungal pathogens with about 8000 fungal species causing almost 100,000 diseases in plants. El Hussein *et al.* [107] stated that plant pathogenic fungi are collectively responsible for about 80% of plant diseases. All fungal taxonomic groups have soil-borne fungal plant diseases, and they differ in terms of their host preference. Ghosh *et al.* [89] reported that there are an estimated 1.5 million species of fungi, among which more than 8,000 are recognized to be pathogens of plants. They can affect crops during the developmental stage from the sprouting stage to the seed maturing stage [108]. However, most of these pathogens infect plant root causing various detrimental effects such as root rot, wilting, yellowing, stunting, and dieback which can eventually result in the eventual death of the crop-plant. These pathogens employ diverse strategies including mechanical pressure, chemical action, or entry via wounds or stomata to infiltrate and infest their hosts. Sanchez-Vallet *et al.* [109], Rovenich *et al.* [110], Ryder and Talbot [111], and Rodriguez-Moreno *et al.* [105] reported various host penetration strategies employed by plant pathogenic fungal species. These include forcing a cell wall breach with the use of unique infection structures termed appressoria that produce concentrated turgor pressure. Some pathogenic fungal species combined the application of turgor pressure with the targeted release of enzymes that break down cell wall. In addition, they noted that some pathogenic fungal species had evolved defense mechanisms to subvert immune responses of the host that involve fungal cell walls. These included changes to the composition of cell walls and the release of effectors to either secure cell walls or impede the recognition of cell wall components. Strategies to inhibit host defense responses were also reported in some plant pathogenic fungal species [112–117].

Plant pathogenic fungi are typically divided into biotrophs, necrotrophs, and hemibiotrophs based on their mode of infection [104, 118, 119]. Biotrophs use a variety of techniques to enter their hosts and obtain nutrients from neighboring cells. They cause infections without killing their hosts in order to subsist on living plant tissues. Necrotrophs, on the other hand, can only complete their lifespan on dead tissues, therefore they infect the living host and ultimately kill the afflicted area. These microorganisms continuously release hydrolytic enzymes and poisons that kill plant cells [104, 118]. In between these categories are the hemibiotrophs which exhibit a dual-phase approach, they initiate their interaction with a suitable host in a biotrophic phase but transition to a necrotrophic stage when the infection progresses sufficiently [105].

According to Jain *et al.* [95], blight, damping off, dieback, Anthracnose, canker, gall, leaf spot, rust, powdery mildew, scab, root rot, and wilt are the most prevalent diseases caused by plant pathogenic fungus. Fungal pathogens, including species of *Gaeumannomyces graminis, Fusarium* sp., *Sclerotinia Sclerotium, Cylindrocladium* sp. *Rhizoctonia* sp., *Pythium* sp., *Magnaporthe oryzae, Armillaria melleam, Verticillium* sp., *Phytophthora* sp., *Sclerotinia* sp., and *Macrophomina phaseolina*, cause the loss of billions of dollars each year [3, 38, 81, 89, 102, 103, 108]. In order for a fungal pathogen to initiate infection and induce disease in a plant, the fungal spore must possess the capability to undergo germination on the surface of the host. This germination process is contingent upon the presence of favorable environmental conditions, including adequate humidity, a ready supply of nutrients, and the availability of a suitable host, as highlighted by Shuping and Eloff [106]. In the absence of such conditions, the spores utilize a self-inhibition mechanism to prevent germination and remain viable for years until favorable conditions are present [120].

Fusarium species are among the most important phytopathogenic fungi [32]. They are soil-borne vascular wilt pathogens with a worldwide distribution. They are to blame for the plant illness fusarium wilt, a soil-borne fungal disease in which the xylem vessels (water-conducting channels) are obstructed, causing the plant to wilt. Infected plants' leaves and possibly their stems lose their turgidity, change color from a light green to a greenish-yellow to a brown hue, and eventually collapse and die [95]. The fungal disease is caused by host-specific pathogenic strains of several species of *Fusarium*, most commonly *F. oxysporum*. Though other species *F. eumartii, F. avenaceum, F. solani, F. sulphureum*, and *F. tabacinum* are also culprits [95, 121]. However, the use of chemicals has been reported to be effective in controlling *Fusarium* wilt, breeding for resistant cultivars is the best method of control [32]. In addition, microbial species such as rhizosphere fungi, *Streptomyces griseoviridis, Trichoderma harzianum, T. koningii, T. asperellum, and Penicillium spp.* have been used to combat the disease. Species belonging to the *Pythium* genus have generally been considered virulent plant pathogens [122, 123]. They cause damping off, seed rot, damping-off, collar rot, root rot disease, and stem rot, which rank amidst the most devastating and economically important challenges in agriculture worldwide [124–127]. Blast is the most devastating disease of cultivated rice, it is caused by the filamentous fungus *Magnaporthe oryzae*

[128–130]. The disease destroys up to 30% of world rice harvests each year [131, 132]. *Magnaporthe oryzae* infects the rice plant at all stages of development (Figure 5.4) [133]. It is also an important pathogen of barley and millets.

It is common practice to use chemical fumigants and fungicides on a regular basis to prevent fungal disease outbreaks, despite the fact that doing so can have adverse effects on the environment, human health, aquatic ecosystems, beneficial soil microbes, and even the protective ozone layer [38]. To avoid these issues, new and better fungicides with little adverse effects are needed. In recent times, biological/natural products that are more ecologically friendly are of interest in the development of antifungal agents [134]. The employment of microbial antagonists as biological controls is a common component of these alternative techniques [135]. Biocontrol depends on the interaction between the plant pathogens and the control agent, in all cases pathogens are antagonized by the presence or activity of the control agents [136]. Biological agents are more cost effective and environmentally friendly. Various species of organisms have been used to control plant pathogenic fungi. Thambugala *et al.* [137] listed over 300 species belonging to 13 classes and 113 genera as fungal antagonists used as biological agents. They reported *Trichoderma* as the genus with the greatest potential comprising 25 biocontrol agents. *Streptomyces* bacteria have also been exploited as microbial antagonists for the biocontrol of fungal plant diseases (Law *et al.*, 2017) [131]. Several other antagonistic bacterial

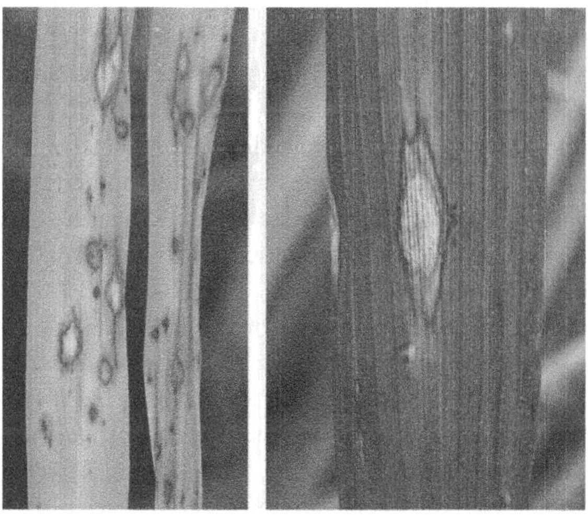

Figure 5.4 Symptoms of rice blast disease caused by the ascomycete fungus *Magnaporthe oryzae*. Source [103].

species including *P. aerofaciens, Bacillus substilis, Pseudomonas flourescens, P. putida, Burkholderia cepacian, B. polmyxa, and B. cereus* were previously reported as soil-borne fungal biocontrol agent (Heydari, and Pessarakli, 2010) [136].

5.3 Mushrooms as Biocontrol Agents

Mushrooms can be considered as biocontrol agents in agriculture and forestry due to their ability to suppress pests, diseases, and weeds naturally. Below is a detailed explanation of how mushrooms can serve as biocontrol agents:

1. Mycoparasitism: Some mushrooms have mycoparasitic properties, which skill they can assault and damage distinctive fungal pathogens. This mechanism controls ailments triggered through a variety of fungi. For example, certain species of *Trichoderma* and *Gliocladium* mushrooms are recognized for their capability to parasitize and inhibit the increase of fungal pathogens like Fusarium and Pythium [138].
2. Antibiosis: Several mushrooms produce secondary metabolites that have antimicrobial properties. These substances can stop bacteria, fungi, and other microbes from multiplying and developing. For instance, species like *Agaricus bisporus* and *Pleurotus ostreatus* (Table 5.1) represents an alternative for parasite management in the livestock sector, as it has nematicidal endeavor in opposition to exceptional taxonomic genera of nematodes that infect farm animals (e.g., the eggs and infectious and histotrophic larvae of *H. contortus*), parasite nematodes of public health significance (e.g., *Ancylostoma caninum*) and free-living nematodes (e.g., *Panagrellus redivivus*) [45].
3. Competition: Mushrooms as biocontrol retailers can compete with pests and weeds for assets such as nutrients, space, and light. This opposition can lead to reduced increase and survival of the pests or weeds, thereby providing manipulation and management. For instance, *Agaricus spp.* mushrooms have been shown to reduce germination and increase weed seeds. However, other species of fungi, such

Table 5.1 Isolated compounds from different species of *Pleurotus* genus.

Species	Nematicide compound	Concentration	Reference
P. ostreatus	Trans-2-decenedioic acid	95 %	[138]
P. pulmonarius	p-anisaldehyde, p-anisyl alcohol	75%	[139]
P. ferulae	Cheimonophyllon E, 5-hydroxymethyl-furancarbaldehyde	50%	[140]
P. pulmonarius	1-(4-methoxyphenyl)-1,2-propanediol	75%	[141]
P. ostreatus	2-hydroxy-(4′-methoxy)-propiophenone	87-94 %	[141]
P. eryngii	S-coriolic acid	95%	[142]

as *Lentinula edodes* and *Neolentinus ponderosus* [143], additionally have fantastic possibilities in the search for bioactive products with nematicidal undertaking in the cattle arena.

4. Pest repellence: Certain mushrooms, such as *Trametes versicolor*, produce unstable compounds that can repel insects and arthropods. These compounds act as herbal pesticides and can deter pests from infesting agricultural vegetation or forestry sites. Additionally, toxocysts, which are structures produced by edible mushrooms, primarily *P. ostreatus*, are used in their mode of action. These structures release NRRL 3526, a nematotoxin that works against the bacteriophage nematode P. redivivus.

5.3.1 Mushrooms as Natural Biopesticide

Mushrooms can certainly serve as herbal biopesticides in agriculture and pest management. Root-knot nematodes, *Meloidogyne spp*., are serious pests that motive wide-ranging losses to special vegetation worldwide and are ranked among the most tough agricultural pests to control due to their large host range, brief existence cycles, and excessive reproduction charge [144]. Biological control is a sustainable choice to chemical control which includes the use of natural and environment-friendly factors to limit plant-parasitic nematodes [145, 146]. According to Gupta and Mishra [147], mushrooms are abundant sources of substances that have been shown to have cancer-fighting, anti-oxidants, hypocholesterolemic, and wound-healing properties. Details on how mushrooms can be used as biopesticides are explained as follows:

1. Insect pathogenic fungi: Some mushrooms, such as species in the genus *Beauveria* and *Metarhizium*, are recognized as entomopathogenic fungi. These fungi have the capability to infect and kill insects. They produce spores that connect to the insect's cuticle and germinate, penetrating the insect's body and causing inner infections [148]. This ultimately leads to the death of the pest. Entomopathogenic fungi can be used for the biological management of agricultural pests, inclusive of aphids, whiteflies, thrips, and beetles [149].
2. Nematode control: Certain mushrooms, like species in the genera *Pochonia* and *Paecilomyces*, have nematophagous properties. They can parasitize and infect plant-parasitic nematodes that motive damage to crops. These mushrooms produce specialized structures, like trapping devices or adhesive networks, which can seize and kill nematodes. By making use of these mushrooms in the soil, farmers can decrease nematode populations naturally and protect their crops [145].
3. Antifeedant and repellent properties: Some mushrooms produce compounds that deter pests or make them much less possibly to feed on crops. These compounds can act as antifeedants or repellents, decreasing the damage caused through pests. For example, extracts from sure mushrooms have proven repellent effects toward bugs like mosquitoes and beetles [150].
4. Indirect consequences on pests: Mushrooms can in a roundabout way affect pest populations by means of enhancing the common health and resistance of plants. By mycorrhizal associations, mushrooms shape jointly recommended relationships with plant roots, improving nutrient uptake and water availability for the plants. This expanded plant energy can make vegetation extra resilient to pests and diseases. Additionally, mushrooms that degrade organic count make contributions to nutrient cycling, which cannot directly

affect pest populations by way of affecting the availability of sources [31, 151].

5. Compatibility with different control methods: Biopesticide mushrooms can be used in conjunction with different pest administration strategies, such as biological manipulate marketers or natural insecticides. Integrated pest administration (IPM) procedures combining more than one manipulate strategy provide a holistic and sustainable pest management system for farmers [152].

When thinking about the use of mushrooms as biopesticides, it is vital to apprehend the specific mushroom species, goal pests, application methods, and environmental conditions [154, 155].

5.3.2 Mechanisms of Mushroom-Mediated Biocontrol

Mushrooms possess various mechanisms that contribute to their effectiveness as natural biopesticides. Here are some key mechanisms:

1. Production of Bioactive Compounds: Mushrooms produce a wide range of bioactive substances, including secondary metabolites, enzymes, and proteins with pesticidal properties. These compounds can function as toxins, repellents, growth inhibitors, or disruptors of pests' physiological strategies [45].
2. Entomopathogenic Activity: Certain mushroom species, such as *Beauveria bassiana* and *Metarhizium anisopliae*, exhibit entomopathogenic activity. They release fungal spores that adhere to insects' exoskeletons, penetrate, germinate, and release enzymes that break down the insect's tissues, ultimately leading to its demise [156].
3. Nematode Parasitism: Some mushroom species display nematophagous activity, meaning they parasitize and infect plant-parasitic nematodes. These mushrooms employ adhesive traps like nets or rings to capture nematodes. Once ensnared, the mushrooms penetrate the nematodes' cuticles, develop inside them, and ultimately kill the nematodes [157].
4. Induced Resistance: Mushrooms can activate plant defense responses. Certain mushroom extracts have been shown to enhance the production of defensive compounds in plants, making them more resistant to pests and diseases.

This mechanism indirectly controls pests by strengthening the plants' innate defense mechanisms [45].

5. Microbial Interactions: Mushrooms can engage in synergistic interactions with other microorganisms, including beneficial bacteria and fungi. These interactions help suppress or control pests indirectly by promoting the growth of antagonistic microorganisms that can inhibit pest development or infection [155].

It is important to note that the effectiveness and mechanisms of mushrooms as biopesticides can vary depending on the specific mushroom species, target pests, and environmental conditions. Proper application and methods are also crucial for maximizing their effectiveness.

With the advancements in genetics and biotechnology in recent decades, it is now more feasible to manipulate biocontrol-related aspects in fungi and host plants. This includes incorporating fungal genes that encode antimicrobial substances (e.g., chitinases) into plant genomes, developing avirulent or hypovirulent strains for fungal pathogens using hypovirulence-associated mycoviruses, overexpressing or disrupting specific fungal genes to enhance biocontrol efficacy or expand the range of targeted pathogens, and conferring induced systemic resistance (ISR) to plants through fungus-based elicitors [13].

5.4 Mushroom Species and Biocontrol Potential

Mushroom species have the ability to combat various pests and pathogens in agriculture and forestry, by producing compounds that can inhibit the growth or activity of harmful organisms. Some mushroom species with notable biocontrol potential and their mechanisms of action are discussed.

***Volvariella* spp:** *Volvariella* is a genus of mushrooms, and different species within this genus are commonly referred to as "volvarias" or "straw mushrooms" (Figure 5.5). These mushrooms are edible and are cultivated for culinary purposes in various parts of the world. They typically have a characteristic bulbous or volva-shaped base and can be found growing on substrates like straw, which is why they are often called straw mushrooms [158]. The biocontrol potential of *Volvariella* spp., specifically as a natural enemy of certain pests and as a promoter of soil health, has garnered interest in agricultural and ecological contexts. Certain species of *Volvariella*,

Figure 5.5 *Volvariella* spp. (Bull.; Fr.) Singer is an edible mushroom also known as the straw mushroom [161].

particularly *Volvariella volvacea*, exhibit entomopathogenic properties [159]. This means they can parasitize or kill insect pests. The fungal spores of *Volvariella* attach to the bodies of insects, germinate, and then penetrate the insect's exoskeleton, leading to the eventual death of the pest. This mushroom can be used to control soil-borne pests, such as nematodes and certain insect larvae. When incorporated into the soil, the fungal mycelium colonizes the root zone, making it an effective biological control agent against these pests [160].

Volvariella spp. have shown promise in suppressing plant-parasitic nematodes, such as root-knot nematodes (*Meloidogyne* spp.) and cyst nematodes (*Heterodera* spp.). These nematodes can be detrimental to crop plants by causing root damage and reducing yields. *Volvariella* spp. can act as a nematode antagonist by competing for resources and releasing compounds that affect nematode development and reproduction.

Volvariella spp. are employed as integrated pest management (IPM) strategies, which reduces the reliance on synthetic chemical pesticides, promoting more sustainable agricultural practices. This aligns with the principles of organic farming and environmentally friendly pest control. *Volvariella* spp. are biodegradable and pose minimal risks to non-target organisms and the environment compared to chemical pesticides, making them an eco-friendly choice for pest management [162]. While *Volvariella* spp. have biocontrol potential, their effectiveness may vary depending on environmental conditions, target pests, and specific species or strains used. Research is ongoing to optimize their use and compatibility with different crops and ecosystems.

Figure 5.6 *Agaricus bisporus* [167].

***Agaricus bisporus*:** Commonly known as the white button mushroom, is a widely cultivated mushroom species that is primarily grown for its culinary uses (Figure 5.6) [163]. However, it also has potential applications in biocontrol [164], which refers to the use of living organisms to control pests, diseases, or invasive species. In the case of *Agaricus bisporus*, its role in biocontrol is mainly associated with its ability to suppress certain plant pathogens and pests through various mechanisms [165]. The common button mushroom has shown biocontrol potential against nematode pests. *A. bisporus* produces metabolites that are toxic to nematodes, making it a promising candidate for biological control in agriculture [166].

Agaricus bisporus mycelium, the vegetative part of the mushroom, can host and support beneficial microorganisms that antagonize plant pathogens [168]. These microorganisms may produce antimicrobial compounds or compete with pathogens for resources, ultimately reducing their populations in the soil or on plant surfaces. Agaricus bisporus produces a variety of enzymes as they grow and decompose organic matter [169]. Some of these enzymes, such as chitinases and glucanases, can break down the cell walls of certain plant pathogens, rendering them ineffective.

***Pleurotus ostreatus*:** *Pleurotus ostreatus* usually known as the oyster mushroom, is a fascinating fungal species that is primarily recognized for its culinary and medicinal properties (Figure 5.7). However, it also holds potential in the field of biocontrol, which involves using living organisms

Figure 5.7 *Pleurotus ostreatus* [172].

to manage pests and diseases in agriculture and forestry. Here, we will discuss in detail how *Pleurotus ostreatus* can be utilized in biocontrol [170, 171].

Oyster mushrooms have been explored for their ability to control various pests and pathogens through different mechanisms [173]. They can produce secondary metabolites with antimicrobial properties that affect the growth of certain plant pathogens and pests. These metabolites can disrupt the life cycle of the target organisms, reducing their populations. *Pleurotus ostreatus* plays a crucial role in breaking down organic matter, such as dead plant material. By decomposing this matter, it can eliminate habitats and food sources for certain pests and pathogens. They can compete with other fungi and microorganisms for resources like nutrients and space. This competition can limit the growth and spread of pathogenic organisms [174].

In agriculture, Oyster mushrooms can be cultivated in agricultural fields as part of integrated pest management (IPM) strategies and can be used to control soil-borne pathogens, such as nematodes and fungi, that affect crop plants. By growing oyster mushrooms alongside crops, these pathogens may be suppressed. These mushrooms have demonstrated biocontrol potential against several plant pathogens. They can suppress the growth of pathogenic fungi through mechanisms such as mycoparasitism, antibiosis (production of antifungal compounds), and competition for resources [175, 176].

Hypsizygus marmoreus: *Hypsizygus marmoreus*, commonly called the "white beech mushroom" or "marbled mushroom," is primarily cultivated for its culinary uses rather than its biocontrol potential. While it is not typically used as a biological control agent for pests or pathogens, it is important to note that mushrooms like *Hypsizygus marmoreus* can indirectly contribute to sustainable and ecologically responsible agricultural practices in several ways [177]. Mushroom cultivation often involves the use of organic substrates such as agricultural waste, sawdust, or wood chips. These substrates provide a food source for mushroom mycelium, which breaks down organic matter during the growing process. As a result, mushroom cultivation can contribute to the decomposition of organic materials, improving soil structure and nutrient cycling.

Mushrooms sequester carbon from the environment during growth and nutrition, which can help mitigate climate change by reducing atmospheric carbon dioxide levels. The carbon-rich mycelium of *Hypsizygus marmoreus* contributes to this carbon sequestration. Mushroom cultivation often utilizes agricultural residues and by-products, reducing waste in the agricultural industry. This practice aligns with the principles of sustainability and resource efficiency. The cultivation of mushrooms like *Hypsizygus marmoreus* can promote biodiversity in agricultural systems. Mushrooms and their mycelium can support a variety of beneficial microorganisms and insects in the soil, which can contribute to overall ecosystem health and balance.

***Trametes versicolor* (Turkey Tail):** This mushroom produces compounds with antiviral and antibacterial properties [178]. It has been studied for its potential in controlling viral and bacterial diseases in plants, making it a candidate for biocontrol in agriculture. *Trametes versicolor*, also known as the turkey tail mushroom, has various promising applications in biocontrol. *Trametes versicolor* produces bioactive compounds that can inhibit the growth of pests, nematodes, and plant diseases [179]. These compounds can be used to create biopesticides, especially effective against soil-borne plant pathogens like Rhizoctonia spp.

Extracts from this mushroom exhibit antimicrobial properties against bacteria and fungi. They can be used to develop natural antimicrobial agents for agriculture and medicine, potentially controlling diseases in crops and other applications [179]. Extracts from the mushroom may help control pathogenic bacteria and fungi in aquaculture, reducing the need for chemical treatments and promoting fish and shellfish health. It is adept at breaking down environmental pollutants like pesticides and industrial chemicals. It can be used in bioremediation efforts to clean up contaminated soil and water, thanks to its ligninolytic enzymes. The fungus can serve as a biosensor, reacting to environmental changes and pollutants. Its growth and metabolic activity can be used to monitor environmental conditions effectively.

***Pleurotus tuber-regium*:** This mushroom has shown potential as a biological control agent against certain weed species. It produces allelopathic compounds that inhibit weed germination and growth [180].

***Ganoderma lucidum* (Reishi Mushroom):** Research suggests that *Ganoderma* species can produce antifungal compounds and enzymes that halt the development of plant pathogens. They may have the potential to control soil-borne fungal diseases in crops [181].

***Lentinula edodes* (Shiitake Mushroom):** Shiitake mushrooms have been studied for their ability to abate the development of various plant pathogens, including certain bacteria and fungi. They produce antimicrobial compounds that can inhibit these pathogens [182, 183].

Mushrooms with biocontrol potential offer a sustainable and ecologically friendly approach to managing pests and diseases in agriculture and forestry. Their mechanisms of action can vary, including the production of antifungal, antibacterial, and antinematode compounds, as well as mycoparasitism and competition. Continued research and development of these fungi as biocontrol agents hold promise for reducing the reliance on chemical pesticides and promoting more sustainable agricultural practices [184].

5.5 Advantages of Mushroom Biocontrol

Due to their capacity to eliminate some of these pathogens, many mushrooms play a crucial role in the biological control of plant-parasitic pathogens as an alternative to chemical methods. This method has advantages since it is easy to obtain and it is also environmentally beneficial [1, 185]. These mushrooms generate bioactive substances that have the potential to protect non-targeted species, have selective effects, and lead to the creation of environmentally friendly products [141, 186]. *C. militaris* is one of these mushrooms that can be artificially grown and yield large amounts of bioactive chemicals [187]. Another well-researched fungus is *Hericium erinaceus*, which has been shown to have biological effects like antibacterial and anti-proliferative properties [153, 188]. This method is risk-free and leaves no residue behind. In certain cases where minor adjustments may be made, it can be affordable. It is a long-lasting method that prevents potential reapplication, making it simple to establish [189].

5.6 Challenges and Limitations of Mushroom Bio-Control

The efficiency of this biocontrol approach has been examined; however, the majority of the studies were conducted in greenhouses and laboratories. Therefore, it is unknown how effective they are in the field. It is extremely difficult to translate laboratory results into similarly successful field applications [190]. According to reports, these bio-controls are frequently applied slowly since there is a lag between the growth of the fungus and the pathogen population. When used in high level of pathogens, it is less effective and more susceptible to pathogen contamination. In some cases, the field application may need extensive research and be challenging [189]. One of the main causes of bio-control agents' decreased effectiveness in the field is their capacity to adapt to specific local biotic and abiotic environmental variables. Therefore, to completely comprehend this phenomenon, it is necessary to study their spatial distribution pattern in the soil [191]. The efficiency of these agents usually includes trade-offs with other inherent characteristics of the agents, such as specificity and environmental persistence [165, 192], and is closely related to their modes of action [193]. This biological control method may be expensive to commercialize because it requires a number of steps, including characterization, formulation and substrate development, mass production, product efficacy testing, storage stability inspection, industrial partner identification, consideration of human and environmental safety issues, registration, and marketing [137].

5.7 Conclusion and Future Outlook

The climatic and other ecological circumstances, physicochemical characteristics of the soil, and subsequent organic or chemical amendments over and above indigenous microorganisms present in the soil are the chief factors that influence crop production and also the soil-borne plant pathogens that may affect crops. The application of mushrooms as management tools against soil-borne pathogens is rapidly gaining global acceptance for sustainable agriculture. This is against the use of chemical compounds which have various deleterious short- and long-term effects on the environment. Mushrooms offer a harmless and eco-friendly green substitute for preventing and controlling diseases of plants caused by soil-borne pathogens of plants such as parasitic nematodes, bacteria, and fungi. This chapter therefore highlights the types of soil-borne pathogenic nematodes, bacteria, and fungi that affect plants; the potential applications of various types of mushrooms as bio-control agents and their mechanisms of action against these pathogens. The advantages, limitations, and challenges of these mushrooms as bio-control agents and their ecological impacts for sustainable agriculture were also stressed. However, more research is required to isolate, characterizem, and identify new mushrooms with antimicrobial potentials against soil-borne pathogens of plant and to also determine the effect of the combination of different species of mushrooms against these soil-borne plant pathogens. In the same vein, the consortium of mushrooms and other groups of beneficial (micro) organisms especially bacteria and fungi as well as plants with antimicrobial activities requires investigation. Finally, the use of molecular techniques especially omics technology such as metagenomics, proteomics, metabolomics, and celiomics is a matter of urgency needed to provide an advanced understanding of the action mechanisms of mushroom against these plant pathogens and their interactions with the environment, other mushrooms, other groups of micro and macro-organism.

References

1. Elkhateeb, W.A., Daba, G.M., Soliman, G.M., The anti-nemic potential of mushroom against plant-parasitic nematodes. *J. Microbiol. Biotechnol.*, 6, 1, 1–6, 2021.
2. Gupta, G., Parihar, S., Ahirwar, N., Snehi, S., Singh, V., Plant growth promoting rhizobacteria (PGPR): Current and future prospects for development of sustainable agriculture. *J. Microbiol. Biochem. Technol.*, 7, 2, 96–102, 2015.
3. Shafique, H.A., Sultana, V., Ehteshamul-Haque, S., Athar, M., Management of soil-borne diseases of organic vegetables. *J. Plant Prot. Res.*, 56, 3, 221–230, 2016.
4. Khan, M.R., Altaf, S., Mohiddin, F.A., Khan, U., Anwer, A., Biological control of plant nematodes with phosphate solubilizing microorganisms, in: *Phosphate Solubilizing Microbes for Crop Improvement*, M.S. Khan and A. Zaidi (Eds.), pp. 395–426, Nova Science Publishers, New York, NY, USA, 2009, 451 pp.
5. Baysal-Gurel, F., Gardener, B.M., Miller, S.A., Soil-borne disease management in organic vegetable production, 2012, Available on: www.extension.org/pages/64951.

6. Ferreira, J.M., Carreira, D.N., Braga, F.R., Soares, F.E.F., First report of the nematicidal activity of *Flammulina velutipes*, its spent mushroom compost and metabolites. *3 Biotech*, 9, 11, 410, 2019, https://doi.org/10.1007/s13205-019-1951-x.
7. Youssef, M.M.A. and Lashein, A.M.S., Effect of cabbage (Brassica oleracea) leaf residue as a biofumigant, on root knot nematode, *Meloidogyne incognita* infecting tomato. *J. Plant Prot. Res.*, 53, 3, 271–274, 2013.
8. Bernard, G.C., Egnin, M., Bonsi, C., The Impact of Plant-Parasitic Nematodes on Agriculture and Methods of Control, in: *Nematology: Concepts, Diagnosis and Control*. InTech, London, 2017, doi: 10.5772/intechopen.68958.
9. Gaziea, S.M., Shereen, M.A.H., Laila, H.F., El-Hady, S.E., Efficiency of biological control of root-knot nematodes in infected grapevines seedling by genetic improved bacteria. *Plant Arch.*, 20, 1, 951–961, 2020.
10. Eisenback, J.D. and Triantaphyllou, H.H., Root-knot nematodes: *Meloidogyne* species and races, in: *Manual of Agricultural Nematology*, pp. 281–286, Marcel Dekker, New York, 1991.
11. Karssen, G. and Moens, M., Taxonomy and Principal General Root-Knot Nematodes, in: *Plant Nematology*, R. Perry and M. Moens (Eds.), pp. 60–90, CAB International, Wallingford, UK, 2006.
12. Van der Heijden, M.G., Martin, F.M., Selosse, M.A., Sanders, I.R., Mycorrhizal ecology and evolution: the past, the present, and the future. *New Phytol.*, 205, 4, 1406–1423, 2015.
13. Elkhateeb, W.A., Daba, G.M., Wen, T.C., Thomas, P.W., Medicinal mushrooms as a new source of natural therapeutic bioactive compounds. *Egypt. Pharma. J.*, 18, 88–101, 2019.
14. Jing, P., Zhao, S.J., Lu, M.M., Cai, Z., Pang, J., Song, L.H., Multiplefingerprint analysis for investigating quality control of Flammulina velutipes fruiting body polysaccharides. *J. Agric. Food. Chem.*, 62, 50, 12128–12133, 2014.
15. Dong, Y.R., Cheng, S.J., Qi, G.H., Yang, Z.P., Yin, S.Y., Chen, G.T., Antimicrobial and antioxidant activities of Flammulina velutipes polysaccharides and polysaccharide-iron(III) complex. *Carbohydr. Polym.*, 161, 26–32, 2017.
16. Soares, F.E., Sufate, B.L., Queiroz, J.H., Nematophagous fungi: far beyond the endoparasite, predator and ovicidal groups. *ANRES*, 52, 1, 1–8, 2018.
17. Briones, M.J., The serendipitous value of soil fauna in ecosystem functioning: The unexplained explained. *Front. Environ. Sci.*, 6, 149, 2018.
18. Ehrenfeld, J.G., Plant–Soil Interactions, in: *Encyclopedia of Biodiversity*, pp. 109–128, 2013, doi:10.1016/b978-0-12-384719-5.00179-9.
19. Balestrini, R., Lumini, E., Borriello, R., Bianciotto, V., Plant-Soil Biota Interactions, in: *Soil Microbiology, Ecology and Biochemistry*, pp. 311–338, 2015, doi:10.1016/b978-0-12-415955-6.00011-6.
20. Mihajlović, M., Rekanović, E., Hrustić, J., Grahovac, M., Tanović, B., Methods for management of soilborne plant pathogens. *Pesticidi Fitomedicina, 32*, 1, 9–24, 2017.

21. Brevik, E.C., Slaughter, L., Singh, B.R., Steffan, J.J., Collier, D., Barnhart, P., Pereira, P., Soil and human health: current status and future needs. *Air Soil Water Res.*, *13*, 1178622120934441, 2020.
22. Khan, M.R. and Anwer, M.A., Fungal bioinoculants for plant disease management, in: *Microbes and Microbial Technology: Agricultural and Environmental Applications*, pp. 447–488, Springer New York, New York, NY, 2011.
23. Poveda, J., Abril-Urias, P., Escobar, C., Biological control of plant-parasitic nematodes by filamentous fungi inducers of resistance: Trichoderma, mycorrhizal and endophytic fungi. *Front. Microbiol.*, *11*, 992, 2020.
24. Fiers, M., Edel-Hermann, V., Chatot, C., Le Hingrat, Y., Alabouvette, C., Steinberg, C., Potato soil-borne diseases. A review. *Agron. Sustain. Dev.*, *32*, 1, 93–132, 2012.
25. Mansfield, J., Genin, S., Magori, S., Citovsky, V., Sriariyanum, M., Ronald, P., Foster, G.D., Top 10 plant pathogenic bacteria in molecular plant pathology. *Mol. Plant Pathol.*, *13*, 6, 614–629, 2012.
26. Jones, J.T., Haegeman, A., Danchin, E.G., Gaur, H.S., Helder, J., Jones, M.G., Perry, R.N., Top 10 plant-parasitic nematodes in molecular plant pathology. *Mol. Plant Pathol.*, *14*, 9, 946–961, 2013.
27. Peeters, N., Guidot, A., Vailleau, F., Valls, M., R alstonia solanacearum, a widespread bacterial plant pathogen in the post-genomic era. *Mol. Plant Pathol.*, *14*, 7, 651–662, 2013.
28. Andika, I.B., Kondo, H., Sun, L., Interplays between soil-borne plant viruses and RNA silencing-mediated antiviral defense in roots. *Front. Microbiol.*, *7*, 1458, 2016.
29. Mitiku, M., Plant-parasitic nematodes and their management: A review. *Agric. Res. Technol.*, *8*, 30–38, 2018.
30. Sarwar, M., Aslam, M., Sarwar, S., Iftikhar, R., Different nematodes and plasmodiophorids as vectors of plant viruses, in: *Applied Plant Virology*, pp. 275–290, Academic Press, Cambridge, Massachusetts, 2020.
31. Kumar, Y. and Yadav, B.C., Plant-parasitic nematodes: Nature's most successful plant parasite. *Int. J. Res. Rev.*, *7*, 3, 379–386, 2020.
32. Okungbowa, F.I. and Shittu, H.O., Fusarium wilts: An overview. *Environ. Res. J.*, *6*, 2, 83–102, 2012.
33. Ravichandra, N.G. and Ravichandra, N.G., Symptoms Caused by Phytonematodes and Disease Diagnosis, in: *Horticultural Nematology*, pp. 89–100, 2014.
34. Seethapathy, P., Kurusamy, R., Kuppusamy, P., Soil borne diseases of major pulses and their biological management. *Innov. Farm.*, *2*, 1, 01–11, 2017.
35. Tronsmo, A.M., Munk, L., Djurle, A., Yuen, J., Collinge, D.B., What is a plant disease, in: *Plant pathology and plant diseases*, pp. 3–10, 2020.
36. Sodikov, B., Khakimov, A., Rakhmonov, U., Omonlikov, A., Gulmatov, R., Utaganov, S., Soil-borne plant pathogenic fungi biodiversity of sunflower. *IOP Conf. Ser.: Earth Environ. Sci.*, 1068, 1, 012018, 2022.

37. Dutta, P., Kumari, A., Mahanta, M., Upamanya, G.K., Heisnam, P., Borua, S., Vijayreddy, D., Nanotechnological approaches for management of soil-borne plant pathogens. *Front. Plant Sci., 14,* 1136233, 2023.
38. Panth, M., Hassler, S.C., Baysal-Gurel, F., Methods for management of soil-borne diseases in crop production. *Agriculture, 10,* 1, 16, 2020.
39. Seesao, Y., Gay, M., Merlin, S., Viscogliosi, E., Aliouat-Denis, C.M., Audebert, C., A review of methods for nematode identification. *J. Microbiol. Methods,* 138, 37–49, 2017, doi: 10.1016/j.mimet.2016.05.030.
40. Jalali Sendi, J. and Khosravi, R., Recent developments in controlling insect, acari, nematode, and plant pathogens of agricultural and medical importance by Artemisia annua L.(Asteraceae), in: *Artemisia annua-pharmacology and biotechnology*, pp. 229–247, 2014.
41. Feyisa, B., Survey and identification of plant parasitic nematodes on faba bean crop in Ethiopia. *J. Plant Pathol. Microbiol.*, 12, 261, 2021.
42. Smythe, A.B., Holovachov, O., Kocot, K.M., Improved phylogenomic sampling of free-living nematodes enhances resolution of higher-level nematode phylogeny. *BMC Evol. Biol.*, 19, 1, 1–15, 2019.
43. Mendoza-de Gives, P., Soil-Borne nematodes: Impact in agriculture and livestock and sustainable strategies of prevention and control with special reference to the use of nematode natural enemies. *Pathogens*, 11, 6, 640, 2022.
44. Pulavarty, A., Egan, A., Karpinska, A., Horgan, K., Kakouli-Duarte, T., Plant parasitic nematodes: A review on their behaviour, host interaction, management approaches and their occurrence in two sites in the Republic of Ireland. *Plants*, 10, 11, 2352, 2021.
45. Garcia, N., Grenier, E., Buisson, A., Folcher, L., Diversity of plant parasitic nematodes characterized from fields of the French national monitoring programme for the Columbia root-knot nematode. *PLoS One*, 17, 3, e0265070, 2022.
46. Coyne, D.L., Cortada, L., Dalzell, J.J., Claudius-Cole, A.O., Haukeland, S., Luambano, N., Talwana, H., Plant-parasitic nematodes and food security in Sub-Saharan Africa. *Annu. Rev. Phytopathol.*, 56, 381–403, 2018.
47. Ali, M.A., Azeem, F., Abbas, A., Joyia, F.A., Li, H., Dababat, A.A., Transgenic strategies for enhancement of nematode resistance in plants. *Front. Plant Sci.*, 8, 750, 2017.
48. Singh, S., Singh, B., Singh, A.P., Nematodes: A threat to sustainability of agriculture. *Procedia Environ. Sci.*, 29, 215–216, 2015, doi: 10.1016/j.proenv.2015.07.270.
49. Carneiro, R.M.D.G., Lima, F.S.O., Correia, V.R., Methods and Tools Currently Used for the Identification of Plant Parasitic Nematodes, in: *Nematology: Concepts, Diagnosis and Control*, InTech, London, 2017, doi: 10.5772/intechopen.69403.
50. Jibril, S.M., Jakada, B.H., Kutama, A.S., Umar, H.Y., Plant and pathogens: pathogen recognision, invasion and plant defense mechanism. *Int. J. Curr. Microbiol. Appl. Sci.*, 5, 6, 247–257, 2016.

51. Palomares-Rius, J.E., Escobar, C., Cabrera, J., Vovlas, A., Castillo, P., Anatomical alterations in plant tissues induced by plant-parasitic nematodes. *Front. Plant Sci.*, 8, 1987, 2017.
52. Thakur, S., Duggal, P., Mann, S.S., A review: Sustainable approaches for the nematode management in organic farming. *Pharma Innov. J.*, 12, 3, 1556–1559, 2023.
53. Siddique, S. and Grundler, F.M., Parasitic nematodes manipulate plant development to establish feeding sites. *Curr. Opin. Microbiol.*, 46, 102–108, 2018, doi: 10.1016/j.mib.2018.09.004.
54. Kyndt, T., Vieira, P., Gheysen, G., de Almeida-Engler, J., Nematode feeding sites: unique organs in plant roots. *Planta*, 238, 5, 807–818, 2013, doi: 10.1007/s00425-013-1923-z.
55. Holbein, J., Grundler, F.M., Siddique, S., Plant basal resistance to nematodes: an update. *J. Exp. Bot.*, 67, 7, 2049–2061, 2016.
56. Moens, M. and Perry, R.N., Migratory plant endoparasitic nematodes: a group rich in contrasts and divergence. *Annu. Rev. Phytopathol.*, 47, 313–332, 2009.
57. Perry, R.N. and Moens, M., Introduction to plant-parasitic nematodes; modes of parasitism, in: *Genomics and molecular genetics of plant-nematode interactions*, pp. 3–20, 2011.
58. Holbein, J., Franke, R.B., Marhavý, P., Fujita, S., Górecka, M., Sobczak, M., Siddique, S., Root endodermal barrier system contributes to defence against plant-parasitic cyst and root-knot nematodes. *Plant J.*, *100*, 2, 221–236, 2019.
59. Volcy, C., Past and present of the nematode Radopholus similis (Cobb) Thorne with emphasis on Musa: a review. *Agron. Colomb.*, 29, 3, 433–440, 2011.
60. Yu, Y.T., Liu, H.L., Zhu, A.G., Zhang, G., Zeng, L.B., Xue, S.D., A review of root lesion nematode: identification and plant resistance. *Adv. Microbiol.*, 2, 411–416, 2012, http://dx.doi.org/10.4236/aim.2012.24052.
61. Lamovšek, J., Urek, G., Trdan, S., Biological control of root-knot nematodes (*Meloidogyne* spp.): microbes against the pests. *Acta Agric. Slov.*, 101, 2, 263–275, 2013.
62. Ali, N., Chapuis, E., Tavoillot, J., Mateille, T., Plant-parasitic nematodes associated with olive tree (Olea europaea L.) with a focus on the Mediterranean Basin: A review. *C.R. Biol.*, 337, 7–8, 423–442, 2014.
63. Onkendi, E.M., Kariuki, G.M., Marais, M., Moleleki, L.N., The threat of root-knot nematodes (*Meloidogyne* spp.) in Africa: a review. *Plant Pathol.*, 63, 4, 727–737, 2014.
64. Ralmi, N.H.A.A., Khandaker, M.M., Mat, N., Occurrence and control of root knot nematode in crops: a review. *Aust. J. Crop Sci.*, 11, 12, 1649, 2016.
65. Lopes-Caitar, V.S., Pinheiro, J.B., Marcelino-Guimaraes, F.C., Nematodes in horticulture: An overview. *J. Hortic. Sci. Crop Res.*, 1, 1, 106, 2019.
66. Abd-Elgawad, M.M., Managing nematodes in Egyptian citrus orchards. *Bull. Natl. Res. Cent.*, 44, 1–15, 2020.

67. Abd-Elgawad, M.M., Biological control of nematodes infecting eggplant in Egypt. *Bull. Natl. Res. Cent.*, 45, 1–9, 2021.
68. Migunova, V.D. and Sasanelli, N., Bacteria as biocontrol tool against phytoparasitic nematodes. *Plants*, 10, 2, 389, 2021.
69. Reddy, P.P. and Reddy, P.P., Temperate Fruit Crops, in: *Nematode Diseases of Crops and their Management*, pp. 203–219, 2021.
70. Arjoune, Y., Sugunaraj, N., Peri, S., Nair, S.V., Skurdal, A., Ranganathan, P., Johnson, B., Soybean cyst nematode detection and management: a review. *Plant Methods*, 18, 1, 1–39, 2022.
71. Saucet, S.B., Van Ghelder, C., Abad, P., Duval, H., Esmenjaud, D., Resistance to root-knot nematodes *Meloidogyne* spp. in woody plants. *New Phytol.*, 211, 1, 41–56, 2016.
72. Subedi, S., Thapa, B., Shrestha, J., Overview of root-knot nematode (*Meloidogyne incognita*) and control management. *J. Agric. Nat. Resour.*, 3, 2, 21–31, 2020, DOI: https://doi.org/10.3126/janr.v3i2.32298.
73. Bhat, A.A., Shakeel, A., Waqar, S., Handoo, Z.A., Khan, A.A., Microbes vs. nematodes: insights into biocontrol through antagonistic organisms to control root-knot nematodes. *Plants*, 12, 3, 451, 2023.
74. Postnikova, O.A., Hult, M., Shao, J., Skantar, A., Nemchinov, L.G., Transcriptome analysis of resistant and susceptible alfalfa cultivars infected with root-knot nematode *Meloidogyne incognita*. *PLoS One*, 10, 2, e0118269, 2015.
75. Khan, A., Aiman, S.I., Baber, Y., Hassan, F., Usman, H.M., Sohail, M.A., Abbas, A., An overview of root-knot nematodes and their management. *J. Entomol. Zool. Stud.*, 9, 1, 35–40, 2021.
76. Yigezu Wendimu, G., Biology, taxonomy, and management of the root-knot nematode (*Meloidogyne incognita*) in sweet potato. *Adv. Agric.*, 2021, 1–13, 2021.
77. Bohlmann, H. and Sobczak, M., The plant cell wall in the feeding sites of cyst nematodes. *Front. Plant Sci.*, 5, 89, 2014.
78. Gartner, U., Hein, I., Brown, L.H., Chen, X., Mantelin, S., Sharma, S.K., Blok, V.C., Resisting potato cyst nematodes with resistance. *Front. Plant Sci.*, 12, 661194, 2021.
79. Timper, P., Conserving and enhancing biological control of nematodes. *J. Nematol.*, 46, 2, 75, 2014.
80. Ali, M.A., Azeem, F., Abbas, A., Joyia, F. A., Li, A., Dababat, A.A., Transgenic strategies for enhancement of nematode reistance in plants. *Front. Plant Sci.*, 8, 750, 2017.
81. Ayala-Doñas, A., Cara-García, M.D., Talavera-Rubia, M., Verdejo-Lucas, S., Management of soil-borne fungi and root-knot nematodes in cucurbits through breeding for resistance and grafting. *Agronomy*, 10, 11, 1641, 2020.
82. Abd-Elgawad, M.M. and Askary, T.H., Fungal and bacterial nematicides in integrated nematode management strategies. *Egypt. J. Biol. Pest Control*, 28, 1, 1–24, 2018.

83. Sharma, K.K., Trichoderma in agriculture: An overview of global scenario on research and its application. *Int. J. Curr. Microbiol. Appl. Sci.*, 7, 1922–1933, 2018.
84. Isah, M., Bello, Z.H., Labbo, A.M., Ibrahim, A., Roles of Trichoderma species as a Potent Biochemical Agent for Sustainable Farming Practice: A Mini Review. *Caliphate J. Sci. Technol. (CaJoST)*, 2, 196–205, 2021.
85. Luo, N., Jiao, Y., Mao, Z.C., Li, H.X., Xie, B.Y., Advances of Trichoderma in Controlling Root Knot Nematodes and Cyst Nematodes. *Biotechnol. Bull.*, 39, 2, 35, 2023.
86. Yao, X., Guo, H., Zhang, K., Zhao, M., Ruan, J., Chen, J., Trichoderma and its role in biological control of plant fungal and nematode disease. *Front. Microbiol.*, 14, 1160551, 2023.
87. Tariq, M., Khan, A., Asif, M., Khan, F., Ansari, T., Shariq, M., Siddiqui, M.A., Biological control: a sustainable and practical approach for plant disease management. *Acta Agric. Scand. Sect B*, 70, 6, 507–524, 2020.
88. Buttimer, C., McAuliffe, O., Ross, R.P., Hill, C., O'Mahony, J., Coffey, A., Bacteriophages and bacterial plant diseases. *Front. Microbiol.*, 8, 34, 2017.
89. Ghosh, R., Tarafdar, A., Chobe, D.R., Sharath Chandran, U.S., Rani, S., Sharma, M., Diagnostic techniques of soil borne plant diseases: recent advances and next generation evolutionary trends. *Biol. Forum-Int. J.*, 11, 2, 1–13, 2019.
90. Sundin, G.W., Castiblanco, L.F., Yuan, X., Zeng, Q., Yang, C.H., Bacterial disease management: challenges, experience, innovation and future prospects: challenges in bacterial molecular plant pathology. *Mol. Plant Pathol.*, 17, 9, 1506–1518, 2016.
91. Tahat, M.M. and Sijam, K., Ralstoina solanacearum: The bacterial wilt causal agent. *Asian J. Plant Sci.*, 9, 7, 385, 2010.
92. Zulperi, D. and Sijam, K., First report of Ralstonia solanacearum race 2 biovar 1 causing Moko disease of banana in Malaysia. *Plant Dis.*, 98, 2, 275–275, 2014.
93. Vrancken, K., Holtappels, M., Schoofs, H., Deckers, T., Valcke, R., Pathogenicity and infection strategies of the fire blight pathogen Erwinia amylovora in Rosaceae: state of the art. *Microbiology*, 159, Pt_5, 823–832, 2013.
94. Aćimović, S.G., Zeng, Q., McGhee, G.C., Sundin, G.W., Wise, J.C., Control of fire blight (Erwinia amylovora) on apple trees with trunk-injected plant resistance inducers and antibiotics and assessment of induction of pathogenesis-related protein genes. *Front. Plant Sci.*, 6, 16, 2015.
95. Jain, A., Sarsaiya, S., Wu, Q., Lu, Y., Shi, J., A review of plant leaf fungal diseases and its environment speciation. *Bioengineered*, 10, 1, 409–424, 2019.
96. Koczan, J.M., McGrath, M.J., Zhao, Y., Sundin, G.W., Contribution of Erwinia amylovora exopolysaccharides amylovoran and levan to biofilm formation: implications in pathogenicity. *Phytopathology*, 99, 11, 1237–1244, 2009.

97. McGhee, G.C. and Sundin, G.W., Evaluation of kasugamycin for fire blight management, effect on nontarget bacteria, and assessment of kasugamycin resistance potential in Erwinia amylovora. *Phytopathology*, 101, 2, 192–204, 2011.
98. Johnson, K.B. and Temple, T.N., Evaluation of strategies for fire blight control in organic pome fruit without antibiotics. *Plant Dis.*, 97, 3, 402–409, 2013.
99. Toth, I.K., Van Der Wolf, J.M., Saddler, G., LOjkowska, E., Hélias, V., Pirhonen, M., Elphinstone, J.G., Dickeya species: an emerging problem for potato production in Europe. *Plant Pathol.*, 60, 3, 385–399, 2011.
100. Khayi, S., Chan, K.G., Faure, D., Patterns of Genomic Variations in the Plant Pathogen Dickeya solani. *Microorganisms*, 10, 11, 2254, 2022.
101. Hugouvieux-Cotte-Pattat, N., Condemine, G., Gueguen, E., Shevchik, V.E., Dickeya plant pathogens. eLS, 1–10, 2020.
102. Ritz, K., FUNGI, in: *Encyclopedia of Soils in the Environment*, pp. 110–119, 2005, doi:10.1016/b0-12-348530-4/00147-8.
103. Godfray, H.C.J., Mason-D'Croz, D., Robinson, S., Food system consequences of a fungal disease epidemic in a major crop. *Philos. Trans. R. Soc. B Biol. Sci.*, 371, 1709, 20150467, 2016.
104. Doehlemann, G., Ökmen, B., Zhu, W., Sharon, A., Plant pathogenic fungi. *Microbiol. Spectrum*, 5, 1, 5–1, 2017.
105. Rodriguez-Moreno, L., Ebert, M.K., Bolton, M.D., Thomma, B.P., Tools of the crook-infection strategies of fungal plant pathogens. *Plant J.*, 93, 4, 664–674, 2018.
106. Shuping, D.S.S. and Eloff, J.N., The use of plants to protect plants and food against fungal pathogens: A review. *Afr. J. Tradit. Complementary Altern. Med.*, 14, 4, 120–127, 2017.
107. El Hussein, A.A., Alhasan, R.E., Abdelwahab, S.A., El Siddig, M.A., Isolation and identification of Streptomyces rochei strain active against Phytopathogenic Fungi. *Br. Microbiol. Res. J.*, 4, 10, 1057–1068, 2014.
108. Hariharan, G. and Prasannath, K., Recent advances in molecular diagnostics of fungal plant pathogens: A mini review. *Front. Cell. Infect. Microbiol.*, 10, 600234, 2021.
109. Sánchez-Vallet, A., Saleem-Batcha, R., Kombrink, A., Hansen, G., Valkenburg, D.J., Thomma, B.P., Mesters, J.R., Fungal effector Ecp6 outcompetes host immune receptor for chitin binding through intrachain LysM dimerization. *elife*, 2, e00790, 2013 Jul 2.
110. Rovenich, H., Boshoven, J.C., Thomma, B.P., Filamentous pathogen effector functions: of pathogens, hosts and microbiomes. *Curr. Opin. Plant Biol.*, 20, 96–103, 2014.
111. Ryder, L.S. and Talbot, N.J., Regulation of appressorium development in pathogenic fungi. *Curr. Opin. Plant Biol.*, 26, 8–13, 2015.
112. Hemetsberger, C., Herrberger, C., Zechmann, B., Hillmer, M., Doehlemann, G., The Ustilago maydis effector Pep1 suppresses plant immunity by inhibition of host peroxidase activity. *PLoS Pathog.*, 8, 5, e1002684, 2012.

113. Plett, J.M., Daguerre, Y., Wittulsky, S., Vayssières, A., Deveau, A., Melton, S.J., Martin, F., Effector MiSSP7 of the mutualistic fungus Laccaria bicolor stabilizes the Populus JAZ6 protein and represses jasmonic acid (JA) responsive genes. *Proc. Natl. Acad. Sci.*, 111, 22, 8299–8304, 2014.
114. Jashni, M.K., Mehrabi, R., Collemare, J., Mesarich, C.H., de Wit, P.J., The battle in the apoplast: further insights into the roles of proteases and their inhibitors in plant–pathogen interactions. *Front. Plant Sci.*, 6, 584, 2015.
115. Masachis, S., Segorbe, D., Turrà, D., Leon-Ruiz, M., Fürst, U., El Ghalid, M., Di Pietro, A., A fungal pathogen secretes plant alkalinizing peptides to increase infection. *Nat. Microbiol.*, 1, 16043, 2016.
116. Thynne, E., Saur, I.M., Simbaqueba, J., Ogilvie, H.A., Gonzalez-Cendales, Y., Mead, O., Solomon, P.S., Fungal phytopathogens encode functional homologues of plant rapid alkalinization factor (RALF) peptides. *Mol. Plant Pathol.*, 18, 6, 811–824, 2017.
117. Sørensen, J.L., Benfield, A.H., Wollenberg, R.D., Westphal, K., Wimmer, R., Nielsen, M.R., Gardiner, D.M., The cereal pathogen Fusarium pseudograminearum produces a new class of active cytokinins during infection. *Mol. Plant Pathol.*, 19, 5, 1140–1154, 2018.
118. Wen, L., Cell death in plant immune response to necrotrophs. *J. Plant Biochem. Physiol.*, 1, 1–3, 2013.
119. De Silva, N.I., Lumyong, S., Hyde, K.D., Bulgakov, T., Phillips, A.J.L., Yan, J.Y., Mycosphere essays 9: defining biotrophs and hemibiotrophs. *Mycosphere*, 7, 5, 545–559, 2016.
120. Chitarra, G.S., Abee, T., Rombouts, F.M., Posthumus, M.A., Dijksterhuis, J., Germination of Penicillium paneum conidia is regulated by 1-octen-3-ol, a volatile self-inhibitor. *Appl. Environ. Microbiol.*, 70, 5, 2823–2829, 2004.
121. Arie, T., Fusarium diseases of cultivated plants, control, diagnosis, and molecular and genetic studies. *J. Pestic. Sci.*, 44, 4, 275–281, 2019.
122. Van West, P., Saprolegnia parasitica, an oomycete pathogen with a fishy appetite: new challenges for an old problem. *Mycologist*, 20, 3, 99–104, 2006.
123. Parveen, T. and Sharma, K., Pythium diseases, control and management strategies: A review. *Int. J. Plant Anim. Environ. Sci.*, 5, 1, 244–257, 2015.
124. Alhussaen, K., Hussein, E.I., Al-Batayneh, K.M., Al-Khatib, M., Al-Khateeb, W., Jacob, J.H., Hegazy, M.I., Identification and controlling Pythium sp. infecting tomato seedlings cultivated in Jordan Valley using garlic extract. *Asian J. Plant Pathol.*, 5, 2, 84–92, 2011.
125. Redekar, N.R., Eberhart, J.L., Parke, J.L., Diversity of Phytophthora, Pythium, and Phytopythium species in recycled irrigation water in a container nursery. *Phytobiomes J.*, 3, 1, 31–45, 2019.
126. Weiland, J.E., Scagel, C.F., Grünwald, N.J., Davis, E.A., Beck, B.R., Foster, Z.S., Fieland, V.J., Soilborne Phytophthora and Pythium diversity from rhododendron in propagation, container, and field production systems of the Pacific Northwest. *Plant Dis.*, 104, 6, 1841–1850, 2020.

127. Arora, H., Sharma, A., Sharma, S., Haron, F.F., Gafur, A., Sayyed, R.Z., Datta, R., Pythium damping-off and root rot of Capsicum annuum L.: impacts, diagnosis, and management. *Microorganisms*, 9, 4, 823, 2021.
128. Dean, R., Van Kan, J.A., Pretorius, Z.A., Hammond-Kosack, K.E., Di Pietro, A., Spanu, P.D., Foster, G.D., The Top 10 fungal pathogens in molecular plant pathology. *Mol. Plant Pathol.*, 13, 4, 414–430, 2012.
129. Yan, X. and Talbot, N.J., Investigating the cell biology of plant infection by the rice blast fungus Magnaporthe oryzae. *Curr. Opin. Microbiol.*, 34, 147–153, 2016.
130. Quoc, N.B. and Bao Chau, N.N., The role of cell wall degrading enzymes in pathogenesis of Magnaporthe oryzae. *Curr. Protein Pept. Sci.*, 18, 10, 1019–1034, 2017.
131. Law, J.W.F., Ser, H.L., Khan, T.M., Chuah, L.H., Pusparajah, P., Chan, K.G., Lee, L.H., The potential of Streptomyces as biocontrol agents against the rice blast fungus, Magnaporthe oryzae (Pyricularia oryzae). *Front. Microbiol.*, 8, 3, 2017.
132. Ashkani, S., Yusop, M.R., Shabanimofrad, M., Harun, A.R., Sahebi, M., Latif, M.A., Genetic analysis of resistance to rice blast: a study on the inheritance of resistance to the blast disease pathogen in an F3 population of rice. *J. Phytopathol.*, 163, 4, 300–309, 2015.
133. Wilson, R.A. and Talbot, N.J., Under pressure: investigating the biology of plant infection by Magnaporthe oryzae. *Nat. Rev. Microbiol.*, 7, 3, 185–195, 2009.
134. Yoon, M.Y., Cha, B., Kim, J.C., Recent trends in studies on botanical fungicides in agriculture. *Plant Pathol. J.*, 29, 1, 1, 2013.
135. Suprapta, D.N., Potential of microbial antagonists as biocontrol agents against plant fungal pathogens. *J. ISSAAS*, 18, 2, 1–8, 2012.
136. Heydari, A. and Pessarakli, M., A review on biological control of fungal plant pathogens using microbial antagonists. *J. Biol. Sci.*, 10, 4, 273–290, 2010.
137. Thambugala, K.M., Daranagama, D.A., Phillips, A.J., Kannangara, S.D., Promputtha, I., Fungi vs. fungi in biocontrol: An overview of fungal antagonists applied against fungal plant pathogens. *Front. Cell. Infect. Microbiol.*, 10, 604923, 2020.
138. Armas-Tizapantzi, A., Mata, G., Hernandez-Cuevas, L.V., Montiel-Gonzalez, A.M., Estructuras tipo toxocistos en Pleurotus ostreatus y P. pulmonarius. *Sci. Fungorum*, 49, e1250, 2019, Doi: 10.33885/sf.2019.49.1250.
139. Kwok, O.C., Plattner, R., Weisleder, D., Wicklow, D.T., A nematicidal toxin from Pleurotus ostreatus NRRL 3526. *J. Chem. Ecol.*, 18, 127–136, 1992, Doi: 10.1007/BF00993748.
140. Stadler, M., Mayer, A., Anke, H., Sterner, O., Fatty acids and other compounds with nematicidal activity from cultures of Basidiomycetes. *Planta Med.*, 60, 128–132, 1994, Doi: 10.1055/s-2006-959433.
141. Li, G., Wang, X., Zheng, L., Li, L., Huang, R., Zhang, K., Nematicidal metabolites from the fungus Pleurotus ferulae Lenzi. *Ann. Microbiol.*, 57, 527–529, 2007.

142. Koitabashi, M., Kajitani, Y., Hirashima, K., Antifungal substances produced by fungal strain Kyu-W63 from wheat leaf and its taxonomic position. *J. Gen. Plant Pathol.*, 70, 124–130, 2004, Doi: 10.1007/s10327-003-0095-2.
143. Xiang, H.Q. and Feng, Z.X., The nematicidal toxicity of the fruit bodies of Pleurotus ostreatus. *J. Shenyang Agric. Univ.*, 32, 173–175, 2001.
144. Montañez-Palma, L.F., Efecto del extracto hidroalcohólico del hongo comestible Neolentinus ponderosus sobre el nematodo Haemonchus contortus. Tesis de Maestría, Centro de Investigaciones Biológicas. Universidad Autónoma del Estado de Morelos, México, 2020.
145. Huang, W., Sun, J., Cui, J., Wang, G., Kong, L., Peng, H., Efficacy evaluation of fungus Syncephalastrum racemosum and nematicide avermectin against the root-knot nematode *Meloidogyne incognita* on cucumber. *PLoS One*, 9, e89717, 2014.
146. Qui, S.J., Can, J.Y., Liu, W.P., Becker, J.O., Degradation and absorption of fosthiazate in soil. *J. Agric. Food. Chem.*, 52, 6239–6242, 2004.
147. Gupta, E. and Mishra, P., Functional food with some health benefits, so called superfood: a review. *Curr. Nutr. Food Sci.*, 17, 2, 144–66, 2021 Feb 1.
148. Ramadan, W.A. and Soliman, G.M., Effect of different applications of bioagent Achromobacter xylosoxidans against *Meloidogyne incognita* and gene expression in infected eggplant. *Jordan J. Biol. Sci.*, 13, 363–370, 2020.
149. Le Mire, G., Nguyen, M.L., Fassotte, B., du Jardin, P., Verheggen, F., Delaplace, P., Jijakli, H., Implementing plant biostimulants and biocontrol strategies in the agroecological management of cultivated ecosystems, a review. *Biotechnol. Agron. Soc. Environ.*, 20, 299–313, 2016.
150. Archidona-Yuste, A., Cantalapiedra-Navarrete, C., Lie'banas, G., Rapoport, H.F., Castillo, P., Palomares-Rius, J.E., Diversity of root-knot nematodes of the genus Meloidogyne Goeldi, 1892 (Nematoda: Meloidogynidae) associated with olive plants and environmental cues regarding their distribution in southern Spain. *PLoS One*, 13, e0198236, 2018.
151. Tanweer, A. and Hisamuddin, A., Evaluation of Cassia tora for control of root-knot nematode (*Meloidogyne incognita*) in soybean. *Afr. J. Basic Appl. Sci.*, 4, 21–24, 2012.
152. Zhang, X., Zhang, X., Gu, S., Pan, L., Sun, H., Gong, E., Structure analysis and antioxidant activity of polysaccharide-iron (III) from Cordyceps militaris mycelia. *Int. J. Biol. Macromol.*, 178, 170–179, 2021.
153. Daba, G.M., Elkhateeb, W.A., EL-Dein, A.N., Ahmed, E.F., EL Hagrassi, A.M., Fayad, W., Wen, T.C., Therapeutic potentials of n-hexane extracts of the three medicinal mushrooms regarding their anti-colon cancer, antioxidant, and hypocholesterolemic capabilities. *Biodiversitas*, 21, 2437–2445, 2020.
154. Haydock, P.P., Woods, S.R., Grove, I.G., Hare, M.C., Perry, R.N., Moens, M., Chemical control of nematodes, in: *Plant nematology*, R.N. Perry and M. Moens (Eds.), pp. 392–410, CABI Publishing, Oxford shire, UK, 2006.

155. Dahlin, P. and Hallmann, J., New insights on the role of allyl is thiocyanate in controlling the root knot nematode *Meloidogyne hapla. Plants*, 9, 603, 2020.
156. Islam, W., Adnan, M., Shabbir, A., Naveed, H., Abubakar, Y.S., Qasim, M., Insect- fungal-interactions: a detailed review on entomopathogenic fungi pathogenicity to combat insect pests. *Microb. Pathog.*, 159, 105122, 2021.
157. Wanchoo, A., Lewis, M.W., Keyhani, N.O., Lectin mapping reveals stage-specific display of surface carbohydrates in *in vitro* and haemolymph-derived cells of the entomopathogenic fungus *Beauveria bassiana. Microbiology (Reading)*, 155, Pt 9, 3121–33, 2009.
158. Rashid, S.N., Biodiversity, Distribution and Morphological characterization Of Mushrooms In Southern Region Of Bangladesh (Doctoral Dissertation), Department Of Plant Pathology, Sher-E-Bangla Agricultural University, 2014.
159. Aman, A.S., Kumar, A., Kumar, P., Mishra, P.K., Mohan, M., Implementation of An Effective and Successful Management Tactics against Diamondback Moth (Plutella xylostella L.) on Cabbage. International Year of Millets 2023, 47, 2022.
160. Chaurasia, P.K. and Bharati, S.L., Applicability of fungi in agriculture and environmental sustainability, in: *Microbes in land use change management*, pp. 155–172, Elsevier, Amsterdam, 2021.
161. Thuc, L.V., Corales, R.G., Sajor, J.T., Truc, N.T.T., Hien, P.H., Ramos, R.E., Van Hung, N., Rice-straw mushroom production, in: *Sustainable rice straw management*, pp. 93–109, 2020.
162. Bhumarkar, R., Mahajan, G., Kumar, A., Doubling farmers income and attaining resilience in agriculture through crop diversification, Agriculture & Food: E-Newsletter, 2021.
163. Singh, O. and Kushwaha, K.P.S., Evaluation and studies of different strains of white button mushroom Agaricus bisporus (Lange.) Sing. *Scientist*, 2, 1, 106–116, 2023.
164. Altaf, S., Jan, S.K., Basu, U., Ahanger, S.A., Dave, A., Kakraliya, S.S., Mushtaq, M., Sustainable Management of Green Mold Disease of White Button Mushroom Using Botanicals and Biocontrol Agents under Temperate Conditions. *Horticulturae*, 8, 9, 768, 2022.
165. Lahlali, R., Ezrari, S., Radouane, N., Kenfaoui, J., Esmaeel, Q., El Hamss, H., Belabess, Z., Barka, E.A., Biological Control of Plant Pathogens: A Global Perspective. *Microorganisms*, 10, 3, 596, 2022, https://doi.org/10.3390/microorganisms10030596.
166. Keshari, N. and Kranti, K.V., Integrated management of phytopathogenic nematodes infesting mushroom, in: *Management of Phytonematodes: Recent Advances and Future Challenges*, pp. 141–70, 2020.
167. Saiful, M.D.I., Cultivation techniques of edible mushrooms: Agaricus bisporus, Pleurotus spp, Lentinula edodes and Volvariella volvocea, 2015.

168. Shamugam, S. and Kertesz, M.A., Bacterial interactions with the mycelium of the cultivated edible mushrooms Agaricus bisporus and Pleurotus ostreatus. *J. Appl. Microbiol.*, 134, 1, lxac018, 2023.
169. Chang, W., Feng, W., Yang, Y., Shen, Y., Song, T., Li, Y., Cai, W., Metagenomics analysis of the effects of Agaricus bisporus mycelia on microbial diversity and CAZymes in compost. *PeerJ*, 10, e14426, 2022.
170. Törős, G., El-Ramady, H., Prokisch, J., Velasco, F., Llanaj, X., Nguyen, D.H., Peles, F., Modulation of the Gut Microbiota with Prebiotics and Antimicrobial Agents from *Pleurotus ostreatus* Mushroom. *Foods*, 12, 10, 2010, 2023.
171. Di Francesco, A., Zajc, J., Stenberg, J.A., Aureobasidium spp.: diversity, versatility, and agricultural utility. *Horticulturae*, 9, 1, 59, 2023.
172. Adam, S., Phoenix Oyster Mushroom: What They Are + How to Grow Them, 2022, Retrieved from https://grocycle.com/phoenix-oyster-mushroom/ September 16, 2023.
173. Akyüz, M., İnci, S., Kirbağ, S., Therapeutic properties of oyster mushroom (*P. ostreatus*) cultivation on different cellulosic substrates. *J. Res. Pharm.*, 27, 1, 320, 2023.
174. Lesa, K.N., Khandaker, M.U., Mohammad Rashed Iqbal, F., Sharma, R., Islam, F., Mitra, S., Emran, T.B., Nutritional value, medicinal importance, and health-promoting effects of dietary mushroom (*Pleurotus ostreatus*). *J. Food Qual.*, 1, 2454180, 2022.
175. AgrSciEd, J.A.M.S., Abstracts of the First International Conference on Plant Protection (ICPP). *J. Agric. Mar. Sci. [JAMS]*, 28, 2, 55–162, 2023.
176. Adetunji, C.O., Ahamed, M.I., Altalhi, T. (Eds.), *Microbial Biostimulants for Sustainable Agriculture and Environmental Bioremediation*, CRC Press, Boca Raton, Florida, 2022.
177. Zhao, J., Lin, J., Yan, J., Zhang, C., Wang, T., Gan, B., Evaluation of the nutritional value, umami taste, and volatile organic compounds of Hypsizygus marmoreus by simulated salivary digestion *in vitro*. *Curr. Res. Food Sci.*, 7, 100591, 2023.
178. Lowenthal, R., Taylor, M., Gidden, J.A., Heflin, B., Lay Jr., J.O., Avaritt, N., Urbaniak, A., The mycelium of the Trametes versicolor synn. Coriolus versicolor (Turkey tail mushroom) exhibit anti-melanoma activity *in vitro*. *Biomed. Pharmacother.*, 161, 114424, 2023.
179. WA, E., Elnahas, M.O., Thomas, P.W., Daba, G.M., Trametes versicolor and Dictyophora indusiata champions of medicinal mushrooms. *Pharm. Res.*, 4, 1, 00019, 2020.
180. Miller, S.L. and Sayner, A., Cultivation of Medicinal Mushrooms for a World in Need, in: *Wild Mushrooms and Health*, pp. 95–129, CRC Press, Boca Raton, Florida, 2023.
181. Muniroh, M.S., Nusaibah, S.A., Vadamalai, G., Siddique, Y., Proficiency of biocontrol agents as plant growth promoters and hydrolytic enzyme producers in Ganoderma boninense infected oil palm seedlings. *Curr. Plant Biol.*, 162, 20, 100116, 2019.

182. Moussa, A.Y., Fayez, S., Xiao, H., Xu, B., New insights into antimicrobial and antibiofilm effects of edible mushrooms. *Food Res. Int.*, 162, 111982, 2022.
183. Muto, Y., Fukushima-Sakuno, E., Ishihara, A., Osaki-Oka, K., Antimicrobial activity of octan-3-one released from spent mushroom substrate of shiitake (Lentinula edodes) and its inhibitory effects on plant diseases. *J. Gen. Plant Pathol.*, 89, 2, 122–131, 2023.
184. Mattedi, A., Sabbi, E., Farda, B., Djebaili, R., Mitra, D., Ercole, C., Pellegrini, M., Solid-State Fermentation: Applications and Future Perspectives for Biostimulant and Biopesticides Production. *Microorganisms*, 11, 6, 1408, 2023.
185. Youssef, M.M.A. and El-Nagdi, W.M.A., New approach for biocontroling root-knot nematode, *Meloidogyne incognita* on cowpea by commercial fresh oyster mushroom (*Pleurotus ostreatus*). *Jordan J. Biol. Sci.*, 14, 173–177, 2021.
186. Bua-art, S., Saksirirat, W., Hiransalee, A., Kanokmedhaku, S., Lekphrom, R., Effect of bioactive compound from luminescent mushroom (Neonothopanus nambi Speg.) on root-knot nematode (*Meloidogyne incognita* Chitwood) and non-target organisms. *KKU Res. J.*, 16, 331–341, 2011.
187. Yang, C.H., Kao, Y.H., Huang, K.S., Wang, C.Y., Lin, L.W., Cordyceps militaris and mycelial fermentation induced apoptosis and autophagy of human glioblastoma cells. *Cell Death Dis.*, 3, e431–e431, 2012.
188. Soliman, G., Elkhateeb, W., Wen, T.C., Daba, G., Mushrooms as efficient biocontrol agents against the root-knot nematode, *meloidogyne incognita. Egypt. Pharm. J.*, 21, 68–74, 2022.
189. Kwenti, T.E., Biological Control of Parasites, in: *Natural remedies in the fight against parasites*. InTech, London, 2017, doi: 10.5772/68012.
190. Zhang, Y., Li, S., Li, H., Wang, R., Zhang, K.Q., Xu, J., Fungi-Nematode Interactions: Diversity, Ecology, and Biocontrol Prospects in Agriculture. *J. Fungi (Basel Switzerland)*, 6, 4, 206, 2020, https://doi.org/10.3390/jof6040206.
191. Nandini, B., Puttaswamy, H., Saini, R.K., Prakash, H.S., Geetha, N., Trichovariability in rhizosphere soil samples and their biocontrol potential against downy mildew pathogen in pearl millet. *Sci. Rep.*, 11, 9517, 2021, doi: 10.1038/s41598-021-89061-2.
192. He, D.C., He, M.H., Amalin, D.M., Liu, W., Alvindia, D.G., Zhan, J., Biological Control of Plant Diseases: An Evolutionary and Eco-Economic Consideration. *Pathogens*, 10, 1311, 2021, doi: 10.3390/pathogens10101311.
193. Hassan, M.N., Afghan, S., Hafeez, F.Y., Biological control of red rot in sugarcane by native pyoluteorin-producing Pseudomonas putida strain NH-50 under field conditions and its potential modes of action. *Pest Manage. Sci.*, 67, 1147–1154, 2011, doi: 10.1002/ps.2165.

6

Production of Stable Enzymes from Mushrooms with Numerous Biomedical Applications

Idris Abdullahi Dabban[1,2*], Olorunshola Lucky Opeyemi[1], Mustapha Adamu[2], Bashaku Yakubu Dabban[3], Auwalu Bala[4], Oyegbade Samuel Adeniyi[5], Oluwafemi Adebayo Oyewole[2] and Charles Oluwaseun Adetunji[6]

[1]Department of Biological Sciences, The Federal Polytechnic Bida, Niger State, Nigeria
[2]Department of Microbiology, Federal University of Technology, Minna, Niger State, Nigeria
[3]Department of Integrated Sciences, Federal College of Education (Technical), Gusau, Zamfara State, Nigeria
[4]Department of Microbiology, Sokoto State University, Sokoto State, Nigeria
[5]Department of Biological Sciences, Covenant University, Ota, Ogun State, Nigeria
[6]Department of Microbiology, Edo State University, Uzairue, Edo State, Nigeria

Abstract
Mankind has faced several challenges to the success of metabolic activities in several biochemical systems and most of these challenges have been trounced using biological catalysts known as enzymes. A plethora of mushrooms have for a very long time been acknowledged as good and reliable sources of medicinal and nutritional compounds. Mushrooms are within the fungal class which produces over 50% of global enzymes. Over the years, critical functions have been played by mushroom-secreted enzyme and their catalytic products to solve various biomedical and biotechnological problems as these enzymes are stable, cheap, efficient, extracellular, eco-friendly, and well-suited with confirmed application in various industries. The majority of mushroom enzymes applied in biomedicine and biotechnology are from the oxidoreductase and hydrolase class. Biomedical application of mushroom enzymes or their catalytic products is mostly in the field of

**Corresponding author*: abdullahi.idris@fedpolybida.edu.ng

Charles Oluwaseun Adetunji, Chukwuebuka Egbuna, Oluwatosin Ademola Ijabadeniyi and Samantha C. Karunarathna (eds.) Mushroom Biotechnology for Improved Agriculture and Human Health, (123–158) © 2025 Scrivener Publishing LLC

antimicrobial therapy, as antioxidants, and in the treatment of cancer and tumors. This chapter therefore highlights the classes/types of mushrooms, the stable enzymes produced by these mushrooms, and the biomedical and biotechnological applications of mushrooms and their catalytic products.

Keywords: Mushroom, enzymes, biomedicine, application, biotechnology

6.1 Introduction

Enzymes are very important biological catalysts and they play significant functions in worldwide medical, food, industrial, environmental, and biotechnological applications. They are mostly proteins that decrease the activation energy (E_a) and increase the rate of biochemical reactions; they however remain unchanged after the metabolic reaction [1]. Enzymes have for a very long time solved many of man's problems associated with metabolic biological reactions ranging from simple fermentation of beer, bread, and wine to more technical and advanced systems of drug production and polymerase chain reactions [2].

The biochemical reactions enzymes catalyze are grouped into seven main classes, i.e., (a) transferases, (b) Oxidoreductases, (c) lyases, (d) hydrolases, (e) isomerases, (f) translocases, and (g) ligases [3]. Enzymes are commonly found in nature as they have been successfully sourced and isolated from plants, animals as well as microorganisms. However, global attention has, for a long time, been shifted to enzymes from microorganisms, especially bacteria and fungi. This is attributed to an increase in industrial demand which cannot be met by plant and animal sources coupled with the fact that the production of biocatalysts from microbial sources is fast, cheap, scalable, and ease of purification and the microorganisms can be genetically manipulated [4, 5].

Out of the microbial sources of enzymes, the fungal group is more preferred than other microbial sources of enzymes [1]. Enzymes from fungal sources have efficient catalysis with thermal and environmental stability, extracellular, high yield, and ease of purification and organisms from this group have the Generally Recognized As Safe (GRAS) status [1, 2]. Over 50% of commercially available enzymes sold in the industrial market are from fungal sources, especially from *Trichoderma, Aspergillus, Penicillium, Rhizopus, Saccharomyces*, *Kluveromyces,* and *Candida* species which fulfill the requirements for commercial scale production of enzymes. However, the continuous and ever-growing market need for industrially important enzymes has increased the search for producers of novel enzymes, especially from local sources of microorganisms with desirable commercial and industrial traits [1].

Mushrooms are macroscopic, filamentous, saprophytic higher Basidiomycetes, and Ascomycetes fungi that grow apically and are composed of the chitino-cellulosic cell wall [6–8]. Their mode of reproduction is asexual (sporulation) and sexual [8]. Mushrooms grow to produce fruiting bodies and they are divided into edible and poisonous mushrooms [6]. The edible mushrooms (e.g., *Pleurotus ostreatus*) have nutritional properties and they contain a lot of amino acids, vitamins, minerals, antioxidants, fibre, fatty acids, and several secondary metabolites [9–11]. They have therefore been used for centuries as traditional medicine and food due to their different ecological, morphological, and physiological distinctiveness [8].

The cultivation of edible mushrooms has in recent times revealed to be a promising alternative source of biomedically important enzymes. This could be due to the high population (more than 14,000 known species), low cost of cultivation, high rates of production globally, and their nutritional benefits to human wellness [8, 12].

Mushrooms are saprophytes by obtaining their nutrients mostly from ligno-cellulosic waste matter, and they therefore are known to produce ligno-cellulytic enzymes such as laccases, cellulases, lignin peroxidases, and beta-gluocsidases, which possess industrial and biomedical applications [13, 14]. Given that mushrooms are known innate bio-decomposers; they therefore produce both intracellular and extracellular biocatalysts needed for catalysis of a plethora of substrates [15]. The biochemical processes that take place during growing and maturation of mushroom mycelia and sporophores (fruiting bodies) cause the release of extracellular enzymes that break down complex insoluble materials to soluble and more simple compounds, which are then processed by mushrooms' intracellular enzymes [16].

Globally, the button mushroom (*Agaricus bisporus*) is a widely cultivated mushroom. Others include shiitake oyster mushrooms (*Pleurotus* spp), (*Lentinus edodes*), straw mushrooms (*Volvariella volvacea*), winter mushrooms (*Flamulina velutipes*), and wood ear mushroom (*Auricula auricula*) [17]. However, several medicinal mushrooms have also been reported to produce enzymes with biomedical (antimicrobial, anticancer, antioxidants) and industrial applications. This includes *Tuber maculatum* and *Tuber aestivum* (Truffles) which produce biomedically important enzymes including xylanase, amylase, laccase, lipase, catalase, cellulase, peroxidase, etc. [18]. This chapter therefore elucidates the types and classes of mushrooms,

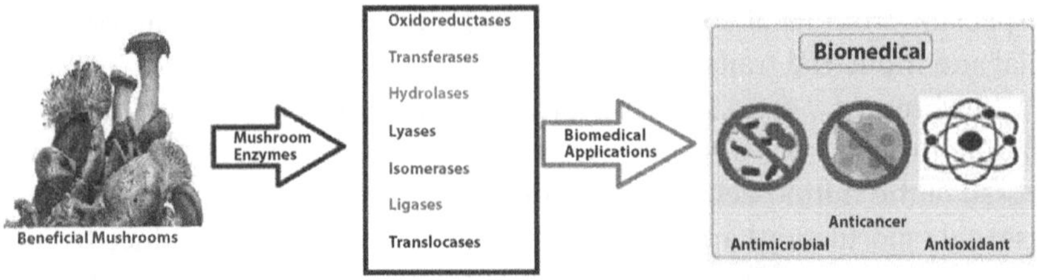

Figure 6.1 Pictorial introduction on biomedical applications of beneficial mushroom enzymes.

the stable enzymes they produce, industrial, environmental, and biomedical applications of enzymes produced by mushrooms (Figure 6.1).

6.2 Classes/Types of Mushrooms

The classification of mushrooms is generally into a few major groups; this is because these groups extend beyond one group since types of mushrooms in one group may fall often into another group. A typical example is the oyster mushroom which falls under wild group of mushrooms as well as cultivated mushrooms. Generally, mushrooms are classified as wild, cultivated, medicinal, edible (useful), poisonous, and psychoactive mushrooms (Figure 6.2) [19]. Some examples of mushrooms (Figure 6.3) and their characteristics are described below.

6.2.1 Button Mushroom (*Agaricus bisporus*)

The button mushroom, scientifically known as *Agaricus bisporus*, is widely recognised as one of the most commonly available mushroom varieties in the food market. Nevertheless, it is noteworthy that these organisms exhibit highly toxic characteristics when found in their natural habitats [20].

6.2.2 Oyster (*Pleurotus ostreatus*)

The *Pleurotus ostreatus*, also known as oyster mushrooms, possess a visually ornate appearance and are characterised by a moderate yet tasty taste. These items can be easily obtained from retail establishments and are straightforward to grow in domestic settings. Although *Pleurotus pulmonarias* and *P. populinus*are not widely recognized as the most popular oyster mushroom species globally, they are nonetheless classified as oyster mushrooms [21].

Wild

- found naturally in undomesticated areas
- difficult/impossible to cultivate especially in large scale
- most are poisonous but look edible

Cultivated

- grown commercially in large scale
- requires a diverse technique for cultivation
- Examples include oyster mushrooms, shiitake, portabello, enoki, etc.

Medicinal

- have applications in health and used as medicine
- some cannot be easily consumed due to hardiness and/or bitter taste
- they are converted into teas or capsules and are also used as supplements
- e.g shiitake, reishi, lion's mane, turkey tail and porcini

Edible/Useful

- some are edible while other are not edible
- non edible types are used for other application such as environmental bioremediation of pollutants
- some are used in composting
- other potential applications include production of biofuels, cleaning goods packaging etc.

Poisonous

- includes several wild varieties of mushrooms
- they are highly poisonuos
- requires proper identification before ingestion or other usage.
- many species can be fatal if ingested

Psychoactive

- are mushrooms that can induce hallucinations
- exhibit psychotropic impact
- psilocybin (the hallucinogen c) is found in most mushroom in this group.
- cultivation is banned in most countries due to its harmful nature.

Figure 6.2 Characteristics of the mushroom classification.

6.2.3 Portabello

The mushroom is characterized by its substantial size and robust structure, with the capacity to expand to a diameter of up to 6 inches. Plant-based alternatives are commonly utilized as a substitute for meat and exhibit a highly favourable flavour profile when subjected to grilling techniques [22].

6.2.4 Morel (*Morchella*)

The Morel mushroom, scientifically known as *Morchella*, often undergoes its blooming process in forest regions throughout the period ranging from

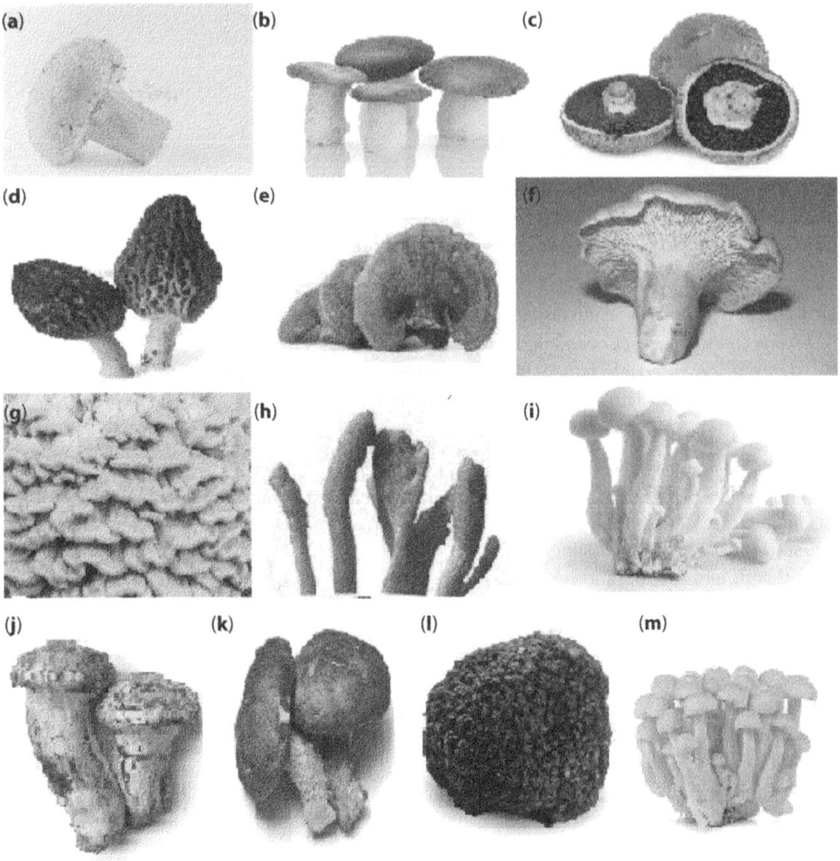

Figure 6.3 Pictorial examples of some edible/beneficial mushrooms.
(a) – Button mushroom (*Agaricus bisporus*); (b) – Oyster (*Pleurotus ostreatus*); (c) – Portabello; (d) – Morel (*Morchella*); (e) – Reishi (*Ganoderma lucidum*); (f) – Chanterelle (*Cantharellus*); (g) – *Laetiporus sulphureus*; (h) – Cordyceps (*Cordyceps militaris*); (i) – Enokitake (*Flammulina filiformis*); (j) – Matsutake mushroom; (k) – Shiitake mushroom (*Lentinula edodes*); (l) – Truffles; (m) – Shimeji

early to late April. This particular species of fungus is highly regarded for its palatability and suitability for consumption. The object in question exhibits characteristics consistent with that of a sponge. Despite their resemblance to true morels, fake morels are toxic [23].

6.2.5 Reishi (*Ganoderma lucidum*)

Reishi (*Ganoderma lucidum*) is widely recognised as a medicinal fungus that possesses immune-enhancing properties and exhibits antioxidant activity [24].

6.2.6 Burnt Matches (*Eutypellascorpia*)

The species *Eutypellascorpia,* commonly known as burnt matches, bears a resemblance to a cluster of charred paper matches and is commonly observed throughout the winter season on twigs and trees [25].

6.2.7 Chanterelle (*Cantharellus*)

The Chanterelle mushroom, scientifically known as *Cantharellus,* is commonly referred to as a highly esteemed fungal species. The food item possesses a robust taste profile, exhibiting vibrant hues of orange or yellow, and displaying a pleasantly supple consistency. The collection of chanterelle mushrooms typically occurs during the autumn season, specifically from the soil located beneath deciduous trees such as oak trees. It is essential to know that some species of chanterelle mushrooms possess hazardous properties [26].

6.2.8 *Laetiporus sulphureus*

Laetiporus sulphureus, commonly referred to as the chicken mushroom, exhibits a flavour profile that is notably reminiscent of chicken. The occurrence of these mushrooms is species-dependent, often observed in large, fan-shaped clusters that may be associated with a tree, whether it is living or deceased [27].

6.2.9 Cordyceps (*Cordyceps militaris*)

Cordyceps militaris (Cordyceps) is a species of fit-to-be-eaten fungus that has gained significant recognition for its medicinal properties. Several investigations have been carried out to investigate the possible anti-inflammatory and anti-cancer qualities of this important mushroom [28].

6.2.10 Enokitake (*Flammulina velutipes*)

Enokitake mushrooms, also known as *Flammulina velutipes,* are readily available for purchase in many retail establishments. These mushrooms are characterized by their slender, needle-like shape and distinctive white hue. Enoki mushrooms are employed in several culinary applications, like pickling or rapid frying, in order to impart a distinctive texture to any given dish [29].

6.2.11 Giraffe Spots (*Endophora albobadia*)

A species of crust fungi known as *Peniophora albobadia* belongs to the Peniophoraceae family. The organism, which grows expanding crusts on the bark of decaying twigs and fallen branches from various hardwood species, is saprophytic. The colors white and reddish-brown, which accurately reflect the stark contrast between the fertile part and the margin, were used to create the species epithet. It is frequent to find sticks and branches on the ground during the winter that resemble the giraffe's coat pattern [30].

6.2.12 Destroying Angel (*Amanita sp.*)

Amanita sp., also referred to as the destroying angel or the European destroying angel amanita, is a highly toxic basidiomycete fungus belonging to the genus *Amanita. Amanita virosa* is often found throughout Europe and is known to form associations with a diverse range of deciduous and coniferous plants [31]. The fruiting bodies, often known as mushrooms, exhibit a substantial size and are observed during the summer and autumn seasons. Notably, the caps, stipes, and gills of these mushrooms all display a uniform white coloration. During the transition from summer to autumn, there is a notable proliferation of highly lethal mushroom species in forested areas. The visual characteristics of the object in question closely mimic those of commercially available button mushrooms commonly found in supermarkets. The term "deadly" is employed in accordance with its semantic connotation, as indicated by Alvarado *et al.* [32].

6.2.13 Green-Spored Lepiota (*Chloropyhllum*)

The *Chlorophyllum*, commonly known as the Green-spored Lepiota, is a white, erect mushroom that typically inhabits grassy areas. It has the potential to reach a diameter of up to 8 inches. The residual substance left by the object in question consists of spores that exhibit a coloration ranging from grey to green. It is advisable to refrain from engaging in these perilous species [33].

6.2.14 Matsutake Mushroom

The matsutake mushroom, often known as the pine mushroom, is significantly more prevalent in Asia, Northern Europe, and North American cultures. Due to its distinctive growth pattern beneath specific trees and occasional concealment by foliage on the forest floor, the matsutake mushroom is a rare discovery in its natural habitat [34]. Pine mushrooms are sought after as food by several animal species, including rabbits, deer, and squirrels, among others, so indicating that humans are not the sole organisms expressing a need for these mushrooms. The matsukate cultivar is known to be characterized by a pronounced spicy flavour and scent [34].

6.2.15 Shiitake Mushroom (*Lentinula edodes*)

The Shiitake mushroom (*Lentinula edodes*) is a highly esteemed Asian fungus, second most consumed globally, known for its wide range of culinary applications and therapeutic attributes. The mushroom is believed to contain immune-stimulating chemicals prominent of which is Lentinen [35].

6.2.16 Truffles

Truffles, renowned for their exceptional flavour, are highly coveted due to their scarcity. These subterranean fungi are employed by canines that have undergone specialised training and possess highly developed olfactory abilities. Truffles have a subterranean reproductive structure which makes it a challenge for visual detection of the fungus due to the absence of an outward stalk or cap that would emerge above the soil surface. Truffles exhibit a visually unattractive and irregular appearance; however, they are in high demand worldwide due to their opulent, nutty, and earthy taste [36].

6.2.17 Shimeji

The discernible attributes of these diminutive edible fungi encompass their elongated stalks and inwardly curved, compacted crowns. Consuming these mushrooms in their raw state is not advisable due to their potential difficulty in digestion, which can be mitigated by subjecting them to roasting for a sufficient duration. They exhibit favourable reactions to a wide range of culinary techniques, regardless of whether they are subjected to elevated, reduced, rapid, or gradual heat levels. One may proceed to either fry or braise the shimeji mushrooms in order to savour their delicate and earthy flavour [37].

6.3 Stable Enzymes Produced by Mushrooms

One of the essential current components to be obtained from therapeutic mushrooms (truffles and various other eatable mushroom genera) is stable enzymes. The fact that fungal cells act as mini-factories with a wide range of synergist activities gives fungal catalysts (enzymes) an advantage over those derived from other sources in general [38]. They also produce their products effectively using cost-effective media. According to Field *et al.* (1993), the enzymes released by mushrooms are ambiguous (nonspecific), and they can dissolve a wide range of fundamentally different combinations and substances. The various drugs, nutraceuticals, and extracellular enzymes that can be made from mushrooms are well known. According to the growth conditions and substrates used, mushrooms have a large collection of enzymes including lipases, lignin debasing enzymes, oxidases, cellulases, proteolytic catalysts, and invertase [39, 40]. A few truffle groups have been examined for catalyst (enzyme) creation, for example, *Tuber maculatum*, which produces xylanase, lipase, amylase, cellulase, laccase, peroxidase, catalase, and *Tuber aestivum* which produces peroxidase, catalase, and amylase [38]. Palatable mushrooms are supposedly rich wellsprings of different lipolytic, cellulolytic, ligninolytic, proteolytic, and stress enzymes which have different biotechnological and biomedical applications [38].

6.3.1 Ligninolytic Enzymes

Ligninolytic enzymes are extracellular, oxidative, and vague (nonspecific) proteins that produce temperamental items by means of oxidative responses [12].

6.3.1.1 Lignin Peroxidase

Lignin peroxidases have hemi-containing iron particles in their dynamic site and they catalyze hydrogen peroxide subordinate oxidation of nonphenolic fragrant mixtures (compounds) into cation revolutionaries to decay them [12]. Lignin depolymerization by means of lignin peroxidase is very productive due to its high decrease potential [12]. A mushroom *Pleurotus sajor-caju* developed on coir pith was uncovered to show the greatest lignin and manganese peroxidase activity of around 16.25 U/ml and 20.2 U/ml separately on the twentieth day of fermentation with Azolla and Soya structures as supplement [41]. The development of fungi in enhanced media brought about a decrease in organic carbon and lignin content consistently during aging. The deterioration began from the fifth day of innoculation and reached the peak somewhere in the range of the fifteenth and twenty-fourth day with a slow reduction from that point. Between 25 to 30 days, lignin content was diminished from 38.5% to 16.15% [41].

6.3.1.2 *Manganese Peroxidase*

This class of ligninolytic catalyst has been supposedly delivered by a progression of fungi. This enzyme oxidizes phenolic mixtures to phenoxy revolutionary by oxidation of Mn (II) to Mn (III) particles within the sight of Hydrogen peroxide (H_2O_2) as a cofactor. Beginning depolymerization of lignin requires lignin peroxidase and manganese peroxidase, so the presence of the enzymes in the response combination is accounted for to accelerate the depolymerization cycle [42].

6.3.1.3 *Versatile Peroxidase*

The heme access channel, the reactant tryptophan, and the Mn oxidation site are all features of this unusual enzyme with a wide substrate preference. It is known as hybrid peroxidase because adaptable peroxidase has similar synergist tools to those of both lignin peroxidase and manganese peroxidase. The ability of adaptive peroxidase to degrade substrates with high decrease potential without the need for veratryl liquid and to oxidize Mn^{2+} on its own is what sets it apart the most. Because of its dual functions, it has shown to be a desirable choice for biotechnological applications [42]. The main enzyme involved in lignin modification without Mn^{2+} is adaptable peroxidase [43].

6.3.1.4 *Laccases*

Laccases are blue multi-copper oxidases that catalyze the reduction of atomic oxygen to water along with the oxidation of other aromatic substrates. With oxygen serving as the sole electron acceptor, this enzyme can degrade both phenolic and non-phenolic substances [42]. Although many Basidiomycetes, Ascomycetes, and Deuteromycetes fungi produce laccase, the white decay organisms are particularly effective at producing laccase [44]. Laccase from fungal sources has amazing utility in working with lignin debasement. In any event, the lower fungi (Chytridiomycetes and Zygomycetes), which have not yet received sufficient attention [45], have not yet been reported to produce laccase.

Wild consumable mushrooms *Panus* sp, *Pleurotus giganticus, and Lentinus sajor-caju*, from Tripura were distinguished as laccase makers utilizing 0.02% guaiacol as substrate. Additionally, *Hypsizygus marmoreus, Agaricus bisporus, Pleurotus sajors-caju, Agrocybe aegerita, Lentinula edodes, Pleurotus eryngii,* and *Pleurotus ostreatus* were viewed as strong producers of laccase in extracellular medium containing copper sulfate alongside Tween 80 and 4, 6-dimethyl-2-mercaptopyrimidine [42]. The consumable and medicinal significant mushroom *Hericium erinaceus* was grown for the creation of laccase in the growth medium arranged by malt concentrate and peptone [46]. Laccases are fundamental for detoxification and sterilization of waste, decolorizing colors (dyes), and degenerating polyaromatic hydrocarbons in refuse management [46]. For removing lignin from wood and bio-blanching wood mash, laccases are used in the paper industry. Laccases are used in the food sector to modify wine, process and handle fruit liquids, and enhance the surface and texture of dough when baking. According to Chen *et al.* [47], laccases are used for discovering, enzymatic and immunological tests, hair coloring, and reducing the effects of poison ivy dermatitis. Lacasses participate in the cycles of wound healing and defense against external stresses [48]. However, they do not pursue incredibly intricate designs, such as phenolic compounds with a few fragrant rings [49].

6.3.1.5 Tyrosinase

Tyrosinase, also known as phenolase, o-diphenyl oxidase, polyphenol oxidase, catechol oxidase, monophenol oxidase, or cresolase, belongs to a class of type-3 copper proteins that are involved in the synthesis of melanin. Ensuing oxidation and other nonenzymatic responses results to the arrangement of dim earthy colored shades [50]. They additionally assume huge parts in the spore development, guard, and destructiveness component. Among various phases of development, the most extreme tyrosinase is accounted for to be gathered during the untimely phase of development for eatable mushrooms. Tyrosinase is the principal enzyme that is liable for the enzymatic browning of fruits after harvest [51].

6.3.2 Hydrolases

6.3.2.1 Cellulases

Cellulases are multi-compound (enzyme) complex made out of endo-β-1,4-glucanase, exo-β-1,4-glucanase I and II, and β-glucosidase or cellobiase. The enzymes act consecutively to break the straight polymer into monosaccharide glucose unit. The palatable mushrooms enzyme creation fluctuates extraordinarily relying upon the idea of the lignocellulosic materials. In *Pleurotus eryngii* grown on kenaf stalks, bulrush stalks, cotton seed, and ramie stalks body showed ramie stalks and kenaf stalks were best appropriate to cultivate the mushroom with organic productivity of 55% and 57%, separately [52].

Both the substrates initiated the creation of cellulase enzymes. A few cellulases created by helpful mushrooms have tracked down applications in the textile factories for the biostoning of denim pieces of clothing, working on the effectiveness of cleansers such as detergent and cleaning tiles. In addition, these catalysts (enzymes) are fundamental for planning effectively degradable delicate cardboard, and paper, improved lignocellulosic biomass for livestock feed, buildups for the development of soil richness, and for getting ready definitions to eliminate modern ooze [52]. In food and brewery industries, cellulases are used to extract and purify leafy food juices, produce natural product nectars and puree, synthesize olive oil utilized in carotenoid extractions for the creation of colouring agents for food specialists and the plant-added substances for the food industries as a thickener [53]. Mushroom cellulases likewise have a few biomedical applications as they can be utilized as an intense antitumor agent.

6.3.2.2 *Xylanase*

Xylanases are fundamental for the total degradation of plant biomass involving cellulose, hemicellulose, and lignin to create biofuels and other value added compounds [54]. It is recommended that xylanase assume a part in cell lengthening and processing of fruit juices [55]. In the paper and mash industry, xylanase is expected for blanching kraft mash (pulp) and deinking printed papers [56].

6.3.2.3 *Pectinases*

Pleurotus sajor-caju grown on an onion squander produced a thermostable pectinase through solid-state fermentation [57]. Furthermore, pectinase was additionally reported to be synthesized by *Lentinula boryana* on wheat straw dextrose agar at 25°C for 7 days [58]. Pectinases are broadly utilized in the natural product (fruit) juice extraction, juice purification, degumming of regular strands, production of pectin-free starch, treatment of wastewater or sewage, and in cocoa as well as the tobacco industry. Pectinase treatment expands the pace of fermentation of tea and eliminates its frothing propensity and the enzyme likewise eliminates the adhesive coat from coffee beans [59].

6.3.2.4 *Amylases*

Enzymes that take part in the hydrolytic breakdown of starch are cooperatively referred to as amylolytic compounds (enzymes) or amylases. Explicit catalysts (enzymes) within this group includes β-amylase, pullulanase, α-amylase, gluco-amylase (otherwise called amyloglucosidase), and insoamylase. Amylases are grouped into two classes endoamylases and exoamylases [60]. Endoamylases catalyze hydrolysis in an irregular way on the inside of the starch particle. Different chain-length oligosaccharides are arranged straight and expanded as a result of this activity. Exoamylases hydrolyze from the non-decreasing end, which efficiently produces quick outcomes. The complete breakdown of starch is being carried out by a massive array of amylases. The complex collection of enzymes known as amylases hydrolyzes polysaccharides like starch to glucose and hence has a wide range of biotechnological applications in the food industry, fermentation, textile, and paper industries [61].

6.3.2.5 *Proteases*

Proteases synthesized by beneficial mushrooms have been used for different biotechnological and biomedical applications. Pleureryn is a little and novel aspartic protease with a sub-atomic mass of 11.5 kDa from the consumable mushroom *P. eryngii* [60]. In any case, its N-terminal arrangement looks like DNA replication authorizing factors more than aspartic proteases. It likewise shows some inhibitory action against HIV-1 opposite transcriptase. Pleureryn displays a pH ideal of 5 and an ideal temperature of 45°C with significant action staying at high temperatures and at pH 4 and 12. Pleureryn is exceptional in that it is somewhat steady to changes in pH or temperature [61]. Other mushroom proteases will quite often be less steady. Proteases are key in different industries like dairy, cowhide (leather), cleansers such as detergent, fermenting, meat, and visual ventures due to the hydrolytic capacities of amide linkage. Mushrooms like *Lepista nuda* and *Termitomyces clypeatus* have also been reported to sufficiently produce proteases [62].

6.3.2.6 *Esterase*

A few biomedical applications of mushroom esterases have been reported including a combination of chiral drugs, therapy for some cancerous growths, neuropathy, manufacture of biosensors, therapy of innate angioedema, therapy of hypercholesterolemia, and other quality treatment procedures [63].

6.3.3 Stress Enzymes

Higher mineral concentrations in different parts of the fruiting body of edible mushrooms in the wild may frequently cause a pressure reaction in the structure that activates responsive oxygen species. This in reverse instigates a mushroom cell reinforcement safeguard framework to create cancer prevention enzyme [63, 64]. Palatable mushrooms, for example, *Agaricus bisporus, Pleurotus sajor-kaju, Pleurotus ostreatus, and Volvariella volvaceae,* have been tried for their cancer prevention potentials, ascorbate oxidase, peroxidase, and catalase productions [65]. *Agaricus bisporus* contained the most elevated peroxidase and catalase action contrasting with others. Ascorbate oxidate movement was highest in *Agaricus bisporus* (16U) trailed by *Volvariella volvaceae* with 7U. These catalysts (enzymes) are known for their different biomedical applications [65].

6.4 Biomedical and Biotechnological Applications of Stable Mushroom Enzymes

Biomedical and industrial biotechnology as well as other endeavors/fields have been adorned and enlightened by enzymes since they are one of the most effective and secured biotechnological tool that are globally appreciated and regarded important [66]. Enzymes of various classes reported from different species of mushrooms are widely utilized and used in different biomedical and biotechnological industries such as antimicrobial, antioxidants, dairy, treatments of cancers and tumors, food and feed processing, paper, textile, brewery, leather, pharmaceuticals, and agrochemicals [65]. Old processes such as the use of detrimental chemicals having diverse adverse effects on the environment have been substituted by various industries through the use of new approaches using enzymes, especially from mushrooms [67]. The megahit of mushroom for biomedical and industrial production of enzymes is widely due to their abilities and adaptabilities. A number of enzymes, antibiotics, mycotoxins, and organic acids have been produced by a lot of well-known beneficial mushrooms [67].

Other macrofungi and mushrooms have globally been accepted across the world not just for their nutritional properties and values but also for their medicinal values and benefits [17, 68–71]. In the aspect of traditional medicine, mushrooms and other macrofungi have been utilized for over 3000 years for preventing and treating varieties of diseases and infections [53]. The most important biomedical usage of enzymes produced from mushrooms includes antitumors, anticancers, antimicrobials, and antioxidants. Furthermore, in pharmaceutical industries, the products of the various activities of some of the enzymes produced from mushrooms have been applied specifically in the production of drugs [1].

6.4.1 Mushroom Enzymes as Antimicrobial Agents

Production of significant biomolecules which includes enzymes, edible biomass, and antimicrobial compounds, mushrooms have developed an innate ability that is well conceded for their future effectiveness which can be employed in antimicrobial therapy [1]. The usage of enzymes gotten from beneficial mushrooms as an antimicrobial agent can be widely grouped into direct and indirect. The indirect group is based on the ability of the biocatalysts synthesized from these mushrooms to yield middling biomolecules which has stationary qualities against microorganisms [79] while in the direct group, there is a direct distinctive antimicrobial virtue by the enzymes yielded by these mushrooms [80] such as the release of important organelles from microbial cells as a result of lysis, discontinuance of the production of DNA, RNA, and vital enzymes as well as the hindrance of crucial metabolic route and pathways and essential microbial receptors [81, 82].

The oyster mushroom *Pleurotus ostreatus* is important and has the potential to synthesize several exceptional enzymes that are useful biomedically. This mushroom has been reported to produce pure laccases and ribonucleases that possess an effective and potent activities against different viruses like hepatitis C (HCV) [83], hepatitis B (HBV) [84], and human immunodeficiency virus (HIV) [135, 136]. Laccases produced from *Tricholoma mongolicum* have been studied to have antiviral values [84]. Reports have shown that *Pleurotus ostreatus* also has the ability to synthesize vital enzymes with antioxidant properties like peroxidase and superoxide dismutase and catalase [80]. One of the various biomedical significant applications of these enzymes is the reduction of oxidative detriment in humans [80, 85, 86].

Aside creation of biomedically significant chemicals by mushrooms, other important protein compounds with antimicrobial activities have been accounted for in mushrooms, of which the most examined is the antiviral property of mushroom lectin particularly against Human Immunodeficiency Infection (HIV) [1]. This compound is discovered to be manufactured by the mushrooms *Russula delica* and *Agrocybe aegerita*, it frustrates the penetration of a few infections for example HIV by joining with the glycoproteins of the viral encompass, causing an adjustment of the viral design and game plan, thus hindering viral connection and resulting blockage of passage into the cells [87, 88]. Moreover, refined lectin is accounted to be non-poisonous and has found application as an important flu immunization adjuvant against H9N2 viral infections in a review utilizing contaminated mice [89]. One more comparable protein is the Trichogin discharged by *Tricholoma giganteum* which was portrayed to repress the opposite transcriptase in HIV [1].

6.4.2 Mushroom Enzymes as Anticancer and Antitumor Agents

Aside from biomedical and biotechnologically significant enzymes, mushrooms likewise secrete an extensive variety of biomolecules including non-ribosomal peptides, polyketides, terpenes, polyketide-terpene half breeds, chitosan-oligosaccharides as optional/secondary metabolites which have extraordinary application in therapeutics for production of immune suppressants, cholesterol-lessening compounds (for example lovastatin) as well as anti-infection agents [89]. The non-specific and high oxidation capacity of laccase, which operates without the aid of any cofactors, makes it an effective biocatalyst in established and developing biotechnological applications [72]. This significant enzyme is utilized in the readiness of a few anticancer medications and in the detoxification of cosmetic items [72]. Besides, it is likewise utilized as the center parts of numerous economically significant polymers [73, 74].

The first chemical substance of high pharmaceutical value produced utilizing a laccase enzyme was actinocin, which was gotten from 4-methyl-3-hydroxyanthranilic acid. Actinocin has anticancer properties and blocks DNA record in cancer cells [75]. Another anticancer medication discovered utilizing laccase was vinblastine, which is utilized for treating leukemia. A few palatable mushrooms, for example, *Pleurotus ostreatus* (shellfish mushroom), *Agaricus bisporus* (champignon), and *Lentinula edodes* (rice mushroom) are proficient laccase synthesizers [76, 77]. The adequacy of laccase as a catalyst was first shown in exudates from *Rhus vernicifera*, the Japanese veneer tree in 1883 [76] and subsequently in specific fungi as revealed by Aisemberg *et al.* [44].

Mushrooms like *Lepista nuda* and *Termitomyces clypeatus* are rich manufacturers of proteases. Additionally, proteases from *Lepista nuda* had shown antiproliferative action on human hepatoma cells. Moreover, protease from *Termitomyces clypeatus* was accounted for to be equipped for killing malignant growth cells, re-establishing p53 (cancer protein of which this cancer protein goes about as a cancer silencer) level and cutting surface proteoglycans [40]. Cellulase from mushrooms, for example, *Pleurotus eryngii* which act successively to break down direct polymers into monosaccharide glucose units can likewise be valuable as an intense antitumor agent [47, 62].

6.4.3 Mushroom Enzymes as Antioxidants

Reactive oxygen species (or straightforward oxidants) are compounds whose presence particularly in high concentrations in the human system causes detrimental impacts through following up on and changing the conformational design of many significant biomolecules like proteins, carbohydrates, lipids, and nucleic acids [78]. Cell reinforcements again are fundamental biomolecules that upgrade human well-being by following up on and inactivating oxidants, in this way forestalling their malicious consequences on the human system [1, 78]. A few types of mushrooms have been accounted for to yield flavonoids, steroids, beta-glucans, phenols, and other significant cell reinforcements [90]. The cell reinforcement property of consumable mushrooms is famous as a result of their ability to produce antioxidative pressure enzymes, for example, superoxide dismutase, peroxidase, transferases, and catalases [91]. Consumable mushrooms, for example, *Agaricus bisporus, Pleurotus ostreatus, Pleurotus sajor-kaju*, and *Volvariella volvaceae*, have been reported for their cell reinforcement (antioxidant) ability, ascorbate oxidase, peroxidase, and catalase production [65, 91].

Maturing is portrayed by diminished degrees of ascorbate and ascorbate peroxidase movement [92]. Extra-mitochondrial frameworks like microsomal electron transport and other enzyme frameworks like xanthine oxidase, aldehyde oxidase, dihydroorotic dehydrogenase, and a class of flavoproteins dehydrogenase are answerable for oxidative pressure, and expanded concentration of responsive oxygen species [92]. Stress enhancement or bearableness is accomplished by a bunch of rummaging mushroom enzymes, for example, catalase, glutathione reductase, ascorbate synthase, and superoxide dismutase [93].

Matured Wistar rodents (rats) treated with *Pleurotus ostreatus* showed a huge decline in the degree of xanthine dehydrogenase in the kidneys, heart, and cerebrum when contrasted with untreated matured rodents (rats) [38]. Derivatives from *P. ostreatus* treated matured rodents and showed fundamentally expanded degrees of glucose-6-phosphate dehydrogenase in every one of the previously mentioned organs including the liver. Comparative perception was seen when the mushroom derivative was administered to mature rodents (rats) in the event of an ascorbate peroxidase catalyst. Differential instances were discovered by comparing the isoforms of the enzyme found in different organs of treated and untreated mice. According to Thomas and Jayakumar [38], adult untreated rats' brain tissues contained less ascorbate peroxidase isozyme, indicating that the organ is very vulnerable to stress. Consequently, mushroom derivatives (catalysts) were viewed to be powerful as helpful enhancement in lessening age-related oxidative pressure in matured rodents by expanding auxiliary cell reinforcement proteins like glucose-6-phosphate dehydrogenase, ascorbate peroxidase while decreasing oxidative marker xanthine dehydrogenase in significant organs like liver, kidney, heart, and cerebrum [38].

6.4.4 Other Applications of Mushroom Enzymes

Notwithstanding the biomedical applications, enzymes derived from mushrooms have a few other modern and biotechnological possible usages. Different uses of mushroom catalysts are summed up in Table 6.1. Mushroom laccases possess wide applications in the pharmaceutical endeavour because of their capacity to blend a few complex medically significant mixtures including anesthetics, anti-microbials, mitigating agents, tranquilizers, and sedatives like mitomycin, dimerized vindoline, cephalosporins, triazolo (benzo) cycloalkyl thiadiazines, and vinblastine [45, 94]. Laccases are additionally used in the analysis, enzymatic and immunological measures and tests, hair passing on, and furthermore in decreasing impacts of toxin ivy dermatitis [95]. It is additionally vital to take note that mushroom laccases have been examined on the grounds that they can possibly repress HIV switch transcriptase [96] and antitumor impact [97]. Aliphatic and fragrant (aromatic) amines might be changed over completely to 3-(3, 4-dihydroxyphenyl)-propionic acid with the assistance of laccase-based oxidation [75, 94] and this multitude of items has massive remedial importance.

In the interim, the enzyme xylanases are additionally fundamental for producing chemically significant xylo-oligosaccharides, for example, probiotics, anti-inflammatory, hostile to hyperlipidemic, anti-allergic, anti-cancerous, anti-oxidant, and anti-microbial agents [105].

Different mushrooms have been accounted for to create extracellular asparaginase action, i.e., these dynamic enzymes were separated from culture filtrates [110]. According to Shivashankar and Premkumari [109], the esterase from *Hypsizygus ulmarius* is linked to ester arrangement and transesterification with potential applications in the food, pharmaceutical, and cosmetic industries. According to Ghorai *et al.* [111], these enzymes are connected to the digestion of a variety of medicines, including as the cholesterol-lowering medication lovastatin, the antiviral medication oseltamivir, the narcotic painkiller meperidine (Demerol), cocaine, and heroin.

The uniqueness and disposable nature of enzymes make them an incredibly tempting option for many current biotechnological applications. Their ease of organization, short age, ease of control of associated traits (genes),

Table 6.1 Biomedical and biotechnological application of some mushroom enzymes.

Enzymes class	Application	References
Lignolytic		
Lignin peroxidase E.C.1.11.1.14	Biodegradation of lignin	[98, 99]
Manganese peroxidase E.C.1.11.1.13	Bioremediation, biopulping, biobleaching	[100]
Versatile peroxidase EC 1.11.1.16	Double capacity to oxidize Mn(II) and phenolic and non-phenolic compounds, Better than both lignin peroxidase and manganese peroxidase	[42, 101]
Tyrosinase E.C.1.14.18.1	Food industry, biosensor, bioremediation	[42, 102]
Laccase E.C. 1.10.3.2	Paper mash (pulp) bleaching, material colors, engineered color decolouration, biosensors, detoxification of contaminated water, bioremediation, synthetic combination	[46, 103]
Cellulolytic enzymes		
Endo-β-1,4-glucanase EC 3.2.1.74 Exo-β-1,4-glucanase EC 3.2.1.91	Textile industry, food and feed industry, pulp and paper industry, bioethanol production	[52, 104]
β-glucosidase or Cellobiase E.C. 3.2.1.21	Food handling industry, drug industry (agrochemicals and drugs), biofuel	[104, 105]
Xylanase EC 3.2.1.8	Food and feed industry, paper and fiber industry, biofuel	[61]

(Continued)

Table 6.1 Biomedical and biotechnological application of some mushroom enzymes. (*Continued*)

Enzymes class	Application	References
Amylase EC 3.2.1.1	Food, cleansers, material, drink, creature feed, paper creation, fermentation	[57, 106]
Invertase (sucrase) E.C. 3.2.1.26	Food industry, liquor industry	[107]
Pectinase EC 3.2.1.15	Extraction, purification, filtration, and depectinization of organic product juices, wines Maceration of natural products, vegetables, and evacuation of internal mass of lotus seed, garlic, almond, and nut	[52, 59]
Other enzymes		
Protease Many under EC 3.4.21.	Brewing, dairy, meat, detergent, leather, and photographic industries	[39, 108]
Esterase E.C. 3.1.1.1	Delicate and expert medication plan, Union of short chain esters, digestion, and detoxification of agrochemicals and drugs	[109]

and extensive availability of mushrooms in nature, enzymes from this microbial source are chosen over those from animal or plant sources. All of these could be attributed to the reason for the massive increase in demand for mushroom in the global market [106].

The presence of different enzymes in mushrooms additionally adds to their nutritive worth, flavor, and timeframe of realistic usability [94, 106]. There is solid proof proposing that significant cell reinforcement and cytoprotective catalysts are available in different consumable mushrooms [112]. This emphasizes the importance of a therapeutic approach in light of healthy mediations with biomass mushroom supplementation to prevent and limit the harmful effects in many problems (such as those connected with high oxidation harm like coronary illness, asthma, provocative circumstances, and neurodegenerative issues) [113]. The oral intake of specific enzymes produced from mushrooms decreases the persistent harmful health challenges [112]. maintain pH equilibrium levels; increase blood flow and other intracellular network; expulsion of poisonous substances; recuperation of gastrointestinal bacterial equilibrium (biota balance); upgrade and equilibrium of the safe framework; and improvement in cell digestion [1].

Stomach-related enzymes produced by mushrooms generally possess the capacity to work over a wide scope of acidic conditions (pH 3.5 to 8.5). This implies that they can endure the stomach digestive process, the acidity of the stomach, and the alkalinity of the small digestive tract [113]. Mushroom-based stomach-related enzymes can consequently help with processing food in the stomach and afterward keep on aiding the cycle after the food enters the small digestive tract [114]. This guarantees the complete digestion of food [115].

Mushroom proteases have been displayed to lead the course of proteolysis, debase malignant growth cells as well as poisons. Mushroom proteases can help with eliminating abundant protein from the circulatory framework [112]. Glucoamylase catalyst separates huge starch atoms in the human body into glucose which is the helpful principal energy compound [115]. Phenoloxidases incorporate laccases and tyrosinases. These enjoy the benefit that they can respond with atomic oxygen without the requirement for remotely provided co-substrates which prompts lower costs. Moreover, when tyrosinases oxidize phenols and other sweet-smelling compounds in wastewater regularly, the oxidized item will polymerize into insoluble mixtures that can be eliminated by filtration or precipitation [116]. Mushroom enzyme, such as laccase, has found horticultural applications and has been utilized to remediate soil contaminated by herbicides, pesticides, and certain explosives [45, 114, 115].

Mushroom tyrosinases have been investigated for phenol expulsion from wastewater. Edwards *et al.* [117] immobilized business mushroom tyrosinase on chitosan-covered polysulphone narrow layers inside a slim bioreactor and utilized this framework to oxidize a scope of phenols present in manufactured and modern effluents. The expulsion was productive even at low phenol concentration which commonly cause issues for physical and synthetic strategies. Further, the chitosan-adsorbed o-quinones produced during the response accordingly assist with lessening enzyme inactivation. In another model, as revealed by Ensuncho *et al.* [118], utilized *Agaricus bisporus* tyrosinase immobilized on a chitosan-alginate support in both seat scale and pilot-scale bioreactors in which air was risen through the framework. Over 92% of the phenol was eliminated in the seat-scale reactor and 60 % in the pilot-scale reactor.

Mushroom manganese peroxidase (MnP) and laccase are the most dynamic of mushroom enzymes used in biodegradation and other environmental remediation processes [1]. They are suggested in the staining of a few engineered colors like amarant (monoazo color) and RB5 (diazo color) through the movement of MnP; and of RBBR (anthraquinone color) which is deteriorated by laccase [119, 120]. Additionally, the enzymatic action of immobilized mushroom laccases can catalyze the oxidation of anthracene and benzene (through a circuitous component that suggests the investment of an oxidizing arbiter like 2, 2-azino-bis (3-ethylbenzthiazoline-6-sulfonic corrosive) and of chlorophenols, including PCPs the most hard-headed pentachlorophenols into less harmful mixtures [121].

Mushroom enzymes have also been applied in biosensor improvement, material biofinishing, refreshment processing, diminishing mixture extensibility in flour, livestock feed, beauty care products, clinical conclusion, enzyme immunoassays, and detergent production [122–125]. Tony *et al.*, [126] dealt with the development of white-decay growth *Pleurotus eryngii* F032 in a reasonable medium and revealed that an azo color Responsive Dark 5 (RB5) is debased in view of its capacity to deliver ligninolytic proteins like lignin peroxidase (LiP), manganese peroxidase (MnP), and laccase. These proteins can corrupt and change the intricate construction of the color into a less harmful compound.

6.5 Some Limitations of Mushroom Enzymes

Mushrooms are known for their different possibilities to manufacture some drugs, nutraceuticals, and extracellular enzymes yet report with respect to specific chemicals like IAA creation, and extracellular natural acid creation are extremely less. Since ancient times people have vigorously consumed mushrooms for dietary and helpful use. Mushroom harmfulness has likewise been known since man began consumption and utilization of mushrooms and has been the reason for the death of numerous verifiable figures including the Roman Sovereign Claudius [127, 128].

Investigations on tyrosinases from mushrooms were carried out to comprehend and to forestall the enzymatic browning that happens within the sight of air when mushrooms and natural products like fruits or vegetables are cut or wounded [129]. This peculiarity is personally connected to the activity of tyrosinase and causes serious monetary losses in the food industries [129].

In the year 1967, the amino acid l-3,4-dihydroxyphenylalanine (also known as L-DOPA or levodopa) had been used as a medication for the treatment of Parkinson's sickness. It is additionally utilized for controlling the myocardium following neurogenic injury. The world market for L-DOPA is around 250 tons each year as reported by Koyanagi *et al.* [130]. The vast majority of the L-DOPA sold financially is delivered from vanillin and hydantoin by a compound cycle that includes eight response steps, one of which is an optical goal. A few specialists have read up organic courses for the creation of L-DOPA. For instance, frameworks including entire cells of *Escherichia coli* as found by Lee *et al.* [131]. *Erwinia herbicola* by Koyanagi *et al.* [130] or *Aspergillus oryzae* [132] have been utilized to change L-tyrosine into L-DOPA in the research center. Nonetheless, the particular exercises in these cycles are low, and both the underlying substrate (L-tyrosine) and the ideal item (L-DOPA) experience side responses that produce different metabolites. The terrible showing of entire cell frameworks triggered analysts to utilize tyrosinases in cell-free frameworks. Up to this point, all reviews have involved mushroom tyrosinase got from Sigma. The productivities of L-DOPA that have been acquired in clump reactors utilizing immobilized tyrosinase go from 1.44 to 54 mg [133]. Sadly, these productivities are still generally low, which is because of two elements. First is the change of the additional L-tyrosine, which is deficient, with under 30% of it being consumed during the interaction. Furthermore, albeit side responses with L-tyrosine have compromised these enzymatic frameworks, a portion of the L-DOPA that is created is lost in a moment tyrosinase-catalyzed response that produces dopaquinone. In spite of the advancement in tyrosinase-based creation of L-DOPA, this cycle would not presumably ever be popularized since an enzymatic cycle including another compound tyrosine phenol-lyase (TPL) appears to seem to have better potential.

The enzyme laccase synthesized from mushroom is restricted because of its low security and high aversion to the cycle conditions yet can be overwhelmed by immobilization strategy which works on the movement and soundness of the biocatalyst in both watery and natural stages offered that help allow the dissemination of the substrate to the dynamic site of the catalyst [134].

6.6 Conclusion and Future Perspectives

Lately, the interest and demand for enzymes from microbial sources especially from mushrooms and other groups of fungi is fast and continuously on the increase. This is because the application of enzymes in biochemical processes is by far better than the use of chemical catalysts, owing to their catalytic potentials in several metabolic reactions under serene conditions coupled with their ease of production and handling. The demand therefore for microbial enzymes is continuously on the rise. However, the enzymes from mushrooms and other related fungi are taking the lead in application and demand due to the ability of mushrooms to secrete in large quantities various types of inexpensive biomedical and industrially important enzymes. This chapter therefore explains the classes/types of mushrooms. The stable enzymes produced by beneficial mushrooms were also elucidated as well as their biomedical and industrial applications. Additional researches are required to search for new species of mushrooms with abilities to synthesize biomedically important enzymes. In the same vein, currently known mushrooms should be continually screened for their abilities to produce new enzymes. In addition, formulations of adequate media compositions and other fermentation conditions for industrial-scale production of these important enzymes by mushrooms should also be investigated. In the current global market, commercially available biomedical enzymes are synthesized from few bacterial and fungal species due to the GRAS (Generally Recognized As Safe) status. Therefore efforts must be put in place for industrial-scale production, purification, characterization, and sale of enzymes from mushrooms with biomedical and biotechnological applications to compliment the already existing commercial enzymes. Meanwhile, recent advancements in the field of biotechnology, genetic engineering, and omics technology especially metagenomics, proteomics, celiomics, and metabolomics can be employed to research and make novel discoveries associated with the production of mushroom enzymes such as to augment recovery of target genes for enzyme production by mushrooms and other fungal groups and decrease enzyme production time and cost by increasing the rate of growth of host mushrooms.

References

1. El-Gendi, H., Saleh, A.K., Badierah, R., Redwan, E.M., El-Maradny, Y.A., El-Fakharany, E.M., A Comprehensive Insight into Fungal Enzymes: Structure, Classification, and Their Role in Mankind's Challenges. *J. Fungi*, *8*, 1, 23, 2021, MDPI AG, Retrieved from http://dx.doi.org/10.3390/jof8010023.
2. Kango, N., Jana, U.K., Choukade, R., Fungal enzymes: sources and biotechnological applications, in: *Advancing Frontiers in Mycology and Mycotechnology: Basic and Applied Aspects of Fungi*, pp. 515–538, 2019.
3. Jeske, L., Placzek, S., Schomburg, I., Chang, A., Schomburg, D., Brenda in 2019: A European Elixir core data resource. *Nucleic Acids Res.*, *47*, D1, D542–D549, 2019.

4. Guerrand, D., Economics of food and feed enzymes: Status and prospectives, in: *Enzymes in human and animal nutrition*, pp. 487–514, Academic Press, Cambridge, Massachusetts, 2018.
5. Singh, R.S., Singh, T., Pandey, A., Microbial enzymes—an overview, in: *Advances in enzyme technology*, pp. 1–40, 2019.
6. Van der Heijden, M.G., Martin, F.M., Selosse, M.A., Sanders, I.R., Mycorrhizal ecology and evolution: the past, the present, and the future. *New Phytol.*, 205, 4, 1406–1423, 2015.
7. ALKolaibe, A.G., Elkhateeb, W.A., Elnahas, M.O., El-Manawaty, M., Deng, C.Y., Wen, T.C., Daba, G.M., Wound healing, anti-pancreatic cancer, and α-amylase inhibitory potentials of the edible mushroom, *Metacordyceps neogunnii. Res. J. Pharm. Technol.*, 14, 10, 5249–5253, 2021, doi: 10.52711/0974-360X.2021.00914.
8. Elkhateeb, W.A., Daba, G.M., Soliman, G.M., The anti-nemic potential of mushroom against plant-parasitic nematodes. *J. Microbiol. Biotechnol.*, 6, 1, 1–6, 2021.
9. Jing, P., Zhao, S.J., Lu, M.M., Cai, Z., Pang, J., Song, L.H., Multiplefingerprint analysis for investigating quality control of Flammulina velutipes fruiting body polysaccharides. *J. Agric. Food. Chem.*, 62, 50, 12128–12133, 2014.
10. Daba, G.M., Elkhateeb, W., ELDien, A.N., Fadl, E., Elhagrasi, A., Fayad, W., Wen, T.C., Therapeutic potentials of n-hexane extracts of the three medicinal mushrooms regarding their anti-colon cancer, antioxidant, and hypocholesterolemic capabilities. *Biodiversitas J. Biol. Divers.*, 21, 6, 1–10, 2020, doi: 10.13057/biodiv/d210615.
11. El-Hagrassi, A., Daba, G., Elkhateeb, W., Ahmed, E., El-Dein, A.N., Fayad, W., Shaheen, M., Shehata, R., El-Manawaty, M., Wen, T., *In vitro* bioactive potential and chemical analysis of the n-hexane extract of the medicinal mushroom, *Cordyceps militaris. Malays. J. Microbiol.*, 16, 1, 40–48, 2020.
12. Kumla, J., Suwannarach, N., Sujarit, K., Penkhrue, W., Kakumyan, P., Jatuwong, K., Lumyong, S., Cultivation of mushrooms and their lignocellulolytic enzyme production through the utilization of agro-industrial waste. *Molecules, 25*, 12, 2811, 2020.
13. Luz, J.M.R.D., Nunes, M.D., Paes, S.A., Torres, D.P., Silva, M. D. C. S. D., Kasuya, M.C.M., Lignocellulolytic enzyme production of Pleurotus ostreatus growth in agroindustrial wastes. *Braz. J. Microbiol., 43*, 1508–1515, 2012.
14. Fen, L., Xuwei, Z., Nanyi, L., Puyu, Z., Shuang, Z., Xue, Z., Haiping, L., Screening of lignocellulose-degrading superior mushroom strains and determination of their CMCase and laccase activity. *Sci. World J., 2014*, 1–6, 2014.
15. Berbee, M.L., James, T.Y., Strullu-Derrien, C., Early diverging fungi: diversity and impact at the dawn of terrestrial life. *Annu. Rev. Microbiol., 71*, 41–60, 2017.
16. Chandra Mishra, R., Estimation of Enzymatic Activities of Different Species of Mushrooms (Doctoral dissertation), 2011.

17. Chang, S.T. and Wasser, S.P., The cultivation and environmental impact of mushrooms, Oxford research encyclopedia of environmental science-Agriculture and the environment, 2017.
18. Thomas, P., Elkhateeb, W., Daba, G., Industrial applications of truffles and truffle-like fungi, in: *Advances in Macrofungi*, 1st ed, pp. 82–88, Taylor and Francis Group: CRC Press, London, 2021, doi: 10.1201/9781003096818-8.
19. Gupta, A., Bansal, V., Saini, A.J., Classification and Species: Mushroom. *Webology*, 18, 1, 1653–1660, 2021, DOI: 10.29121/WEB/V18I1/51.
20. Usman, M., Murtaza, G., Ditta, A., Nutritional, medicinal, and cosmetic value of bioactive compounds in button mushroom (Agaricus bisporus): A review. *Appl. Sci.*, *11*, 13, 5943, 2021.
21. El-Ramady, H., Abdalla, N., Fawzy, Z., Badgar, K., Llanaj, X., Törős, G., Prokisch, J., Green biotechnology of oyster mushroom (Pleurotus ostreatus L.): A sustainable strategy for myco-remediation and bio-fermentation. *Sustainability*, *14*, 6, 3667, 2022.
22. Dheyab, M.A., Owaid, M.N., Rabeea, M.A., Aziz, A.A., Jameel, M.S., Mycosynthesis of gold nanoparticles by the Portabello mushroom extract, Agaricaceae, and their efficacy for decolorization of Azo dye. *Environ. Nanotechnol. Monit. Manage.*, *14*, 100312, 2020.
23. Du, X.H. and Yang, Z.L., Mating systems in true morels (Morchella). *Microbiol. Mol. Biol. Rev.*, *85*, 3, e00220–20, 2021.
24. El Sheikha, A.F., Nutritional profile and health benefits of Ganoderma lucidum "Lingzhi, Reishi, or Mannentake" as functional foods: Current scenario and future perspectives. *Foods*, *11*, 7, 1030, 2022.
25. Moyo, P., Identification and characterisation of diatrypaceae species associeated with declining grapevines and alternative hosts in South Africa (Doctoral dissertation), Stellenbosch University, Stellenbosch, 2017.
26. Drewnowska, M. and Falandysz, J., Investigation on mineral composition and accumulation by popular edible mushroom common chanterelle (Cantharellus cibarius). *Ecotoxicol. Environ. Saf.*, *113*, 9–17, 2015.
27. Khatua, S., Ghosh, S., Acharya, K., Laetiporus sulphureus (Bull.: Fr.) Murr. as food as medicine. *Pharmacogn. J.*, *9*, 6s, s1–s15, 2017.
28. Nxumalo, W., Elateeq, A.A., Sun, Y., Can Cordyceps cicadae be used as an alternative to Cordyceps militaris and Cordyceps sinensis?–A review. *J. Ethnopharmacol.*, *257*, 112879, 2020.
29. Attaran Dowom, S., Rezaeian, S., Pourianfar, H.R., Agronomic and environmental factors affecting cultivation of the winter mushroom or Enokitake: Achievements and prospects. *Appl. Microbiol. Biotechnol.*, *103*, 6, 2469–2481, 2019.
30. Ritter, N., Vermeulen, P., Beatty, M., Rivellini, A., Hodowanec, A., Annotated Checklist of the Macrofungi Observed at the Gordon Natural Area (West Chester University, PA)-Version III, 2020.
31. Pulman, J.A., Childs, K.L., Sgambelluri, R.M., Walton, J.D., Expansion and diversification of the MSDIN family of cyclic peptide genes in the poisonous

agarics Amanita phalloides and A. bisporigera. *BMC Genomics, 17*, 1, 1–14, 2016.
32. Alvarado, P., Gasch-Illescas, A., Morel, S., Dagher-Kharrat, M.B., Moreno, G., Manjón, J.L., Moreau, P.A., Amanita Section Phalloideae Species in the Mediterranean Basin: Destroying Angels Reviewed. *Biology, 11*, 5, 770, 2022.
33. Espinoza, L. and Smith, M.E., The Green-Spore Poison Parasol Mushroom, Chlorophyllum molybdites: PP324, 1/2016. *EDIS, 2016*, 2, 4–4, 2020.
34. Pueschel, V.A. and Schieberle, P., Changes in the key aroma compounds of matsutake mushroom (Tricholoma matsutake Sing.) from Canada during pan-frying elucidated by application of the sensomics approach. *Eur. Food Res. Technol., 247*, 51–65, 2021.
35. Ahmad, I., Arif, M., Mimi, X., Zhang, J., Ding, Y., Lyu, F., Therapeutic values and nutraceutical properties of shiitake mushroom (Lentinula edodes): A review. *Trends Food Sci. Technol.*, 134, 123–135, 2023.
36. Chauhan, O.P., Vijay, V., Pandey, A.K., Semwal, A.D., Biochemical and Health Properties of Truffles. *Def. Life Sci. J., 6*, 251–258, 2021.
37. Ito, T., Kobayashi, T., Egusa, C., Maeda, T., Abe, N., Okubo, Y., Niitsuma, T., A case of food allergy due to three different mushroom species. *Allergol. Int., 69*, 1, 152–153, 2020.
38. Thomas, P.A., Geraldine, P., Jayakumar, T., *Pleurotus ostreatus*, an edible mushroom, enhances glucose-6-phosphate dehydrogenase, ascorbate peroxidase and reduces xanthine dehydrogenase in major organs of aged rats. *Pharm. Biol.*, 52, 5, 646–654, 2014.
39. Majumder, R., Banik, S.P., Khowala, S., AkP from mushroom *Termitomyces clypeatus is a* proteoglycan specific protease with apoptotic effect on HepG2. *Int. J. Biol. Macromol.*, 91, 198–207, 2016.
40. Zhang, X., Zhang, X., Gu, S., Pan, L., Sun, H., Gong, E., Zhua, Z., Wen, T., Daba, G.M., Elkhateeb, W.A., Structure analysis and antioxidant activity of polysaccharide-iron (III) from Cordyceps militaris mycelia. *Int. J. Biol. Macromol.*, 178, 170–179, 2021, doi: 10.1016/j.ijbiomac.2021.02.163.
41. Radhakrishnan, S., Ravindranath, A.D., Hanosh, M.S., Sarma, U.S., Jayakumaran, N.A., Quantitative evaluation of the production of ligninolytic enzymes lignin peroxidase and manganese peroxidase by *P. sajor-kaju* during coir pith composting. *Cord*, 28, 1, 24–33, 2012.
42. Weng, C., Peng, X., Han, Y., Depolymerization and conversion of lignin to value added biproducts by microbial and enzymatic catalysis. *Biotechnol. Biofuels*, 14, 84, 2021.
43. Feldman, D., Kowbel, D.J., Glass, N.L., Yarden, O., Hadar, Y., A role for small secreted proteins (SSPs) in a saprophytic fungal life style: Ligninolytic enzyme regulation in *Pleurotus ostreatus*. *Sci. Rep.*, 7, 14553, 2017.
44. Aisemberg, G.O., Grotewold, E., Taccioli, G.E., Judewicz, N.D., A Major Transcript in the response of *Neurospora crassa* to Protein Synthesis Inhibition by Cycloheximide. *Exp. Mycol., 13*, 2, 121–128, 1989.

45. Arora, D.S. and Sharma, R.K., Ligninolytic fungal laccases and their biotechnological applications. *Biotechnol. Appl. Biochem.*, 160, 1760–1788, 2010.
46. Karittapattawan, P. and Benchawattananon, R., Evaluation of laccase production by monokaryotic strains of edible mushrooms. *Pak. J. Biol. Sci.*, 24, 4, 454–460, 2021, doi: 10.3923/pjbs.2021.454.460, PMID: 34486304.
47. Chen, S., Ma, D., Ge, W., Buswell, J.A., Induction of laccase activity in the edible straw mushroom *Volvariella volvacea. FEMS Microbiol. Lett.*, 218, 1, 143–148, 2003, doi: 10.1111/j.1574-6968.2003.tb11510.x, PMID: 12583910.
48. Dwivedi, A., Melville, B.W., Shamseldin, A.Y., Guha, T.K., Analysis of hydrodynamic lift on a bed sediment particle. *J. Geophys. Res.: Earth Surf.*, *116*, F2, 1–19, 2011.
49. Barbosa, M.O., Moreira, N.F., Ribeiro, A.R., Pereira, M.F., Silva, A.M., Occurrence and removal of organic micropollutants: An overview of the watch list of EU Decision 2015/495. *Water Res.*, *94*, 257–279, 2016.
50. Halaouli, S., Asther, M., Sigoillot, J.C., Hamdi, M., Lomascolo, A., Fungal Tyrosinases: New Prospects in Molecular Characteristics, Bioengineering and Biotechnological Applications. *J. Appl. Microbiol.*, 100, 219–232, 2006.
51. Fernandes, M.S. and Kerkar, S., Microorganisms as a source of tyrosinase inhibitors: a review. *Ann. Microbiol.*, 67, 4, 343–358, 2017.
52. Ghorai, S., Mukherjee, S., Khowala, S., Improved production and properties of β-glucosidase influenced by 2-deoxy-D-glucose in the culture medium of Termitomyces clypeatus. *Biotechnol. Bioprocess Eng.*, 6, 297–304, 2011, doi. 10.1007/s12257- 010-0236-3.
53. Chen, H., Tian, T., Miao, H., Zhao, Y.Y., Traditional uses, fermentation, phytochemistry and pharmacology of Phellinus linteus: a review. *Fitoterapia*, 113, 6–26, 2016.
54. Mata, G., Salmones, D., Pérez-Merlo, R., Hydrolytic enzyme activities in shiitake mushroom (*Lentinula edodes*) strains cultivated on coffee pulp. *Rev. Argent. Microbiol.*, 48, 3, 191–195, 2016, doi: 10.1016/j.ram.2016.05.008, Epub 2016 Sep 7, PMID: 27614795.
55. Bapat, V.A., Trivedi, P.K., Ghosh, A., Sane, V.A., Ganapathi, T.R., Nath, P., Ripening of fleshy fruit: molecular insight and the role of ethylene. *Biotechnol. Adv.*, 28, 94–107, 2010.
56. Xie, C., Yan, L., Gong, W., Zhu, Z., Tan, S., Chen, D., Hu, Z., Peng, Y., Effects of Different Substrates on Lignocellulosic Enzyme Expression, Enzyme Activity, Substrate Utilization and Biological Efficiency of *Pleurotus Eryngii. Cell. Physiol. Biochem.*, 39, 4, 1479–94, 2016, doi: 10.1159/000447851, Epub 2016 Sep 9, PMID: 27607466.
57. Ghorai, S., Banik, S.P., Verma, D., Chowdhury, S., Mukherjee, S., Khowala, S., Fungal biotechnology in food and feed processing. *Food Res. Int.*, 42, 5-6, 577–587, 2009.
58. Bharadwaj, P. and Udupa, P.M., Isolation, purifi cation and characterization of pectinase enzyme from *Streptomyces thermocarboxydus. J. Clin. Microbiol. Biochem. Technol.*, 5, 1, 001–006, 2019, doi: 10.17352/jcmbt.000031.

59. Pereira, G.S., Cipriani, M., Wisbeck, E., Souza, O., Strapazzon, J.O., Regina, M.G., Onion Juice Waste Production of *Pleurotus sajor-caju* and Pectinases. *Food Bioprod. Process.*, 106, 11–18, 2017, doi: 10.1016/j.fbp.2017.08.006.
60. Gupta, R., Gigras, P., Mohapatra, H., Goswami, V.K., Chauhan, B., Microbial α-amylase: a b α-amylase biotechnological perspective. *J. Process Biochem.*, 20, 1–18, 2003.
61. Simair, A.A., Mangrio, G.S., Nasreen, T.K., Umar, D.M., Sher, M.M., Noor-e-Saba, K.M., Zhong, J.-J., Characterization of crude xylanase produced by edible mushroom *Pleurotus eryngii*. *J. Biotechnol. Biomater.*, 5, 2, 6, 2015.
62. Elkhateeb, W.A., EL-Ghwas, D.E., Daba, G.M., Mushrooms as efficient enzymatic machinery. *J. Biomed. Res. Environ. Sci.*, 3, 4, 423–428, 2022.
63. Gunjal, A.B., Esterase Enzyme Production using Solid Sate Fermentation with Spent Mushroom as a Substrate, in: *Microbes for Clean and Green Environment*, Chapter – 9 (ISBN: 978-81-945631-2-9) First Edition, pp. 111 –117, 2020.
64. Georgescu, A.A., Danet, A.F., Radulescu, C., Stihi, C., Dulama, I.D., Chelarescue, D.E., Determination of several elements in edible mushrooms using ICP-MS. *Rom. J. Phys.*, 61, 5-6, 1087–1097, 2016.
65. Surekha, C., D.S.V.G.K., K., Srikakarlapudi J. R., R., Haseena, Evaluation of antioxidant and antimicrobial potentiality of some edible mushrooms. *Int. J. Adv. Biotechnol. Res.*, 2, 1, 130–134, 2011.
66. Bull, A.T., Ward, A.C., Goodfellow, M., Search and discovery strategies for biotechnology: the paradigm shift. *Microbiol. Mol. Biol. Rev.*, 64, 3, 573–606, 2000.
67. Kirk, O., Borchert, T.V., Fuglsang, C.C., Industrial enzyme applications. *Curr. Opin. Biotechnol.*, 13, 4, 345–351, 2002.
68. Bandara, A.R., Rapior, S., Bhat, D.J., Kakumyan, P., Chamyuang, S., Xu, J., Hyde, K.D., Polyporus umbellatus, an edible-medicinal cultivated mushroom with multiple developed health-care products as food, medicine and cosmetics: a review. *Cryptogam. Mycol.*, *36*, 1, 3–42, 2015.
69. Bandara, A.R., Rapior, S., Mortimer, P.E., Kakumyan, P., Hyde, K.D., Xu, J., A review of the polysaccharide, protein and selected nutrient content of Auricularia, and their potential pharmacological value. *Mycosphere J.*, *10*, 1, 579–607, 2019.
70. Valverde, M.E., Hernández-Pérez, T., Paredes-López, O., Edible mushrooms: improving human health and promoting quality life. *Int. J. Microbiol.*, *2015*, 1–14, 2015.
71. Gupta, S., Summuna, B., Gupta, M., Annepu, S.K., Edible mushrooms: cultivation, bioactive molecules, and health benefits, in: *Bioactive molecules in food*, vol. *1*, pp. 1–33, 2018.
72. Chen, C.Y., Huang, Y.C., Wei, C.M., Meng, M., Liu, W.H., Yang, C.H., Properties of the newly isolated extracellular thermo-alkali-stable laccase from thermophilic actinomycetes, Thermobifida fusca and its application in dye intermediates oxidation. *AMB Express*, *3*, 1–9, 2013.

73. Couto, S.R. and Herrera, J.L.T., Industrial and biotechnological applications of laccases: a review. *Biotechnol. Adv.*, *24*, 5, 500–513, 2006.
74. Dana, M., Bakhshi Khaniki, G., Mokhtarieh, A.A., Davarpanah, S.J., Biotechnological and industrial applications of laccase: A review. *J. Appl. Biotechnol. Rep.*, *4*, 4, 675–679, 2017.
75. Burton, S.G., Laccases and phenol oxidases in organic synthesis-a review. *Curr. Org. Chem.*, *7*, 13, 1317–1331, 2003.
76. Bento, I., Silva, C.S., Chen, Z., Martins, L.O., Lindley, P.F., Soares, C.M., Mechanisms underlying dioxygen reduction in laccases. Structural and modelling studies focusing on proton transfer. *BMC Struct. Biol.*, *10*, 1, 1–14, 2010.
77. Ansari, M.K.A., Khatib, U.M., Owens, G., Fatma, T., Evaluation of methyl red tolerant cyanobacteria for simultaneous laccase production and dye decolorization. *Int. J. Waste Resour.*, *6*, 2252–5211, 2016.
78. Aklakur, M., Natural antioxidants from sea: A potential industrial perspective in aquafeed formulation. *Rev. Aquacult.*, *10*, 2, 385–399, 2018.
79. Rafael, O.H.D., Fernándo, Z.G.L., Abraham, P.T., Alberto, V.L.P., Guadalupe, G.S., Pablo, P.J., Production of chitosan-oligosaccharides by the chitin-hydrolytic system of *Trichoderma harzianum* and their antimicrobial and anticancer effects. *Carbohydr. Res.*, *486*, 107836, 2019.
80. Liu, J., Li, H., Luo, X., Ma, L., Li, C., Qu, S., A lectin gene is involved in the defense of Pleurotus ostreatus against the mite predator Tyrophagus putrescentiae. *Front. Microbiol.*, *14*, 1191500, 2023.
81. Fuglsang, C.C., Johansen, C., Christgau, S., Adler-Nissen, J., Antimicrobial enzymes: applications and future potential in the food industry. *Trends Food Sci. Technol.*, *6*, 12, 390–396, 1995.
82. Li, K., Guan, G., Zhu, J., Wu, H., Sun, Q., Antibacterial activity and mechanism of a laccase-catalyzed chitosan–gallic acid derivative against Escherichia coli and Staphylococcus aureus. *Food Control*, *96*, 234–243, 2019.
83. EL-Fakharany, E., Haroun, M.B., Ng, T., Redwan, M.E.R., Oyster mushroom laccase inhibits hepatitis C virus entry into peripheral blood cells and hepatoma cells. *Protein Pept. Lett.*, *17*, 8, 1031–1039, 2010.
84. Li, M., Zhang, G., Wang, H., Ng, T., Purification and characterization of a laccase from the edible wild mushroom Tricholoma mongolicum. *J. Microbiol. Biotechnol.*, *20*, 7, 1069–1076, 2010.
85. Yang, J.H., Lin, H.C., Mau, J.L., Antioxidant properties of several commercial mushrooms. *Food Chem.*, 77, 229–235, 2002.
86. Keyhani, J., Keyhani, E., Arzi, L., Anti-oxidative stress enzymes in *Pleurotus ostreatus*. *Acta Hort.*, 739, 420–427, 2007.
87. Barrientos, L.G., Matei, E., Lasala, F., Delgado, R., Gronenborn, A.M., Dissecting carbohydrate–Cyanovirin-N binding by structure-guided mutagenesis: functional implications for viral entry inhibition. *Protein Eng. Des. Sel.*, *19*, 12, 525–535, 2006.

88. Mathieu, Y., Prosper, P., Favier, F., Harvengt, L., Didierjean, C., Jacquot, J.P., Rouhier, M.M., Gelhaye, E., Diversification of fungal specific class A glutathione transferases in saprotrophic fungi. *PLoS One*, *8*, 11, e80298, 2013.
89. Ma, L.B., Xu, B.Y., Huang, M., Sun, L.H., Yang, Q., Chen, Y.J., Yin, Y.L., He, Q.G., Sun, H., Adjuvant effects mediated by the carbohydrate recognition domain of Agrocybe aegerita lectin interacting with avian influenza H9N2 viral surface glycosylated proteins. *J. Zhejiang Univ. Sci. B*, *18*, 8, 653, 2017.
90. Butkhup, L., Samappito, W., Jorjong, S., Evaluation of bioactivities and phenolic contents of wild edible mushrooms from northeastern Thailand. *Food Sci. Biotechnol.*, *27*, 193–202, 2018.
91. Kaushal, V., Thakur, V., Kumar, M., Bhatt, A.K., Anti-oxidative stress enzyme from *Pleurotus sp. Adv. Biotechnol. Microbiol.*, 10, 5, 555796, 2018.
92. Ghorai, S., Edible Mushroom: A Potent Producer of Industrial Enzymes. *J. Sci. Res.*, *66*, 1, 148–166, 2022.
93. Yaqoob, U., Jan, N., Raman, P.V., Siddique, K.H., John, R., Crosstalk between brassinosteroid signaling, ROS signaling and phenylpropanoid pathway during abiotic stress in plants: Does it exist? *Plant Stress*, *4*, 100075, 2022.
94. Ansari, M.K.A., Lastochkina, O., Iqbal, M., Ansari, A.A., Fatma, T., Rodriguez-Couto, S., Owens, G., Laccase-The Wonder Enzyme for a Variety of Industries. *Acta Sci. Microbiol.*, *4*, 12, 52–66, 2021, (ISSN: 2581-3226).
95. Chukwuma, O.B., Rafatullah, M., Tajarudin, H.A., Ismail, N., Lignocellulolytic enzymes in biotechnological and industrial processes: a review. *Sustainability*, *12*, 18, 7282, 2020.
96. Zhang, G.Q., Tian, T., Liu, Y.P., Wang, H.X., Chen, Q.J., A laccase with anti-proliferative activity against tumor cells from a white root fungus Abortiporus biennis. *Process Biochem.*, *46*, 12, 2336–2340, 2011.
97. Hu, D.D., Zhang, R.Y., Zhang, G.Q., Wang, H.X., Ng, T.B., A laccase with antiproliferative activity against tumor cells from an edible mushroom, white common Agrocybe cylindracea. *Phytomedicine*, *18*, 5, 374–379, 2011.
98. Fernandez-Fueyo, E., Ruiz-Duenas, F.J., Martinez, M.J., Romero, A., Hammel, K.E., Medrano, F.J., Martinez, A.T., Ligninolytic peroxidase genes in the oyster mushroom genome: heterologous expression, molecular structure, catalytic and stability properties, and lignin-degrading ability. *Biotechnol. Biofuels*, 7, 2, 2014.
99. Gomathi, V., Esakkiammal, M., Thilagavathi, S.S., Ramalakshmi, A., Lignocellulosic enzyme production by *Termitomyces spp* from termite garden. *Univers. J. Agric. Res.*, 7, 2, 100–111, 2019.
100. Hock, O.G., Qin, D.D., Lum, H.W., Hee, C.W., Shing, W.L., Evaluation of plastic degradation ability of edible mushroom species based on their growth and manganese peroxidase activity. *Curr. Top. Toxicol.*, 16, 65–72, 2020.
101. Adebayo, E.A. and Martinez-Carrera, D., Oyster mushrooms (*Pleurotus*) are useful for utilizing lignocellulosic biomass. *Afr. J. Biotechnol.*, 14, 1, 52–67, 2015.

102. Sharma, A., Sharma, J., Thakrele, P., Screening, localization and activity of mushroom tyrosinase from various developmental growth phases of *Pleurotus florida*. *Int. J. Curr. Microbiol. Appl. Sci.*, 5, 1, 565–575, 2016.
103. Khaund, P. and Joshi, P.R., Enzymatic profiling of wild edible mushrooms consumed by the ethnic tribes of India. *J. Korean Soc. Appl. Biol. Chem.*, 57, 2, 263–271, 2014.
104. Roy, S., Ghorai, S., Choudhury, L., Das, A., Ghosh, R.B., Banik, S.P., Advances in cellulosic enzyme technologies for enhanced stability and catalysis. *J. Adv. Sci. Res.*, 12, 2 Suppl. 1, 49–65, 2021.
105. Ghorai, S., Chowdhury, S., Pal, S., Banik, S.P., Mukherjee, S., Khowala, S., Enhanced activity and stability of cellobiase (β-glucosidase: EC 3.2.1.21) produced in the presence of 2-deoxy-D-glucose from the fungus *Termitomyces clypeatus*. *Carbohydr. Res.*, 345, 8, 1015–1022, 2010.
106. Jonathan, S.G. and Adeoyo, O.R., Evaluation of ten wild Nigerian mushrooms for amylase and cellulase activities. *Mycobiology*, 39, 2, 103–108, 2011.
107. Chowdhury, S., Ghorai, S., Banik, S.P., Pal, S., Basak, S., Khowala, S., Characterization of a novel low molecular weight sucrase from filamentous fungus *Termitomyces clypeatus*. *Process Biochem.*, 44, 10, 1075–1082, 2009.
108. Benmrad, M.O., Mechri, S., Jaouadi, N.Z., Elhoul, M.B., Rekik, H., Sayadi, S., Bejar, S., Kechaou, N., Jaouadi, B., Purification and biochemical characterization of a novel thermostable protease from the oyster mushroom *Pleurotus sajor-caju* strain CTM10057 with industrial interest. *BMC Biotech.*, 19, 43, 2019.
109. Shivashankar, M. and Premkumari, B., Characterization of esterolytic activity from edible mushroom *Hypsizygus ulmarius*. *J. Bio Innov.*, 3, 3, 124–134, 2014.
110. Chand, S., Mahajan, R.V., Prasad, J.P., Sahoo, D.K., Mihooliya, K.N., Dhar, M.S., Sharma, G., A comprehensive review on microbial l-asparaginase: Bioprocessing, characterization, and industrial applications. *Biotechnol. Appl. Biochem.*, 67, 4, 619–647, 2020.
111. Ghorai, S., Banik, S., Verma, D., Chowdhury, S., Mukherjee, S., Khowala, S., Industrial Biotechnology and Commodity Products, in: *Comprehensive Biotechnology*, M.M. Young (Ed.), pp. 635–646, Elsevier, New York, 2016.
112. Siu, K.C., Chen, X., Wu, J.Y., Constituents actually responsible for the antioxidant activities of crude polysaccharides isolated from mushrooms. *J. Funct. Foods*, 11, 548–556, 2014.
113. Bell, V., Silva, C.R.P.G., Guina, J., Fernandes, T.H., Mushrooms as future generation healthy foods. *Front. Nutr.*, 9, 1050099, 2022.
114. Tabibian, J.H., *Digestive Problems Solved: A Patient's Guide to Expert Insights and Solutions*, Springer Nature, Switzerland, 2023.
115. Beuth, J., Proteolytic enzyme therapy in evidence-based complementary oncology: fact or fiction? *Integr. Cancer Ther.*, 7, 4, 311–316, 2008.

116. Chiacchierini, E., Restuccia, D., Vinci, G., Bioremediation of food industry effluents: recent applications of free and immobilised polyphenoloxidases. *Food Sci. Technol. Int.*, *10*, 6, 373–382, 2004.
117. Edwards, W., Leukes, W.D., Rose, S.G., Burton, Immobilization of polyphenol oxidase on chitosan-coated polysulphone capillary membranes for improved phenolic effluent bioremediation. *Enzyme Microb. Technol.*, *25*, 769–773, 1999.
118. Ensuncho, L., Alvarez-Cuenca, M., Legge, R.L., Removal of aqueous phenol using immobilized enzymes in a bench scale and pilot scale three-phase fluidized bed reactor. *Bioprocess Biosyst. Eng.*, *27*, 185–191, 2005.
119. Saratale, R., Saratale, G., Chang, J., Govindwar, S., Bacterial decolorization and degradation of azo dyes: A review. *J. Taiwan Inst. Chem. Eng.*, *42*, 1, 138–157, 2011 Jan.
120. Cabana, H., Ahamed, A., Leduc, R., Conjugation of laccase from the white rot fungus *Trametes versicolor to* chitosan and its utilization for the elimination of triclosan. *Bioresour. Technol.*, 102, 2, 1656–1662, 2011 Jan.
121. Gómez-Dorado, C., Martínez-Salgado, M., *et al.*, Estudio del efecto de dos inductores y un protector enzimático sobre la actividad de las enzimas Mn-peroxidasa y lacasa producidas por *Trametesversicolor* y su efecto en la decoloración de efluentes de la industria papelera. *Univ. Sci.*, *10*, 2, 37–45, 2005.
122. Boer, G.C., Obici, L., de Souza, C.G.M., Peralta, R.M., Purification and some properties of Mn peroxidase from *Lentinulaedodes*. *Proc. Biochem.*, 41, 1203–1207, 2006.
123. Minussi, R.C., Pastore, G.M., Duran, N., Laccase indication in fungi and laccase/N-OH mediator systems applied in paper mill effluent. *Bioresour. Technol.*, 98, 158–164, 2007.
124. Papinutti, V.L. and Forchiassin, F., Lignocellulolytic enzymes from *Fomes sclerodermeus* growing in solid-state fermentation. *J. Food Eng.*, *8*, 54–59, 2007.
125. Asgher, M., Bhatti, H.N., Ashraf, M., Legge, R.L., Recent development in biodegradation of industrial pollutants by white rot fungi and their enzyme systems. *Biodegradation*, 19, 6, 771–783, 2008.
126. H., T., Adnan, L.A., Yusoff, A.R.M., Yuniarto, A., Rubiyatno, Zubir, M.M.F.A., Khudhair, A.B., Teh, Z.C., Naser, M.A., Microbial decolorization of an azo dye reactive black 5 using white-rot fungus *Pleurotus eryngii* F032. *Water Air Soil Poll.*, 224, 1595, 2013, DOI 10.1007/s11270-013-1595-0.
127. Venturini, M., Rivera, C., Gonzalez, C., Blanco, D., Antimicrobial activity of extracts of edible wild and cultivated mushrooms against foodborne bacterial strains. *J. Food Prot.*, 71, 1701–1706, 2008, DOI: 10.4315/0362-028X-71.8.1701.
128. Balandaykin, M.E. and Zmitrovich, I.V., Review on Chaga medicinal mushroom, *Inonotus obliquus* (Higher Basidiomycetes): Realm of medicinal

applications and approaches on estimating its resource potential. *Int. J. Med. Mushrooms*, 17, 2, 95–104, 2015.

129. Guo, Y., Chen, X., Gong, P., Wang, R., Han, A., Deng, Z., Qi, Z., Long, H., Wang, J., Yao, W., Yang, W., Wang, J., Li, N., Advances in the role and mechanisms of essential oils and plant extracts as natural preservatives to extend the postharvest shelf life of edible mushrooms. *Foods*, *12*, 4, 801, 2023.
130. Koyanagi, T., Katayama, T., Suzuki, H., Nakazawa, H., Yokozeki, K., Kumagai, H., Effective production of 3, 4-dihydroxyphenyl-L-alanine (L-DOPA) with Erwinia herbicola cells carrying a mutant transcriptional regulator TyrR. *J. Biotechnol.*, *115*, 3, 303–306, 2005.
131. Lee, S.G., Ro, H.S., Hong, S.P., Kim, E.H., Sung, M.H., Production of L-DOPA by thermostable tyrosine phenol-lyase of a thermophilic *Symbiobacterium* species overexpressed in recombinant *Escherichia coli*. *J. Microbiol. Biotechnol.*, 6, 98–102, 1996.
132. Haq, I.U. and Ali, S., Mutation of *Aspergillus oryzae* for improved production of 3,4-dihydroxyphenyl-L-alanine (L-DOPA) from L-tyrosine. *Braz. J. Microbiol.*, *37*, 78–86, 2006.
133. Ates, S., Cortenlioglu, E., Bayraktar, E., Mehmetoglu, U., Production of L-DOPA using Cu-alginate gel immobilized tyrosinase in a batch and packed bed reactor. *Enzyme Microb. Technol.*, *40*, 4, 683–687, 2007.
134. DiCosimo, R., McAuliffe, J., Poulose, A.J., Bohlmann, G., Industrial use of immobilized enzymes. *Chem. Soc. Rev.*, *42*, 15, 6437–6474, 2013.
135. Irfan, H., Saleem, A., Ali, A., Akmal, M., Medicinal Importance of *Pleurotus species* (Oyster Mushroom): A review. *BULLET: Jurnal Multidisiplin Ilmu*, 1, 05, 792–796, 2022.
136. Zhang, R., Zhao, L., Wang, H., Ng, T.B., A novel ribonuclease with antiproliferative activity toward leukemia and lymphoma cells and HIV-1 reverse transcriptase inhibitory activity from the mushroom, Hohenbuehelia serotina. *Int. J. Mol. Med.*, 33, 1, 209–214, 2014.

7

Relevance of Mushrooms for Biological Control of Diverse Biotic Agent Mitigating Against Agricultural Crops

Oluwabukola Atinuke Popoola[1*], Charles Oluwaseun Adetunji[2], Oluwafemi Adebayo Oyewole[3], Abel Inobeme[4] and Adeyomoye Olorunsola Israel[5]

[1]Genetics, Genomics and Bioinformatics Department, National Biotechnology Development Agency, Abuja, FCT, Nigeria
[2]Applied Microbiology, Biotechnology and Nanotechnology Laboratory, Department of Microbiology, Edo State University Uzairue, Iyamho, Edo State, Nigeria
[3]Department of Microbiology, Federal University of Technology, Minna, Nigeria
[4]Department of Chemistry, Edo State University Uzairue, Iyamho, Edo State, Nigeria
[5]Department of Physiology, Ondo Medical University, Ondo State, Nigeria

Abstract

Due to its potential as an eco-friendly pest and disease control tool, the use of mushrooms in crop pest and disease management has attracted substantial attention in recent years and sustainable alternative to conventional chemical-based methods. Traditional methods of controlling pests and diseases based on utilization of chemical pesticides and synthetic fertilizers, raise concerns about environmental damage, human health risks, and the development of pesticide-resistant strains. In tackling these challenges, alternative and eco-friendly approaches have recently gained momentum. Mushrooms, a varied collection of fungi, have emerged as a promising aid in the management of agricultural pests and diseases. Several mushroom species, including *Trichoderma* spp., *Beauveria bassiana,* and *Metarhizium* spp. have shown promise in controlling a variety of agricultural pests and pathogens. To efficiently address agricultural hazards, these fungi use a variety of strategies, including parasitism, mycoparasitism, and the synthesis of

**Corresponding author*: atinuke.solina@gmail.com

Charles Oluwaseun Adetunji, Chukwuebuka Egbuna, Oluwatosin Ademola Ijabadeniyi and Samantha C. Karunarathna (eds.) Mushroom Biotechnology for Improved Agriculture and Human Health, (159–182) © 2025 Scrivener Publishing LLC

secondary metabolites. This chapter provides a thorough discussion of the many methods of harnessing mushrooms for pest and disease management in agriculture. Furthermore, mushrooms offer additional benefits to sustainable agriculture, including their capacity to improve soil health, promote plant development, and improve crop resilience; also the biopesticidal properties of specific mushroom species, including their ability to produce secondary metabolites with pesticidal, bactericidal, and fungicidal properties. These bioactive compounds have demonstrated efficacy against a wide variety of crop pests and diseases, making them valuable tools for comprehensive pest control approach strategies. Utilizing mushrooms for pest and disease control in agricultural systems offers a promising avenue for environmentally friendly and sustainable crop production. It is of significance to further research into mushroom-based solutions, emphasizing their potential to decrease dependence on synthetic pesticides, and enhance soil standards to contribute to the long-term viability of global food systems. As we navigate the contemporary challenges, harnessing mushroom potency in agriculture offers a compelling pathway towards a more resilient and sustainable future for food production.

***Keywords*:** Mushrooms, biopesticides, pests, diseases, agricultural crops

7.1 Introduction

Agricultural crops play a pivotal role in ensuring global food security and sustaining the rapidly growing human population. However, this crucial industry faces persistent challenges in the form of pests and diseases that devastate crops, leading to yield losses, increased production costs, and environmental concerns associated with utilizing synthetic pest control agents (Zaki *et al.*, 2020). The quest for sustainable and environmentally friendly agricultural practices has made researchers and farmers alike turn to innovative solutions, and one of such solutions has emerged from an unlikely source of mushrooms (Das *et al.*, 2021). Mushrooms, a diverse group of fungi, have garnered attention for their potential applications in agriculture, particularly in controlling pests and diseases that plague crops (Sharma *et al.*, 2014; Sitara *et al.*, 2023). This emerging field of study, often referred to as "mycopesticides" or "mycobiology," explores the multifaceted role that mushrooms can play in crop protection and agricultural sustainability (Bamisile *et al.*, 2021).

Over the past few decades, the agricultural industry has witnessed a shift towards increasingly sustainable and environmentally friendly practices. Growing concerns about the detrimental effects of synthetic pesticides on human health, non-target creatures, and the environment, as well as the emergence of pests that are tolerant to pesticides, are driving this trend. Mushrooms have risen as a promising alternative due to their unique attributes (Badshah *et al.*, 2012). Mushrooms, as fungal organisms, possess a remarkable ability to produce bioactive compounds, including secondary metabolites with pesticidal properties (Mau *et al.*, 2022). Some of these compounds exhibit potent insecticidal, fungicidal, nematicidal, and antibacterial activities. Moreover, mushrooms can serve as valuable biocontrol agents, disrupting the life cycles of pests and diseases, while also promoting plant growth through symbiotic relationships with crop plants (Bautista, 2018).

The application of mushrooms in agriculture is not merely theoretical; it is gaining traction in both laboratory research and practical field applications. The diversity of mushroom species, each with its own set of bioactive compounds, offers a rich resource for developing targeted pest and disease management strategies (Stajic *et al.*, 2017). Furthermore, the cultivation of mushrooms can be integrated into agricultural systems, creating a sustainable and holistic approach to crop protection (Pavlova *et al.*, 2023).

Traditional application of synthetic pesticides and fungicides, although effective, raises concerns about environmental pollution and detrimental impacts on unintended organisms including the development of pesticide-resistant pests and pathogens (Ghimire *et al.*, 2021). As a result, there is growing interest in exploring alternative, eco-friendly, and sustainable methods for pest and disease management. One such innovative approach involves the use of mushrooms. Mushrooms have long been recognized for their diverse ecological roles, including their interactions with other organisms such as plants, insects, and microorganisms. Recent research has shown that certain mushroom species possess natural properties that can contribute to pest and disease control in agricultural systems (Poveda, 2021).

This chapter explores the various ways in which mushrooms are applied to combat agricultural pests and diseases while promoting sustainable farming practices. Agricultural productivity and food security are essential to the well-being of the growing global population. However, addressing pests and diseases sustainably in crop management remains a substantial hurdle.

7.1.1 Benefits of Using Mushrooms in Pest and Disease Management

Biological Control: Some mushroom species have been found to exhibit biopesticidal properties and effectively control various pests. These mushrooms contain bioactive compounds that can deter, repel, or even kill pests. For example, the entomopathogenic fungus *Beauveria bassiana* is known for its insect-killing properties and is used to manage a variety of insect pests (Liu *et al.*, 2023).

Disease Suppression: Certain mushrooms contain natural antifungal Substances that hinder the growth and development of plant pathogens. These compounds can be used to suppress fungal diseases in crops, without the need for synthetic fungicides. For example, the medicinal mushroom *Trichoderma spp.* has been shown to have antagonistic effects against various plant pathogens (Tao *et al.*, 2023; Surija and Faustine, 2021).

Environmental Sustainability: The use of mycopesticides aligns with the principles of sustainable agriculture, as they have minimal impact on organisms not intended as the primary target and reduce surrounding chemical load. This approach also reduces the risk of the development of pesticide resistance (Riyaz *et al.*, 2021; Deepshikha *et al.*, 2017).

Biodegradability: Mycopesticides derived from mushrooms are biodegradable and break down into harmless components, unlike many synthetic chemical pesticides, which persist in the environment (Mehra and Avanthi, 2024).

Integrated Pest Management (IPM): Integrating mushrooms into comprehensive pest management strategies can boost the overall efficiency in pest and disease control. Mycopesticides can be used in conjunction with other methods such as cultural practices and biological control agents (Badshah *et al.*, 2021).

7.1.2 Challenges of Mycopesticides

Although the application of mushrooms in pest and disease management holds great promise with several advantages, such as being environmentally friendly and target-specific, they also come with various challenges:

Standardization: The development of consistent and standardized formulations of mycopesticides is essential to ensure their efficacy and practical application (Dar *et al.*, 2014).

Commercialization: The mass production and cost-effective formulation of mycopesticides for large-scale agricultural use require further development (Gao, 2015).

Awareness and Adoption: Farmers, researchers, and policymakers need to be educated about the benefits and proper application techniques of mycopesticides.

Regulations: Clear regulatory frameworks must be established to govern the approval and application of mycopesticides in various countries.

Utilizing mushrooms for pest and disease management represents a promising avenue for environmentally-friendly farming. As the harmful consequences of traditional chemical pesticides become increasingly apparent, adopting eco-friendly alternatives such as mycopesticides can contribute to healthier ecosystems, improved food safety, and enhanced agricultural sustainability. Through continued research, technological advancements, and collaborative efforts, the potential of mushrooms to transform pest and disease management practices can be fully realized, benefitting both agricultural systems and the environment (Lecoq and Cease, 2022).

7.2 Fungal Biopesticides

Mushrooms belong to the fungal kingdom, and some species have been recognized for their potential as biopesticides. Certain mushroom species contain natural compounds with insecticidal, nematicidal, and fungicidal properties. An established instance is the entomopathogenic fungus *Beauveria bassiana*, which targets and eliminates a diverse array of insect pests. (Drobnjaković *et al.*, 2019). Researchers have harnessed fungal biocontrol potential to develop environmentally friendly alternatives to chemical pesticides. Fungal biopesticides are a type of biological control agent used in agriculture and pest management. They use various species of fungi to control or suppress pests that damage crops, forestry, or other plants. Unlike chemical pesticides, fungal biopesticides offer a more environmentally friendly and sustainable approach to pest management (Kim *et al.*, 2019). The following is an overview of how they work and their advantages.

7.2.1 How Fungal Biopesticides Work

Fungal biopesticides work by infecting and killing target pests, such as insects or other harmful organisms, through various mechanisms. Fungi can be formulated into different products, such as spore formulations or mycelial suspensions, which are then applied to plants or soil (Liu *et al.*, 2023; Constantin *et al.*, 2019; Sinha *et al.*, 2016).

Mechanisms of Action

Infection: The fungal spores attach to pests and then penetrate their bodies. Once inside, the fungus proliferates, ultimately leading to pest death.

Toxin Production: Some fungi produce toxins that are harmful to the pests. These toxins can disrupt biological processes in pests and contribute to their death.

Parasitism: Certain fungi establish parasitic relationships with pests. They might use the pest's resources to grow and reproduce, eventually killing the pest.

7.2.2 Advantages of Fungal Biopesticides

Fungal biopesticides are a type of biological control agent used in agriculture and pest management. They are derived from various species of fungi and offer several advantages over chemical pesticides and other pest control methods. Below are some of the advantages of fungal biopesticides (Bhuvaneswari and Subashini, 2018).

Environmental Safety: Fungal biopesticides are generally considered safe for the environment, non-toxic to humans and animals, and do not persist in the ecosystem (Gayane *et al.*, 2016).

Target Specificity: Fungal biopesticides are often specific to particular pests, minimizing harm to beneficial organisms, such as pollinators and natural predators.

Reduced Resistance: Pests are less likely to develop resistance to fungal biopesticides compared to chemical pesticides, as the intricate mechanism of operation hinders pests' ability to develop resistance.

Sustainability: Fungal biopesticides fit well within comprehensive pest control strategies, promoting a comprehensive strategy to pest control that combines various techniques to achieve long-term sustainability.

Minimal Residue: Unlike chemical pesticides, fungal biopesticides leave little to no residue on crops, making them suitable for organic and residue-conscious farming.

7.2.3 Challenges and Considerations of Fungal Biopesticides

Fungal biopesticides offer an encouraging substitute for synthetic pesticides in the context of controlling various agricultural pests and diseases. However, they also come with their own set of challenges and considerations that need to be addressed for successful implementation (Leathers *et al.*, 1993). Here are some of the key challenges and considerations associated with fungal biopesticides:

Effectiveness: The effectiveness of fungal biopesticides can vary based on environmental conditions, application methods, and target pest species.

Application Timing: Proper timing of application is crucial to ensure that fungal biopesticides come into contact with pests during vulnerable stages of their life cycles.

Cost: The production and formulation of fungal biopesticides can sometimes be more expensive than chemical pesticides, affecting their widespread adoption.

Research and Development: The development of effective fungal biopesticides for specific pests requires research and optimization to ensure consistent results.

Regulation: Regulatory processes for biopesticides may differ from those for chemical pesticides, requiring appropriate approval for commercial use.

Fungal biopesticides are an exciting area of research and development within sustainable agriculture, offering a promising alternative to conventional chemical pesticides. As with any pest control approach, their successful implementation depends on understanding the target pests and the environment, and integrating them effectively into pest management strategies (Deshpande, 1999). Fungal biopesticides are a type of biological pesticide derived from fungi. These biopesticides utilize living fungal organisms or their byproducts to control pests and diseases that affect plants, animals, and humans. Fungi have been used for pest management for centuries; however, recent advancements in biotechnology have led to the development of more targeted and effective fungal biopesticides (Pala *et al.*, 2018).

7.2.4 Some Key Points Regarding Fungal Biopesticides

Modes of Action: Fungal biopesticides primarily function through several mechanisms, including parasitism, competition, and toxin production. Some fungi directly parasitize target pests by attaching to their bodies, invading their tissues, and ultimately causing death. Others outcompete the pest for resources, indirectly reducing their population. Certain fungi produce toxins or enzymes that are toxic to pests.

Target Pests: Fungal biopesticides have the ability to focus on a broad spectrum of organisms, encompassing insects, mites, nematodes, and other fungi. They have been used to control pests in agriculture, forestry, horticulture, and public health sectors.

Environmental Safety: A primary benefit of fungal biopesticides is their comparably low ecosystem impact and minimal harm to unintending organisms and humans. Unlike chemical pesticides, many fungal biopesticides are highly specific to a target pest and do not pose significant risks to beneficial insects, birds, or other organisms.

Residue Management: Fungal biopesticides often leave little to no residue on crops, making them suitable for integrated pest management (IPM) programs, in which multiple methods are used to control pests while minimizing the use of chemical pesticides.

Application: Fungal biopesticides can be applied using various methods, including foliar sprays, soil drenches, seed treatments, and even formulations that are applied directly to pest habitats. Selection of the utilization method relies on the specific pest, crop variety, and fungal species employed.

Challenges: Despite their benefits, fungal biopesticides face challenges, as environmental factors like temperature and humidity can affect their efficacy. They might have a slower action than some chemical pesticides. Production and formulation processes can also be complex, and maintaining the viability of fungal spores or mycelia in the final product is crucial.

Well-known examples of fungal biopesticides include Beauveria bassiana, Metarhizium anisopliae, and Trichoderma species. These fungi have been commercialized and employed to manage diverse pests across a range of agricultural and horticultural environments. Fungal biopesticides are part of the broader field of biopesticides, which also includes bacterial, viral, and botanical-based products. Their potential to offer effective pest control with reduced environmental impacts has contributed to their growing popularity in sustainable agriculture and pest management strategies (Lengai and Muthomi, 2018).

7.3 Mycoparasitism

Mushrooms are not only effective against insects but can also act as mycoparasites to combat fungi that harm plants. Certain mushroom species, like *Trichoderma* and *Clonostachys*, can establish colonies and suppress the proliferation of pathogenic fungi by out-competing them for nutrients and space. This mycoparasitic activity has been explored as a means of suppressing diseases such as root rot and damping-off in various crops (Brožová, 2024). Mycoparasitism refers to a specific type of parasitism in which one fungus (the mycoparasite) parasitizes and derives nutrients from another fungus. In other words, it is a relationship in which a fungus acts as a parasite on another fungus. This phenomenon is similar to how some animals or plants can be parasites on other animals or plants (Wang *et al.*, 2023).

Mycoparasites exhibit various strategies for attacking and consuming host fungi. These strategies include the following:

Direct Penetration: Some mycoparasites have specialized structures, such as appressoria, which help them physically penetrate the host fungal cells. Once inside, nutrients are derived from the cellular contents of the host.

Extracellular Enzymes: Mycoparasites may secrete enzymes that break down the host cell walls, enabling them to access the nutrients present in the host cells.

Competition for Resources: Mycoparasites can also compete with the host fungus for resources, essentially outcompeting the host for nutrients and space.

Antibiosis: Some mycoparasites can produce toxic compounds that inhibit the growth of their host, similar to how antibiotics work.

Mycoparasitism is of significant interest in agriculture and biological control. Some mycoparasitic fungi can be used as biocontrol agents for plant pathogens. By introducing these beneficial mycoparasites into agricultural systems, the prevalence of harmful plant diseases can be reduced. This method offers an eco-friendly substitute for synthetic pesticides (Rakibu *et al.*, 2022). Mycoparasitism is a fascinating aspect of fungal ecology and has practical applications in various fields, including agriculture, forestry, and environmental management. Mycoparasitism refers to a biological relationship in which one fungus (the mycoparasite) parasitizes another fungus, deriving nutrients and benefits from the host fungus (Yaping *et al.*, 2023). Mycoparasites attack and consume other fungi. This phenomenon is akin to the concept of parasitism observed in the animal kingdom, where one organism (the parasite) benefits at the expense of another (the host) (Atanasova *et al.*, 2013).

Mycoparasitism can involve various mechanisms such as:

Direct Attack and Penetration: Mycoparasites can physically invade the host fungus by growing in their tissues. They may produce specialized structures such as infection pegs or haustoria that penetrate the hyphae of the host (filamentous structures of fungi) and gain access to nutrients.

Competition for Resources: Mycoparasites might compete with the host fungus for nutrients and space. By outcompeting the host, they can suppress its growth and eventually take over resources.

Antibiotic Production: Some mycoparasites secrete antibiotics or other toxic compounds that retard the development of the host fungus, allowing mycoparasites to dominate the environment and consume the weakened host.

Induction of Host Cell Death: Mycoparasites can trigger cell death in host fungi, causing host cells to break down and release nutrients that the mycoparasite can then absorb.

Mycoparasites are ecologically significant because they can affect the dynamics of fungal populations with potential agricultural and biocontrol uses. In agriculture, certain mycoparasitic Fungi serve as biological agents for the management of plant diseases caused by pathogenic fungi. By introducing mycoparasites, farmers have the potential to decrease reliance on synthetic fungicides to promote sustainable pest management strategies (Shimizu *et al.*, 2022).

One example of mycoparasitism is the fungus *Trichoderma*, which is well known for its mycoparasitic abilities. It attacks diverse fungal species, including disease-causing agents in plants. which has been used for various biological control strategies. Mycoparasitism underscores the complexity of interactions in microbial ecosystems and highlights the intricate ways in which organisms adapt to survive and thrive in their environment (Rajani *et al.*, 2021).

Induced Resistance

Certain mushrooms have been shown to stimulate the plant's natural defense mechanisms, thereby enhancing its resistance to pests and diseases. This phenomenon, known as induced systemic resistance, involves the release of compounds by mushrooms that activate plant immune responses. As a result, crops have become more resilient to a range of pathogens and pests. Induced resistance is a phenomenon observed in plants in which the plant's immune system becomes more effective against a broad range of pathogens after exposure to certain types of stress or prior infection. This heightened immune response is a form of defense mechanism that helps the plant to prevent future attacks by pathogens (Sridharan *et al.*, 2021).

There are primary categories of induced resistance:

Systemic Acquired Resistance (SAR): SAR is a type of induced resistance that occurs after localized infection in a plant. When a plant is infected by a pathogen in one part, it can trigger a systemic response throughout the plant. This reaction encompasses the generation of different signaling compounds, such as salicylic acid, which triggers defense mechanisms, including the production of pathogenesis-related (PR) proteins. These proteins are involved in restraining pathogen growth and constitute a component of the plant's immune response (Dutta *et al.*, 2023; de Lamo and Takken, 2020).

Induced Systemic Resistance (ISR): It is a comprehensive term that denotes induced resistance caused by various beneficial microorganisms, such as certain types of bacteria and fungi. These microorganisms can colonize plant roots or leaves and activate systemic defense responses. ISR often involves different signaling pathways than SAR and can provide protection against pathogens and pests (Abbas *et al.*, 2022).

The induced resistance can be triggered by various stress factors and external agents, including:

Pathogen Infection: The presence of certain pathogens can stimulate the plant's immune response, leading to induced resistance.

Physical Damage: Mechanical injuries, such as pruning or herbivore feeding, can trigger defense responses in plants.

Chemical Elicitors: Certain chemicals, such as jasmonic acid and ethylene, can induce resistance when applied to plants.

Beneficial Microorganisms: Some beneficial microorganisms that establish symbiotic relationships with plants can induce resistance as part of their interactions.

The mechanisms underlying induced resistance are complex and involve intricate signaling pathways, hormone interactions, and gene expression changes within the plant. This phenomenon has significant agricultural implications, as it can potentially reduce the need for chemical pesticides and enhance crop protection in a more sustainable manner. Researchers continue to study the molecular and physiological mechanisms of induced resistance to develop strategies for the improvement of plant immunity and crop health (Dutta *et al.*, 2023).

7.4 Nutrient Cycling and Soil Health

Mushrooms have a crucial function in the process of nutrient cycling and the breakdown of organic matter. They break down complex organic materials and release essential nutrients back into soil. Mushrooms indirectly contribute to the overall vigor of crops by improving soil health and fertility. Healthy plants are better equipped to withstand pest and disease conditions. Nutrient cycling is a fundamental ecological process that involves the movement and transformation of nutrients through various components of an ecosystem, including the soil, plants, animals, and atmosphere (Molina *et al.*, 1999).

This process is crucial for maintaining soil fertility, supporting plant growth, and sustaining the overall ecosystem health. Nutrient cycling is tightly linked to soil health as the soil serves as a central hub for nutrient storage, transformation, and exchange within the ecosystem. Primary nutrients contributing to nutrient cycling encompass nitrogen (N), phosphorus (P), and potassium (K), in addition to other vital elements like calcium, magnesium, sulfur, and micronutrients (Schmitz, 2016). Nutrient cycling typically involves the following steps:

Input: Nutrients enter the ecosystem through various pathways such as atmospheric deposition, weathering of rocks and minerals, organic matter decomposition, and biological fixation by certain microbes.

Plant Uptake: Plants take up nutrients from the soil through their root systems. These nutrients are essential for various physiological processes including photosynthesis, growth, and reproduction.

Decomposition: As plants and other organic matter (such as dead animals) die and decay, they contribute organic material to the soil. This organic matter is broken down by decomposers such as bacteria, fungi, and other microorganisms. During decomposition, nutrients are released from the organic matter and become available for uptake by plants.

Mineralization: It constitutes the procedure by which biological material is converted into non-organic forms, releasing nutrients, including nitrogen and phosphorus, in forms readily assimilated by plants.

Leaching and Erosion: Nutrients that are not taken up by plants can be subjected to leaching, where they are washed downward through the soil profile by rainfall or irrigation. Erosion can also transport soil particles and the nutrients attached to them from one location to another.

Nutrient Uptake by Microbes: Soil microorganisms also have a vital role in nutrient cycling. They compete with plants for nutrients and their activities influence nutrient availability and cycling dynamics.

Nutrient Recycling: When plants shed leaves, twigs, and other organic materials, they contribute to the soil organic matter content. This material is broken down by soil organisms, discharging nutrients that can be absorbed by fresh vegetation. Soil health is also closely related to nutrient cycling. Healthy soils support robust nutrient cycling processes, which in turn benefits plant growth and ecosystem productivity. Here, we discuss how soil health and nutrient cycling are related.

Organic Matter: Healthy soils have higher levels of organic matter, which serves as a source of nutrients, enhances soil composition, boosts water retention ability, and fosters beneficial microbial life communities involved in nutrient cycling.

Microbial Activity: Soil microbes are central to nutrient cycling. Healthy soils host diverse microbial populations that break down organic matter, release nutrients, and form symbiotic relationships with plants to enhance their nutrient uptake.

Nutrient Availability: Nutrients released through decomposition and mineralization are more readily available to plants in well-structured soils with optimal pH and moisture levels.

Erosion Control: Healthy soils with a good structure and cover vegetation are less prone to erosion, which helps retain nutrients within the ecosystem.

Balanced Nutrient Management: Maintaining the balance of nutrients in the soil is essential for long-term soil health and sustained productivity. Excessive use of chemical fertilizers can disrupt nutrient cycling and harm soil organisms.

Crop Resilience: Nutrient-rich soils result in healthier plants that are more resistant to pests, diseases, and environmental stressors.

For sustainable agriculture and land management practices, it is important to focus on enhancing nutrient cycling and promoting soil health. This can be achieved through crop rotation, cover cropping, reduced tillage, composting, and the judicious use of organic and synthetic fertilizers to support human needs and ecosystem health (Morugán-Coronado *et al.*, 2022).

7.5 Companion Planting

Incorporating certain mushroom species into companion planting strategies can help deter pests and attract beneficial organisms. For instance, oyster mushrooms (*Pleurotus*) have been used to create "mushroom walls" that act as physical barriers to pests. In addition, the presence of mushrooms can attract predatory insects that feed on crop-damaging pests (Konkolewska *et al.*, 2020).

This approach of planting entails strategic cultivation of various plant species in close proximity to one another to achieve specific benefits.

This practice is based on the concept that certain plant combinations can help each other by promoting growth, deterring pests, enhancing pollination, and improving overall garden health. Companion planting can be used in both traditional and organic gardening methods (Saldanha *et al.*, 2019).

Here are some common examples of companion planting strategies and their benefits:

Pest Control: Some plants release natural compounds or scents that repel pests, thus protecting neighboring plants. For example, marigolds are often planted near vegetables to deter aphids and nematodes.

Enhanced Pollination: Planting flowers that attract pollinators, such as bees and butterflies, can improve the pollination of nearby crops. For instance, planting bee-friendly flowers near fruit-bearing plants can increase fruit production (Matsuoka *et al.*, 2020).

Space Optimization: Companion planting can help maximize the use of garden space. Tall plants, like corn, can provide shade to shorter plants, helping to conserve moisture and prevent weed growth.

Nutrient Enhancement: Certain plant combinations can improve soil fertility. Leguminous plants, like beans and peas, have the ability to fix nitrogen in the soil, benefiting nearby plants that require nitrogen for growth.

Trap Cropping: This involves planting a sacrificial crop that attracts pests away from the main crops. Once the pests are concentrated on the trap crop, they can be more easily controlled.

Disease Prevention: Some plants emit natural compounds that suppress or deter the growth of diseases. For example, onions and garlic are known for their antifungal and antibacterial properties.

Mulching and Disease Suppression

Mushroom-based mulches have gained attention for their potential to suppress soil-borne diseases. When specific mushroom substrates serve as mulch, they emit bioactive compounds capable of suppressing the proliferation of harmful microorganisms. This approach not only contributes to disease management but also enhances soil composition and water retention (Blanco-Pérez *et al.*, 2022; Yordanova and Gerasimova, 2016).

7.6 Challenges and Considerations

While the application of mushrooms in pest and disease management in agriculture has the potential to be a sustainable and environmentally friendly approach, it is not without its challenges. Overcoming these challenges will require ongoing research, investment, and collaboration between researchers, farmers, and regulatory agencies to realize the full potential of this innovative method. The application of mushrooms in the control of pests and diseases impacting farm crops offers an innovative and eco-friendly approach, but it also comes with several challenges and considerations (Sonawane *et al.*, 2022). Here are some of the key ones:

Species Selection: Not all mushroom species are effective in controlling pests and diseases. Selecting the right mushroom species with biopesticidal properties is crucial. Each target pest or disease may require a specific mushroom species, so research and testing are essential.

Efficacy: The effectiveness of mushroom-based biopesticides can exhibit variability that can be influenced by factors such as environmental conditions, the type of crop, and the particular pest or disease in question. Achieving consistent results can be challenging.

Formulation and Application: Developing a suitable formulation for mushroom-based biopesticides and finding efficient application methods can be complex. The formulation should ensure stability, ease of application, and long-lasting effectiveness.

Regulation and Approval: Like chemical pesticides, biopesticides are subject to regulation and approval by government agencies. Meeting these regulatory requirements can be time-consuming and expensive.

Cost: Producing and applying mushroom-based biopesticides can be more expensive than conventional chemical pesticides. This cost factor may limit their adoption, especially in regions with limited resources.

Shelf Life: The durability of mushroom-based biopesticides may be less than that of chemical alternatives, requiring careful storage and handling.

Resistance Management: Just as pests can develop resistance to chemical pesticides, they can also become resistant to biopesticides. Implementing resistance management strategies is essential to maintain their effectiveness.

Education and Adoption: It is essential to provide farmers with knowledge about the advantages and correct application of mushroom-based biopesticides. Changing established practices and convincing farmers to adopt new methods can be challenging.

Environmental Impact: While mushroom-based biopesticides are generally considered environmentally friendly, their production and application may still have environmental impacts. Monitoring and mitigating these impacts are important (Khursheed *et al.*, 2022).

Research and Development: Continuous research is needed to improve the efficacy, cost-effectiveness, and practicality of using mushrooms for the control of pests and diseases in agriculture.

Compatibility with Comprehensive Pest Control: Mushroom-based biopesticides should be integrated into broader comprehensive strategies for maximum effectiveness. This requires coordination and planning.

Safety Concerns: Ensuring the safety of farmers, farmworkers, and consumers when using mushroom-based biopesticides is crucial. Proper training and protective measures are necessary.

The utilization of mushrooms for pest and disease control affecting farm crops represents an innovative and sustainable approach. From biopesticides to mycoparasitism and soil improvement, mushrooms offer a range of mechanisms to enhance crop health and productivity (Sharma *et al.*, 2023. As the demand for sustainable agriculture grows, further research and practical implementation of these mushroom-based strategies have the potential to revolutionize pest and disease management in the farming industry.

7.7 Conclusion and Future Perspectives

The use of mushrooms in controlling pests and diseases that impact agricultural crops holds significant promise as an environmental friendly and sustainable substitute for conventional chemical pest control agents. This approach has gained attention due to its potential to address several key challenges in modern agriculture, including the requirement for pest and disease management practices that are both safer and more eco-friendly. While research in this area is ongoing, there are several noteworthy conclusions that can be drawn:

Effective Pest and Disease Suppression: Mushrooms, particularly species like Trichoderma, Metarhizium, and Beauveria, have demonstrated the ability to effectively suppress a wide range of agricultural pathogens and infestations. They operate using a variety of mechanisms, such as parasitic behavior, antagonistic interactions, and the synthesis of bioactive compounds. (Zhang *et al.*, 2021)

Minimal Environmental Impact: Mushroom-based biopesticides typically exhibit minimal toxicity to humans, animals, and unintended organisms. They possess reduced contamination of soil, water, or food products, reducing the negative environmental impact associated with chemical pesticides.

Enhanced Soil Health: The use of certain mushroom species as biocontrol agents can contribute to improved soil health by promoting beneficial microbial communities, nutrient cycling, and organic matter decomposition.

Resistance Management: Unlike chemical pesticides, which can lead to the development of resistant pests and pathogen populations, mushroom-based biopesticides have shown a lower likelihood of resistance development. This makes them a valuable tool for integrated pest and disease management.

Sustainability: Mushroom cultivation can be integrated into sustainable agricultural practices, as they often thrive on agricultural waste products such as crop residues and straw. This can contribute to waste reduction and resource efficiency.

7.7.1 Future Perspectives

The application of mushrooms in pest and disease management in agriculture is an evolving field with several exciting prospects for the future:

Strain Improvement: Continued research into the genetics and biology of biocontrol mushrooms can result in the production of more efficient strains with enhanced pest and pathogen suppression capabilities.

Formulation and Delivery: Innovations in formulation techniques can improve the shelf life, stability, and application methods of mushroom-based biopesticides, making them more practical for widespread use.

Integration with Other Methods: The integration of mushroom-based biocontrol agents with other pest and disease control approaches, such as crop rotation, the use of biological control, and the cultivation of resistant crop varieties, can lead to more robust and sustainable agricultural systems.

Regulatory Approvals: Governments and regulatory bodies should establish clear guidelines and regulatory frameworks for the registration and use of mushroom-based biopesticides, ensuring their safe and effective adoption.

Education and Adoption: Farmers and agricultural practitioners need education and training on the benefits and proper application of mushroom-based biopesticides to facilitate their widespread adoption.

Research on Novel Species: Exploring and studying lesser-known mushroom species for their biocontrol potential can lead to the discovery of new, effective agents for pathogens and infestation control.

Conclusively, utilizing mushrooms for control of pathogens and infestations impacting farm crops offers a sustainable, environmentally friendly, and promising alternative to conventional chemical pesticides. Continued research, development, and education are key to realizing the full potential of this approach and advancing its integration into modern agriculture. As we face ongoing challenges related to food security, environmental sustainability, and pesticide resistance, mushroom-based solutions have the potential to play a vital role in addressing these issues.

References

Abbas, A., Mubeen, M., Zheng, H., Sohail, M.A., Shakeel, Q., Solanki, M.K., Iftikhar, Y., Sharma, S., Kashyap, B.K., Hussain, S., Maria del Carmen Zuñiga Romano, M.C., Moya-Elizondo, E.A., Zhou, L., *Trichoderma* spp. Genes Involved in the Biocontrol Activity Against *Rhizoctonia solani*. *Front. Microbiol.*, 13, 884469, 2022, DOI: 10.3389/fmicb.2022.884469.

Atanasova, L., Le Crom, S., Gruber, S., Coulpier, F., Seidl-Seiboth, V., Kubicek, C.P., Druzhinina, I.S., Comparative transcriptomics reveals different strategies of Trichoderma mycoparasitism. *BMC Genomics*, 14, 121, 2013, ISSN: 1471-2164. DOI: https://doi.org/10.1186/1471-2164-14-121.

Badshah, H., Qureshi, R.A., Khan, J., Ullah, F., Fahad, S., Ullah, F., Khan, A.M., Hussain, I., Khan, N., Pharmacological screening of Morchella esculenta (L.) Pers., Calvatia gigantea (Batsch ex Pers.) Lloyd and Astraeus hygrometricus Pers., mushroom collected from South Waziristan (FATA). *J. Med. Plant Res.*, 6, 10, 1853–1859, 2012.

Badshah, K., Ullah, F., Ahmad, B., *et al.*, Management of *Lycoriella ingenua* (Diptera: Sciaridae) on oyster mushrooms (*Pleurotus ostreatus*) through different botanicals. *Int. J. Trop. Insect Sci.*, 41, 1435–1440, 2021, https://doi.org/10.1007/s42690-020-00341-2.

Bamisile, B.S., Akuts, K.S., Siddiqui, J.A., Xu, Y., Model Application of Entomopathogenic Fungi as Alternatives to Chemical Pesticides: Prospects, Challenges, and Insights for Next-Generation Sustainable Agriculture. *Front. Plant Sci.*, 12, 741804, 1–21, 2021, https://doi.org/10.3389/fpls.2021.741804.

Bautista, E., Mesa, L., Gómez, M., Alternatives for the production of microbial biopesticides based on fungi: the case of Latin America and the Caribbean. *Sci. Agropecu.*, *9*, 4, 585–604, 2018.

Bhuvaneswari, M.S. and Subashini, G.R., Mycopesticides: Fungal based pesticides for sustainable agriculture, in: *Fungi and their Role in Sustainable Development: Current Perspective*, 2018, DOI: 10.1007/978-981-13-0393-7_11.

Blanco-Pérez, R., Vicente-Díez, I., Pou, A., Pérez-Moreno, I., Marco-Mancebón, V.S., Campos-Herrera, R., Organic mulching modulated native populations of entomopathogenic nematode in vineyard soils differently depending on its potential to control outgrowth of their natural enemies. *J. Invertebr. Pathol.*, 192, 107781, 2022, ISSN: 0022-2011. https://doi.org/10.1016/j.jip.2022.107781.

Constantin, M.E., de Lamo, F.J., Vlieger, B.V., Rep, M., Takken, F.L.W., Endophyte-Mediated Resistance in Tomato to Fusarium oxysporum Is Independent of ET, JA, and SA. *Front. Plant Sci.*, *10*, 979, 2019, https://doi.org/10.3389/fpls.2019.00979.

Dar, S.A., Dar, N.A., Bhat, M.A., Bhat, M.H., Prospects, utilization and challenges of botanical pesticides in sustainable agriculture. *Int. J. Mol. Biol. Biochem.*, *2*, 1, 1–14, 2014.

Das, M.M., Aguilar, C.N., Haridas, M., Sabu, A., Production of bio-fungicide, *Trichoderma harzianum* CH1 under solid-state fermentation using coffee husk. *Bioresour. Technol. Rep.*, 15, 100708, 2021, ISSN 2589-014X. https://doi.org/10.1016/j.biteb.2021.100708.

De Lamo, F.J. and Takken, F.L.W., Biocontrol by Fusarium Oxysporum Using Endophyte-Mediated Resistance. *Front. Plant Sci.*, *11*, 37, 2020, Frontiers Media S.A. doi:10.3389/fpls.2020.00037.

Deepshikha, K.B., Devi, E.P., Sharma, G., Rawat, S., Jaiswal, J.P., Irradiation as an Alternative Method for Post-harvest Disease Management: An Overview. *Int. J. Agric. Environ. Biotechnol.*, *10*, 5, 625, 2017, https://doi.org/10.5958/2230-732x.2017.00077.8.

Deshpande, M.V., Mycopesticide Production by Fermentation: Potential and Challenges. *Crit. Rev. Microbiol.*, 25, 3, 229–243, 1999, DOI: 10.1080/10408419991299220.

Drobnjaković, T., Prijović, M., Milenković, S., Marčić, D., Sublethal effects of a *Beauveria bassiana*-based mycopesticide on Dutch and Serbian populations of *Encarsia formosa* (Hymenoptera: Aphelinidae). *Biocontrol Sci. Technol.*, 29, 10, 991–1008, 2019, DOI: 10.1080/09583157.2019.1635246.

Dutta, P., Mahanta, M., Singh, S.B., Thakuria, D., Deb, L., Kumari, A., Upamanya, G.K., Boruah, S., Dey, U., Mishra, A.K., Vanlaltani, L., VijayReddy, D., Heisnam, P., Pandey, A.K., Molecular interaction between plants and Trichoderma species against soil-borne plant pathogens. *Front. Plant Sci.*, 14, 1145715, 2023, doi: 10.3389/fpls.2023.114571.

Gao, L., Application of Two-Stage Cultivation for Exploring the Nutritional Requirements for Sporulation of Three Biocontrol Fungi. *Biotechnol. Res. Int.*, *2015*, 1, 682839, 2015.

Gayane, S.B., Barazani, A., Wasser, S.P., Medicinal mushrooms with anti-phytopathogenic and insecticidal properties, in: *Mushroom Biotechnology*, pp. 137–153, Academic Press, Cambridge, MA, 2016.

Ghimire, A., Pandey, K.R., Joshi, Y.R., Subedi, S., Major Fungal Contaminants of Mushrooms and Their Management. *Int. J. Appl. Sci. Biotechnol.*, *9*, 2, 80–93, 2021, https://doi.org/10.3126/ijasbt.v9i2.37513.

Khursheed, A., Rather, M.A., Jain, V., Wani, A.R., Rasool, S., Nazir, R., Malik, N.A., Sheikh, S.A., Plant based natural products as potential ecofriendly and safer biopesticides: A comprehensive overview of their advantages over conventional pesticides, limitations and regulatory aspects. *Microb. Pathogen.*, 173, 2022A, 105854, 2022, ISSN: 0882-4010. https://doi.org/10.1016/j.micpath.2022.105854.

Kim, J.C., Lee, M.R., Kim, J.S., Long-term storage stability of Beauveria bassiana ERL836 granules as fungal biopesticide. *J. Asia-Pac. Entomol.*, 22, 2, 537–542, 2019. ISSN: 12268615, DOI: 10.1016/j.aspen.2019.04.001.

Konkolewska, A., Piechalak, A., Ciszewska, L., *et al.*, Combined use of companion planting and PGPR for the assisted phytoextraction of trace metals (Zn, Pb, Cd). *Environ. Sci. Pollut. Res.*, 27, 13809–13825, 2020, https://doi.org/10.1007/s11356-020-07885-3.

Leathers, T.D., Gupta, S.C., Alexander, N.J., Mycopesticides: Status, challenges and potential. *J. Ind. Microbiol.*, *12*, 2, 69–75, 1993. ISSN: 01694146. DOI: 10.1007/BF01569904.

Lecoq, M. and Cease, A.J., Are Mycopesticides the Future of Locust Control? *Agronomy*, 12, 2344, 2022.

Lengai, G.M. and Muthomi, J.W., Biopesticides and their role in sustainable agricultural production. *J. Biosci. Med.*, *6*, 06, 7, 2018.

Liu, D., Smagghe, G., Liu, T.X., Interactions between entomopathogenic fungi and insects and prospects with glycans. *J. Fungi*, 9, 5, 575, 2023, https://doi.org/10.3390/jof9050575.

Matsuoka, T., Tsuchiya, K., Yamada, S., Lundholm, J., Okuro, T., The effects of growth form on the impact of companion planting of nectar-producing plant species with Sedum album for extensive green roofs. *Urban For. Urban Greening*, 56, 126875, 2020, ISSN: 1618-8667. https://doi.org/10.1016/j.ufug.2020.126875.

Mau, Y.S., Prayetno, R.S., Kaka, H., *et al.*, Efficacy of indigenous *Trichoderma* isolates of West Timor, Indonesia, as biocontrol agents of brown spot (*Drechslera oryzae*) on two upland rice varieties. *Egypt. J. Biol. Pest Control*, 32, 62, 2022, https://doi.org/10.1186/s41938-022-00559-x.

Mehra, M.K. and Avanthi, A., Challenges and Safety Regulations for Employing Fungi in Environmental Applications, in: *Bioprospecting of Multi-tasking Fungi for a Sustainable Environment*, vol. I, pp. 315–341, Springer Nature Singapore, Singapore, 2024.

Molina, R., O'Dell, T., Dunham, S., Pilz, D., *Biological diversity and ecosystem functions of forest soil fungi: management implications*, UNITED STATES DEPARTMENT OF AGRICULTURE FOREST SERVICE GENERAL TECHNICAL REPORT PNW, pp. 45–58, 1999.

Morugán-Coronado, A., Pérez-Rodríguez, P., Insolia, E., Soto-Gómez, D., David Fernández-Calviño, D., Zornoza, R., The impact of crop diversification, tillage and fertilization type on soil total microbial, fungal and bacterial abundance: A worldwide meta-analysis of agricultural sites. *Agric. Ecosyst. Environ.*, 329, 107867, 2022, ISSN 0167-8809. https://doi.org/10.1016/j.agee.2022.107867.

Np, M., Naik, G., Chittaragi, A., Naik B, G., Naik, M.K., Km, S., Ms, N., Compatibility of Trichoderma asperellum with fungicides. *Pharma Innov. J.*, 9, 8, 136–140, 2020, Retrieved from http://www.thepharmajournal.com.

Pala, S.A., Wani, A.H., Wani, B.A., Mushroom refinement endeavor auspicate non green revolution in the offing. *Nusantara Biosci.*, 6, 2, 173–185, 2014. ISSN: 2087-3948. DOI: 10.13057/nusbiosci/n060210.

Pavlova, N., Malygin, D., Emelyanov, D., Gusenkov, E., Sokornova, S., Additional Properties of Potential Mycopesticides, in: *International Scientific Conference on Agricultural Machinery Industry "Interagromash""*, 2022, May, Springer International Publishing, Cham, pp. 1846–1852, DOI: 10.1007/978-3-031-21219-2_205.

Poveda, J., Trichoderma as biocontrol agent against pests: New uses for a mycoparasite. *Biol. Control, 159*, 104634, 2021, ISSN 1049-9644. https://doi.org/10.1016/j.biocontrol.2021.104634.

Rajani, P., Rajasekaran, C., Vasanthakumari, M.M., Olsson, S.B., Ravikanth, G., Shaanker, R.U., Inhibition of plant pathogenic fungi by endophytic Trichoderma spp. through mycoparasitism and volatile organic compounds. *Microbiol. Res.*, 242, 126595, 2021. ISSN 0944-5013, https://doi.org/10.1016/j.micres.2020.126595.

Rakibul, H., Lv, B., Uddin, M.J., Chen, Y., Fan, L., Sun, Z., Sun, M., Li., S., Monitoring Mycoparasitism of *Clonostachys rosea* against *Botrytis cinerea* Using GFP. *J. Fungi*, 8, 6, 567, 2022, https://doi.org/10.3390/jof8060567.

Riyaz, M., Mathew, P., Zuber, S.M., Rather, G.A., Botanical pesticides for an eco-friendly and sustainable agriculture: New challenges and prospects, in: *Sustainable Agriculture: Technical Progressions and Transitions*, pp. 69–96, Springer International Publishing, Switzerland AG, 20212022, https://doi.org/10.1007/978-3-030-83066-3_5.

Saldanha, A.V., Gontijo, L.M., Carvalho, R.M.R., Vasconcelos, C.J., Corrêa, A.S., Gandra, R.L.G., Companion planting enhances pest suppression despite reducing parasitoid emergence. *Basic Appl. Ecol.*, 41, 45–55, 2019. ISSN: 1439-1791, https://doi.org/10.1016/j.baae.2019.10.002.

Schmitz, O.J., Biological Diversity and Ecosystem Functions, in: *The New Ecology*, Princeton University Press, New Jersey, 2016, https://doi.org/10.23943/princeton/9780691160566.003.0003.

Sharma, A.K., Jana, M., Srivastav, A., Gupta, M., Sharma, S., Singh Gill, S., Antimicrobial Properties of Some Edible Mushrooms: A Review. *World J. Pharm. Pharm. Sci.*, 3, 5, 1009–1023, 2014.

Sharma, A., Sharma, S., Yadav, P.K., Entomopathogenic fungi and their relevance in sustainable agriculture: A review. *Cogent Food Agric.*, 9, 1, 1–21, 2023.

Shimizu, M., Naznin, H.A., Hieno, A., The Significance of Mycoparasitism by *Streptomyces* sp. MBCN152-1 for Its Biocontrol Activity against *Alternaria brassicicola. Microbes Environ.*, 37, 3, 2022, ME22048. ISSN: 1347-4405. https://doi.org/10.1264/jsme2.ME22048.

Sinha, K.K., Choudhary, A.K., Kumari, P., Chapter 15 - Entomopathogenic Fungi, in: *Ecofriendly Pest Management for Food Security*, Omkar (Ed.), pp. 475–505, Ecofriendly Pest Management for Food Security, Academic Press, Cambridge, MA, 2016, ISBN 9780128032657, https://doi.org/10.1016/B978-0-12-803265-7.00015-4.

Sitara, U., Baloch, P.A., Pathan, A.U.K., Bhatti, M.I., Bhutto, M.A., Ali, Q.M., Iqbal, M., *In Vitro* Studies to Determine Antibacterial and Antifungal Properties of Three *Pleurotus* species (Oyster Mushroom). *Pak. J. Bot.*, *55*, 1, 387–392, 2023, https://doi.org/10.30848/PJB2023-1(14).

Sonawane, H., Shelke, D., Chambhare, M., Dixit, N., Math, S., Sen, S., Sarma, H., Fungi-derived agriculturally important nanoparticles and their application in crop stress management–Prospects and environmental risks. *Environ. Res.*, *212*, 113543, 2022.

Sridharan, A.P., Sugitha, T., Karthikeyan, G., Nakkeeran, S., Sivakumar, U., Metabolites of *Trichoderma longibrachiatum* EF5 inhibits soil borne pathogen, *Macrophomina phaseolina* by triggering amino sugar metabolism. *Microb. Pathogen.*, 150, 104714, 2021, ISSN 0882-4010. https://doi.org/10.1016/j.micpath.2020.104714.

Stajic, M., Cilerdzic, J., Vukojevic, J., Mushrooms as Potent Sources of New Biofungicides. *Current Pharm. Biotechnol.*, 18, 13, 1055–1066, 2017.

Surija, S. and Faustine, M., Basal Stem Rot Disease Eradication and Improving Production with *Ganoderma* Vaccine/Biofungicide CHIPS in Flooded Area of Oil Palm Plantation. *Eur. J. Agric. Food Sci.*, *3*, 3, 101–109, 2021, https://doi.org/10.24018/ejfood.2021.3.3.323.

Tao, C., Wang, Z., Liu, S., Lv, N., Deng, X., Xiong, W., Shen, Z., Zhang, N., Geisen, S., Li, R., Shen, Q., Kowalchuk, G.A., Additive fungal interactions drive biocontrol of Fusarium wilt disease. *New Phytol.*, 238, 3, 1198–1214, 2023, https://doi.org/10.1111/nph.18793.

Waqas, H.M., Akbar, M., Iqbal, M.S., Antibacterial and antioxidant activities of Agaricus bisporus (J.E. Lange) imbach from Pakistan. *Bangladesh J. Bot.*, *48*, 4, 1075–1081, 2019, https://doi.org/10.3329/bjb.v48i4.49055.

Yaping, W., Zhu, X., Wang, J., Shen, C., Wang, W., Identification of Mycoparasitism-Related Genes against the Phytopathogen *Botrytis cinerea* via Transcriptome Analysis of *Trichoderma harzianum* T4. *J. Fungi*, 9, 3, 324, 2023, https://doi.org/10.3390/jof9030324.

Yordanova, M. and Gerasimova, N., Effect of mulching on weed infestation and yield of beetroot *(Beta vulgaris* ssp. *rapaceae atrorubra* Krass). *Org. Agric.*, 6, 133–138, 2016, https://doi.org/10.1007/s13165-015-0122-6.

Zaki, O., Weekers, F., Thonart, P., Tesch, E., Kuenemann, P., Jacques, P., Limiting factors of mycopesticide development. *Biol. Control*, 144, 104220, 2020. ISSN 1049-9644, https://doi.org/10.1016/j.biocontrol.2020.104220.

Zhang, C., Weiwei, W., Ming, X., Zhen, L., Qinman, Z., Jumei, H., Mengyu, X., Rui, W., Tong, L., The Combination of a Biocontrol Agent *Trichoderma asperellum* SC012 and Hymexazol Reduces the Effective Fungicide Dose to Control Fusarium Wilt in Cowpea. *J. Fungi*, 7, 9, 685, 2021, https://doi.org/10.3390/jof7090685.

8

Discovery and Relevance of Novel Pharmacological Substances from Beneficial Mushrooms

Oluwabukola Atinuke Popoola[1*], Charles Oluwaseun Adetunji[2], Oluwafemi Adebayo Oyewole[3], Abel Inobeme[4] and Adeyomoye Olorunsola Israel[5]

[1]Genetics, Genomics and Bioinformatics Department, National Biotechnology Development Agency, Abuja, FCT, Nigeria
[2]Applied Microbiology, Biotechnology and Nanotechnology Laboratory, Department of Microbiology, Edo State University Uzairue, Iyamho, Edo State, Nigeria
[3]Department of Microbiology, Federal University of Technology, Minna, Nigeria
[4]Department of Chemistry, Edo State University Uzairue, Iyamho, Edo State, Nigeria
[5]Department of Physiology, Ondo Medical University, Ondo State, Nigeria

Abstract
Mushrooms have played a significant role in human diets and traditional medicine for centuries, with a growing body of research highlighting their immense potential as sources of novel pharmacological substances. This abstract provides an overview of the discovery and relevance of such substances derived from beneficial mushrooms, shedding light on their diverse therapeutic applications and potential impact on human health. Nourishing fungi, including Reishi (*Ganoderma lucidum*), Shiitake (*Lentinula edodes*), and Cordyceps (*Cordyceps sinensis*) have been widely recognized for their health-promoting properties. Recent advances in scientific research have unveiled the existence of bioactive constituents within these fungi, such as polysaccharides and triterpenoids, with various secondary metabolites, that elicit broad pharmacological effects. A pivotal discovery in this domain is the potent immunomodulatory properties of mushroom-derived

**Corresponding author*: atinuke.solina@gmail.com

Charles Oluwaseun Adetunji, Chukwuebuka Egbuna, Oluwatosin Ademola Ijabadeniyi and Samantha C. Karunarathna (eds.) Mushroom Biotechnology for Improved Agriculture and Human Health, (183–208) © 2025 Scrivener Publishing LLC

compounds. Beta-glucans, a variety of polysaccharides present in mushrooms, have exhibited the capacity to boost the immune system's reaction to infections and illnesses, presenting them as a potential reservoir of innovative immunomodulatory treatments. Additionally, substances like triterpenoids and ergosterol derivatives have exhibited anti-inflammatory and antioxidant effects, with implications for conditions ranging from cancer to neurodegenerative disorders. The discovery of these novel pharmacological substances has opened up exciting avenues for therapeutic development. Moreover, these substances show promise in the management of metabolic disorders, cardiovascular diseases, and even mental health conditions like anxiety and depression. Furthermore, the sustainability and eco-friendliness of mushroom cultivation add an appealing dimension to their relevance in pharmacology. Unlike many traditional sources of pharmaceuticals, mushrooms can be cultivated in controlled environments, minimizing the environmental impact associated with drug production. The discovery and relevance of novel pharmacological substances from beneficial mushrooms are a testament to the untapped potential of nature in improving human health. The broad spectrum of bioactive constituents discovered in these fungi holds promise for a wide array of therapeutic applications, from cancer treatment to immune system enhancement and beyond. With ongoing research efforts and a growing interest in natural remedies, mushroom-derived substances are poised to make a significant impact on the field of pharmacology, offering new hope for improved healthcare and a sustainable future.

***Keywords*:** Mushroom pharmacology, bioactive, substances, beneficial, mushrooms, discovery

8.1 Introduction

Mushrooms, often considered a simple dietary component or a curious natural phenomenon, have recently taken center stage in the world of pharmacology and medicine (Kirdeeva *et al.*, 2022). Beyond their culinary and ecological significance, mushrooms, particularly those deemed "beneficial," have emerged as a rich source of novel pharmacological substances with profound implications for human health. This newfound recognition stems from centuries of traditional use in various cultures and is now supported by rigorous scientific research (Pepe *et al.*, 2023).

Beneficial mushrooms encompass a diverse group of fungal species, several of these have been employed in herbal medicinal practices for countless centuries. While traditional knowledge surrounding these mushrooms laid the foundation for their exploration, it is only in recent decades that modern science has begun to unravel their true potential (Kour *et al.*, 2022).

These mushrooms have garnered attention for their remarkable bioactive compounds and therapeutic properties (Arunachalam *et al.*, 2022; Sharpe *et al.*, 2021).

The discovery of pharmacologically active substances within beneficial mushrooms has opened up new avenues in drug development, nutrition, and complementary medicine (Habtemariam, 2019). These compounds, often referred to as mushroom-derived bioactive molecules, display a diverse array of biological functions, from immune modulation and antioxidant effects to potential anti-cancer properties. Consequently, researchers, healthcare professionals, and the public alike are increasingly recognizing the relevance and significance of these substances (Deshpande and Arya, 2022).

As we embark on this journey through the fungal kingdom's hidden treasures, it becomes clear that the discovery and relevance of novel pharmacological substances from beneficial mushrooms are poised to shape the future of medicine, offering hope and healing in a world faced with ever-evolving health challenges (Saultarelli *et al.*, 2019). This review delves into the fascinating world of beneficial mushrooms, shedding light on the methods employed to discover their pharmacological substances and elucidating the diverse range of therapeutic applications that have emerged. We will explore the mechanisms of action, clinical studies, and future prospects of mushroom-derived bioactive molecules, highlighting their potential to revolutionize healthcare and contribute to the development of novel pharmaceuticals (Hamza *et al.*, 2023).

8.1.1 Brief Overview of Beneficial Mushrooms

Beneficial mushrooms, often referred to as medicinal mushrooms, are a group of fungi that have been used for their potential health benefits for centuries in traditional medicine systems, particularly in Asian cultures like traditional Chinese medicine (TCM) and Ayurveda. These mushrooms contain bioactive compounds that are believed to promote various aspects of health and well-being (Gopal and Muthu, 2023). Here is a brief overview of some popular beneficial mushrooms and their potential benefits:

Reishi Mushroom: This is also known as the "Mushroom of Immortality." In Chinese traditional medicine. Suggested benefits include immune system support, stress reduction, and potential anti-inflammatory effects. Contains compounds like triterpenes and polysaccharides (Kurtipek *et al.*, 2016; Jin *et al.*, 2016).

Lion's Mane Mushroom (*Hericium erinaceus*): Known for its distinctive appearance resembling a lion's mane. Believed to support cognitive health and may have neuroprotective properties. Contains compounds called erinacines and hericenones (Chong *et al.*, 2020; Mendel, 2015).
Cordyceps Mushroom (*Cordyceps sinensis*): Traditionally used for energy and stamina.

Some studies suggest it may improve exercise performance and enhance endurance. Contains cordycepin and adenosine (Joshi *et al.*, 2017).
Chaga Mushroom (*Inonotus obliquus*): Rich in antioxidants, particularly melanin. May support the immune system with potential anti-inflammatory potential. Often consumed as a tea or extract (Chang *et al.*, 2022; Lu *et al.*, 2021; Softa *et al.*, 2019).

Shiitake Mushroom (*Lentinula edodes*): Widely consumed culinary mushroom with potential health benefits. Contains compounds like lentinan, which may support the immune system. Known for its savory flavor (Patel, 2019; Brauer *et al.*, 2002).
Maitake Mushroom (*Grifola frondosa*): Often called the "dancing mushroom." May help support the immune system and exhibit anti-inflammatory properties. Contains beta-glucans (Naguib *et al.*, 2023; Ham *et al.*, 2023; Dissanayake *et al.*, 2018).
Turkey Tail Mushroom (*Trametes versicolor*): Named for its colorful concentric rings resembling a turkey's tail feathers. Rich in polysaccharopeptides (PSP) and polysaccharide-K (PSK), which are studied for their immune-supporting properties. Often used alongside cancer treatments in some cultures (Booi *et al.*, 2023; Lo *et al.*, 2020).
Agaricus Blazei Mushroom (Agaricus blazei): Contains beta-glucans and ergosterol, which may contribute to immune support. Suggested to have potential anticancer properties (Wong *et al.*, 2020; Yasuma *et al.*, 2021; Chugh *et al.*, 2022).
Porcini Mushroom (Boletus edulis): Widely used in European cuisine. Rich in vitamins, minerals, and antioxidants. May have immune-supporting properties due to its beta-glucan content (Tan *et al.*, 2022; Lu *et al.*, 2020; Novakovic *et al.*, 2020).

It is crucial to acknowledge that while extensive research has been conducted on these mushrooms regarding their potential health advantages, further investigation is essential to comprehensively grasp their impacts and modes of operation. Individuals considering the inclusion of these beneficial mushrooms in their dietary regimen should seek guidance from a healthcare expert, especially if they have underlying medical conditions or are taking medications, to ensure safety and suitability. Additionally, these mushrooms are often consumed as supplements, teas, or extracts, as they are not always palatable when eaten in their raw form.

8.1.2 Importance of Discovering Novel Pharmacological Substances in Mushrooms

The discovery of novel pharmacological substances in mushrooms is an important area of research with potential benefits for both medicine and science (Hussein, 2022; Rauf *et al.*, 2017). Here are some key reasons why it is important:

Medicinal Potential: Numerous fungi have been employed in ancient healing practices for centuries. Discovering novel pharmacological substances in mushrooms can elucidate the production of novel drugs and therapies for a diverse array of ailments. For example, compounds found in mushrooms have been explored for their potential to address tumors, neurodegenerative diseases, and also immune system disorders (Mwangi *et al.*, 2022; Rathee *et al.*, 2012; Hrudayanath and Sameer, 2014).
Natural Drug Sources: Mushrooms offer a source of natural compounds that can be used in drug development. These compounds may have unique properties and mechanisms of action that are different from synthetic drugs, potentially leading to more effective and safer treatments (Sommano *et al.*, 2022; Chugh *et al.*, 2022; Pathak *et al.*, 2022; Zeb and Lee, 2021; Gründemann *et al.*, 2020).
Biodiversity Conservation: Studying mushrooms can contribute to the conservation of biodiversity. Many mushrooms are still poorly understood, and discovering their pharmacological properties can help raise awareness about the importance of preserving their habitats (Mjaika, 2022; Bazzicalupo *et al.*, 2022; Stallman and Robinson, 2022; Ekpe *et al.*, 2014).
Antibiotic Resistance: With the rise of antibiotic-resistant bacteria, there is a growing need for new antimicrobial agents. Some mushrooms produce compounds with antibiotic properties that could be used to develop novel antibiotics to combat drug-resistant pathogens (Toros *et al.*, 2023; Huguet *et al.*, 2022; Jiang *et al.*, 2022; Duvenage *et al.*, 2021).
Psychiatry and Neurology: Certain mushrooms, such as *Psilocybe* spp., contain psychoactive compounds like psilocybin and psilocin. These substances have displayed the potential to address psychiatric and neurological conditions such as depression, anxiety, and post-traumatic stress disorder (PTSD) (Rai *et al.*, 2021; Yadav *et al.*, 2020; Khatian and Aslam, 2019; Tyls *et al.*, 2016).

Nutritional Benefits: Certain fungi are abundant in vital nutrients, including vitamins, minerals, and antioxidants. Identifying innovative compounds within mushrooms may pave the way for creating nutritional supplements or functional edibles with prospective health advantages (Ahmed *et al.*, 2023; Sganzerla *et al.*, 2022; Santhiya *et al.*, 2022; Cheung, 2020).
Environmental Applications: Some mushroom compounds have environmental applications, such as bioremediation (using fungi to clean up environmental pollutants) and mycoremediation (using fungi to restore damaged ecosystems). Discovering new compounds can enhance these applications (Hu *et al.*, 2022; Kumar *et al.*, 2021; El-Ramady *et al.*, 2022).
Scientific Knowledge: Studying mushrooms can expand our understanding of fungi and their role in ecosystems. Fungi play vital roles in nutrient cycling and symbiotic relationships with plants, and novel compounds may shed light on these interactions (Pukalski and Latowski, 2022; Chatterjee *et al.*, 2017).
Economic Opportunities: The discovery of valuable pharmacological substances in mushrooms can create economic opportunities for pharmaceutical companies, biotechnology firms, and the mushroom industry, potentially leading to job creation and economic growth (Harisoesyanti, 2023; Moxley *et al.*, 2022).
Global Health: Novel pharmacological substances in mushrooms can have a significant impact on global health by providing treatments for diseases that affect millions of people worldwide (Phillips *et al.*, 2022).

The discovery of novel pharmacological substances in mushrooms has the potential to advance medicine, conservation efforts, and scientific knowledge while addressing pressing health and environmental challenges. This field of study remains highly promising for the future.

8.1.3 Beneficial Mushrooms as a Source of Pharmacological Substances

For generations, mushrooms have been employed in traditional medicinal practices, particularly within Asian societies, for their potential health benefits. Many mushrooms contain pharmacologically active compounds that have garnered interest in modern medicine and research (Zeid *et al.*, 2012; Firenzuoli *et al.*, 2008; Takatu *et al.*, 2001). A few of these beneficial mushrooms and their associated pharmacological substances include:

Reishi Mushroom:

- Polysaccharides: Reishi mushrooms contain beta-glucans and other polysaccharides with immune-modulating and anti-inflammatory properties (Kumar, 2021).
- Triterpenoids: These compounds are believed to have anti-tumor and antioxidant effects.

Lion's Mane Mushroom:

- Hericenones and erinacines: It contains these compounds, which have shown potential for nerve regeneration and cognitive function enhancement.

Cordyceps Mushroom:

- Cordycepin: Recognized for its prospective anti-cancer and anti-inflammatory characteristics.
- Adenosine: Cordyceps contains this nucleoside, which can have various pharmacological effects, including anti-inflammatory and anti-asthmatic actions.

Turkey Tail Mushroom:
Polysaccharide-K (PSK) and polysaccharide-peptide (PSP): These substances have been examined for their immune-modulating impacts, especially in the fields of cancer treatment and immune reinforcement.

Shiitake Mushroom (Lentinula edodes):

- Lentinan: A beta-glucan with potential immune-enhancing properties.
- Eritadenine: May help in reducing cholesterol levels.

Maitake Mushroom (Grifola frondosa):

- Beta-glucans: Maitake contains beta-glucans, which may help support the immune system and have anti-diabetic effects.

Chaga Mushroom (Inonotus obliquus):

- Betulin and betulinic acid: These compounds have shown potential anti-cancer and anti-inflammatory properties.
- Polysaccharides: Chaga mushrooms also contain immune-boosting polysaccharides.

Agaricus Blazei Mushroom (Agaricus subrufescens):

- Beta-glucans: Like many other medicinal mushrooms, *Agaricus blazei* contains immune-enhancing beta-glucans.

Research into the pharmacological properties of mushrooms is ongoing, and some of these substances are being investigated for their capacity to address diverse health conditions, inclusive of cancer, immune disorders, and neurodegenerative diseases. It is essential to note that while these mushrooms show promise, they should be used as complementary treatments and not as a replacement for conventional medical care.

8.1.4 Historical Use of Mushrooms in Traditional Medicine

Fungi possess a rich heritage of utilization in traditional healing practices spanning varied cultures worldwide. While many of their purported medicinal properties have not been extensively studied or validated by modern science, they have been an integral part of traditional healing practices for centuries (González-Mariscal and Sosa-Cortés, 2022; Unlu *et al.*, 2016). Here are some examples of the historical use of mushrooms in traditional medicine:

Traditional Chinese Medicine (TCM):
Reishi (*Ganoderma lucidum*): Reishi mushrooms have been used in TCM for more than 2,000 years to strengthen the immune system and encourage longevity. They are thought to offer a number of health advantages, such as lowering stress and enhancing general well-being.
Shiitake (*Lentinula edodes*): Shiitake mushrooms have been used in TCM to enhance immune function, support cardiovascular health, and promote vitality. They are also believed to possess anti-inflammatory characteristics (Suo *et al.*, 2022; Wu *et al.*, 2020).

Ayurveda (Traditional Indian Medicine):
Lion's Mane (Hericium erinaceus): Lion's Mane mushrooms have been used in Ayurveda to improve cognitive function, memory, and overall mental health. Additionally, it is thought that they possess antioxidant and anti-inflammatory qualities (Singh *et al.*, 2022).

Native American Healing Traditions:
Various types of mushrooms were used by Native American tribes for medicinal purposes. For example, the Amanita muscaria mushroom was used by some tribes for its hallucinogenic and psychoactive effects, as well as for pain relief (Maciejczyk and Kafarski, 2013).

Siberian Folk Medicine:
The Chaga mushroom (*Inonotus obliquus*) has been used for centuries in Siberian traditional medicine due to its possible anti-inflammatory and immune-stimulating effects. It was frequently made into a decoction or tea (Peng and Shahidi, 2020).

Japanese Folk Medicine:
Maitake (*Grifola frondosa*) and Matsutake (*Tricholoma matsutake*) mushrooms have been utilized for their possible health benefits, including cancer prevention and immunological support, in traditional Japanese medicine (Sasaki, 2006).

Indigenous Peoples of the Americas:
Various indigenous cultures in the Americas used psychedelic mushrooms, such as psilocybin-containing species, for spiritual and healing purposes. These mushrooms were often used in shamanic rituals (Goel & Zilate, 2022; Draper, 2020).

It is crucial to remember that although mushrooms have a rich history in traditional medicine, not all claims about their medicinal properties have been scientifically proven, and some can be associated with risks or side effects. Growing interest has been shown in the possible medicinal use of certain mushrooms, and research is ongoing to better understand their properties and applications in modern medicine.

Before using mushrooms or mushroom-based products for medicinal purposes, it's best to speak with a medical expert, particularly if you are taking medication or have any underlying medical concerns, to ascertain their safe and appropriate use.

8.2 Bioactive Compounds in Mushrooms

Since ancient times, people have included mushrooms in their diets because of their distinct flavors and endless culinary possibilities. Beyond their culinary appeal, mushrooms are also gaining recognition for their potential health benefits due to the bioactive substances that are present, including chitin and chitosan, polysaccharides, ergothioneine, antimicrobial compounds, sterols, triterpenoids, vitamins, and minerals (Kour *et al.*, 2022; Venturella *et al.*, 2021; Seo and Choi *et al.*, 2021).

These compounds have been the subject of extensive research, revealing a wide range of therapeutic properties. Some of the prominent bioactive compounds found in mushrooms and their potential health benefits are explored below:

Beta-glucans: The complex polysaccharides found in the cell walls of mushrooms are known as beta-glucans. They are known for their immunomodulatory properties and have been studied for their potential to boost the immune system. Beta-glucans are believed to enhance the body's defense mechanisms, making mushrooms like Shiitake and Reishi popular choices for immune support (Timm *et al.*, 2023; Cerletti and Iacoviello, 2021; Mirończuk-Chodakowska *et al.*, 2021).

Ergosterol: This is a precursor of vitamin D2 which can be synthesized by mushrooms when exposed to ultraviolet (UV) light. This bioactive compound plays a role in bone health and is particularly beneficial for individuals with limited sun exposure. Consuming UV-exposed mushrooms can be a natural way to increase vitamin D intake (Jeong and Park, 2020; Papoutsis *et al.*, 2020).

Polyphenols: Mushrooms contain various polyphenols, including phenolic acids and flavonoids, both of which possess antioxidant qualities. These substances lessen the chance of developing chronic illnesses and fight oxidative stress, including cancer and heart disease. Common culinary mushrooms like Button mushrooms and Portobello mushrooms are sources of these polyphenols (Kozarski *et al.*, 2023; Abdelshafy *et al.*, 2022; Nowacka-Jechalke *et al.*, 2018).

Triterpenes: Certain mushrooms, including the *Ganoderma* species (Reishi mushroom), contain bioactive compounds known as triterpenes. These compounds are linked to potential anti-inflammatory and anticancer effects. Reishi mushroom extracts are often used in traditional medicine for their health-promoting properties (Zhou *et al.*, 2022; Whang *et al.*, 2022).

Lentinan: Lentinan is a polysaccharide found in Shiitake mushrooms and is well-recognized for stimulating the immune system. The potential of this substance is to strengthen the body's defenses against illnesses and may be used as an adjunct therapy in cancer treatment (Alkenany and Khalil, 2022a; Alkenany and Khalil, 2022b).
Selenium: Mushrooms are rich in selenium, a trace element that is crucial for various physiological processes. Selenium functions as an antioxidant and could aid in preventing cell damage. Incorporating selenium-rich mushrooms like Brazil nuts and Shiitake mushrooms into the diet can be beneficial for overall health (Dowley *et al.*, 2023; Xu *et al.*, 2021).

Mushrooms are not only a delicious addition to meals but also a valuable source of bioactive compounds with potential health benefits. These compounds, including beta-glucans, ergosterol, polyphenols, triterpenes, lentinan, and selenium, contribute to the therapeutic properties of mushrooms. Research into these bioactive components and their potential for managing and preventing a wide range of health disorders makes mushrooms an exciting area of study for nutritionists and medical professionals. Consuming a range of mushrooms can be a pleasant and nourishing approach to take advantage of their bioactive ingredients for better health (Suzuki, 2021; El-Ramady *et al.*, 2021).

8.3 Pharmacological Activities of Mushroom-Derived Compounds

Due to their possible health benefits, mushrooms have been utilized in traditional medicine for generations. Many mushroom-derived compounds have been studied for their pharmacological activities, and some have shown promising effects (Venturella *et al.*, 2021; Waktola and Temesgen, 2020). Here are some notable pharmacological activities of mushroom-derived compounds:

Immunomodulatory Effects: Beta-glucans, which are found in mushrooms like shiitake, maitake, and reishi, can strengthen the immune system's defense against infections and may even have anti-cancer effects (Pathak *et al.*, 2022).
Antioxidant Properties: Ergothioneine: Found in various mushrooms, ergothioneine is an antioxidant that can help protect cells from oxidative stress and may have neuroprotective effects (Shaffique *et al.*, 2021; Zhao *et al.*, 2020).

Anti-Inflammatory Effects: Triterpenoids: Some mushrooms, like Ganoderma lucidum (reishi), contain triterpenoids that are anti-inflammatory and may aid in the reduction of illnesses linked to inflammation (Nkadimeng *et al.*, 2021; Nkadimeng *et al.*, 2020; Shahzad *et al.*, 2020).

Antimicrobial Activity: Extracts from a variety of mushrooms, including *Agaricus bisporus* (white button mushroom) and *Pleurotus ostreatus*, have shown antibacterial activity against a range of bacteria and fungi (Mkhize *et al.*, 2022; Gebreyohannes *et al.*, 2019).

Antiviral Activity: Lentinan, a beta-glucan found in shiitake mushrooms, has shown antiviral properties and may help when managing viral infections (Chun *et al.*, 2021; El-Maradny *et al.*, 2021).

Anticancer Potential: The potential of compounds present in mushrooms, such as polysaccharides, lectins, and proteoglycans, to suppress the growth of cancer cells and strengthen the body's defenses against the disease has been investigated (Park, 2022; Mwangi *et al.*, 2022).

Antidiabetic Effects: Some mushroom extracts, such as those from *Agaricus blazei*, have been investigated for their capacity to increase insulin sensitivity and decrease blood sugar (Arunachalam *et al.*, 2022; Martel *et al.*, 2017).

Neuroprotective Effects: Hericenones and erinacines have been reported on the ability of lion's mane mushrooms to stimulate nerve growth factor (NGF) production, which may benefit brain health and potentially help with neurodegenerative conditions.

Cardiovascular Health: Dietary fiber, which can lower blood cholesterol levels, is abundant in mushrooms. Oyster mushrooms contain chemicals called lovastatin that can help decrease cholesterol and promote heart health. Cardiovascular disease is associated with elevated cholesterol levels, however, soluble fiber can help reduce LDL (bad) cholesterol levels (Badalyan Rapior, 2021).

Potassium: Some mushrooms, such as white button mushrooms, include potassium. Maintaining adequate blood pressure is crucial for heart health, and potassium plays a role in this (Ahmad *et al.*, 2023).

Liver Protection: Compounds in certain mushrooms, such as the shiitake mushroom, have been shown to have hepatoprotective effects and may help protect the liver from damage (Wu *et al.*, 2019).

It is crucial to remember that even though these pharmacological actions have been noted in a number of studies, additional investigation is required to completely comprehend the mechanisms and possible therapeutic uses of chemicals obtained from mushrooms. Moreover, the efficacy and safety of mushroom-based treatments may differ, it is therefore imperative to speak with a medical expert prior to using them as complementary or alternative treatments for specific health conditions.

8.4 Clinical Applications and Relevance

Mushrooms possess a broad range of therapeutic uses and relevance in various aspects of medicine. While they are not a replacement for conventional medical treatments, they can complement and support health in several ways (Zhao *et al.*, 2020). Here are some of the clinical applications and relevance of mushrooms:

Immune System Support: Compounds found in mushrooms such as shiitake (*Lentinula edodes*), maitake (*Grifola frondosa*), and reishi (*Ganoderma lucidum*) have the ability to boost immunity. They frequently aid in the body's ability to fight infections and lower the danger of disease (Chen *et al.*, 2020).

Antioxidant Properties: Antioxidants, which are abundant in many mushrooms, can help fight oxidative stress and lower the risk of chronic illnesses. Antioxidants like selenium and ergothioneine are commonly found in mushrooms (Mwangi *et al.*, 2022).

Anti-Inflammatory Effects: Compounds in mushrooms, such as beta-glucans, have qualities that reduce inflammation. This can be for inflammatory diseases like arthritis or inflammatory bowel illness.

Cardiovascular Health: Some mushroom varieties, including oyster mushrooms and shiitake mushrooms, may help lower cholesterol levels and improve heart health. They contain compounds like beta-glucans that can contribute to cardiovascular well-being (Rauf *et al.*, 2023).

Blood Sugar Regulation: Some mushrooms have the potential to help control blood sugar levels, including the lion's mane mushroom (*Hericium erinaceus*). This may be especially important for people who already have diabetes or are at risk of getting it (Asim *et al.*, 2018).

Cancer Support: While not a standalone treatment for cancer, some mushrooms, including turkey tail (*Trametes versicolor*) and maitake, are being studied for their potential to support cancer patients. They might lessen the negative effects of cancer therapy and boost immunity (Nowakowski *et al.*, 2021).

Cognitive Health: The ability of lion's mane mushrooms to enhance cognitive performance has drawn attention, and they may also have uses in neurodegenerative illnesses like Parkinson's and Alzheimer's (Ba *et al.*, 2022).
Gut Health: Mushrooms are a dietary fiber source that can support digestive health. Additionally, the prebiotic properties of certain mushroom varieties can encourage the development of healthy gut flora (Zhao *et al.*, 2023).

Although mushrooms have many health advantages, it is crucial to remember that you should only eat them as part of a balanced diet and after seeing a healthcare provider. Individual responses can differ because some people may be allergic to mushrooms or experience negative side effects. Additionally, the quality and safety of mushroom supplements or extracts can vary, so it is crucial to source them from reputable sources.

Overall, mushrooms have a valuable place in clinical applications and can contribute to the promotion of health and well-being when used appropriately in conjunction with other medical interventions.

8.5 Challenges and Future Directions

In recent times, there has been a growing interest in beneficial mushrooms because of their pharmacological qualities and possible health advantages. Substances derived from these mushrooms, such as polysaccharides, terpenoids, and peptides, have shown promise in various applications, including immunomodulation, cancer treatment, and neuroprotection. However, there are several challenges and future directions in the development and utilization of novel pharmacological substances from beneficial mushrooms; standardization and quality control, safety assessment, and bioavailability: regulatory approval, understanding mechanisms of action, clinical trials, dose optimization, combination therapies, cultivation and sustainability, consumer education, ethical considerations, and global collaboration.

Summarily, the development and utilization of novel pharmacological substances from beneficial mushrooms hold great promise for healthcare and pharmaceutical industries. However, addressing the challenges mentioned above and pursuing these future directions will be essential for realizing their full potential and improving human health.

Conclusion

In conclusion, the discovery and relevance of novel pharmacological substances from mushrooms have great promise for the advancement of healthcare and medicine. Over the years, scientific research has unveiled a wealth of bioactive compounds within these fungi that offer a wide range of potential therapeutic benefits. These substances include polysaccharides, terpenoids, phenolic compounds, and various other secondary metabolites, each with unique properties and the capacity to address various health concerns.

The pharmacological significance of these mushroom-derived compounds cannot be understated. They have shown tremendous promise in the management and treatment of a wide range of chronic illnesses, such as diabetes, cancer, heart disease, and neurological disorders. Additionally, their immunomodulatory properties make them invaluable in boosting the body's natural defense mechanisms, as it is pertinent when considering infectious disorders.

Furthermore, the sustainable and eco-friendly nature of mushroom cultivation makes these compounds even more attractive in an era where environmental concerns are paramount. The ability to harness the therapeutic potential of these organisms while preserving natural ecosystems aligns with our broader goals of sustainability and responsible resource utilization.

However, it is essential to acknowledge that the journey from discovery to clinical application is a complex and rigorous one. Extensive research, clinical trials, and preclinical research are required to determine the safety, efficacy, and optimal dosages for these substances. Regulatory approvals and commercialization also pose challenges that must be overcome.

References

Abdelshafy, A.M., Belwal, T., Liang, Z., Wang, L., Li, D., Luo, Z., Li, L., A comprehensive review on phenolic compounds from edible mushrooms: Occurrence, biological activity, application and future prospective. *Crit. Rev. Food Sci. Nutr.*, *62*, 22, 6204–6224, 2022.

Ahmad, I., Arif, M., Xu, M., Zhang, J., Ding, Y., Lyu, F., Therapeutic values and nutraceutical properties of shiitake mushroom (*Lentinula edodes*): A review. *Trends Food Sci. Technol.*, 134, 123–135, 2023.

Ahmed, A.F., Mahmoud, G.A.E., Hefzy, M., Liu, Z., Ma, C., Overview on the edible mushrooms in Egypt. *J. Future Foods*, 3, 1, 8–15, 2023.

Alkenany, M.R. and Khalil, L.W., The effect of Lentinan administration on some immunological and biochemical parameters in intact rabbits. *Int. J. Health Sci.*, 6, 6362–6381, 2022.

Alkenany, M.R. and Khalil, L.W., The effect of Shiitake mushroom derived Lentinan (β-glucan) administration on some hematological and biochemical parameters in intact rabbits. *Int. J. Health Sci.*, III, 5845–5853.

Arunachalam, K., Sasidharan, S.P., Yang, X., A concise review of mushrooms antiviral and immunomodulatory properties that may combat against COVID-19. *Food Chem. Adv.*, 1, 100023, ISSN 2772-753X. https://doi.org/10.1016/j.focha.2022.100023.

Arunachalam, K., Sreeja, P.S., Yang, X., The antioxidant properties of mushroom polysaccharides can potentially mitigate oxidative stress, beta-cell dysfunction and insulin resistance. *Front. Pharmacol.*, 13, 874474, 2022.

Asim, S., Bacha, S., Ali, S., Farooq, S., Li, Y., Ur Rehman, H., Aslam, S.M., Lion's mane mushroom; new addition to food and natural bounty for human wellness: A review. *Int. J. Biosci., 13*, 4, 396–402, 2018, Retrieved from http://dx.doi.org/10.12692/ijb/13.4.396-402.

Ba, D.M., Gao, X., Al-Shaar, L., Muscat, J., Chinchilli, V.M., Ssentongo, P., Richie, J., Mushroom intake and cognitive performance among US older adults: the National Health and Nutrition Examination Survey, 2011-2014. *Br. J. Nutr., 128*, 11, 2241–2248, 2022, https://doi.org/10.1017/S0007114521005195.

Badalyan, S.M. and Rapior, S., Agaricomycetes mushrooms (Basidiomycota) as potential neuroprotectants. *Ital. J. Mycol.*, 50, 30–43, 2021.

Bazzicalupo, A., Gonçalves, S.C., Hébert, R., Jakob, S., Justo, A., Kernaghan, G., Walker, A.K., Macrofungal conservation in Canada and target species for assessment: a starting point. *Facets*, 7, 1, 448–463, 2022.

Booi, H.N., Lee, M.K., Ting, K.N., Fung, S.Y., Medicinal Mushrooms for Respiratory Health, in: *Mushrooms with Therapeutic Potentials: Recent Advances in Research and Development*, pp. 55–96, Springer Nature Singapore, Singapore, 2023.

Brauer, D., Kimmons, T., Phillips, M., Effects of management on the yield and high-molecular-weight polysaccharide content of shiitake (Lentinula edodes) mushrooms. *J. Agric. Food. Chem., 50*, 19, 5333–5337, 2002, https://doi.org/10.1021/jf020080l.

Cerletti, C., Esposito, S., Iacoviello, L., Edible mushrooms and beta-glucans: Impact on human health. *Nutrients, 13*, 7, 2195, 2021.

Chang, Y., Bai, M., Xue, X.B., Zou, C.X., Huang, X.X., Song, S.J., Isolation of chemical compositions as dietary antioxidant supplements and neuroprotectants from Chaga mushroom (*Inonotus obliquus*). *Food Biosci.*, 47, 101623, 2022.

Chatterjee, S., Sarma, M.K., Deb, U., Steinhauser, G., Walther, C., Gupta, D.K., Mushrooms: from nutrition to mycoremediation. *Environ. Sci. Pollut. Res., 24*, 24, 19480–19493, 2017, https://doi.org/10.1007/s11356-017-9826-3.

Chen, S., Liu, C., Huang, X., Hu, L., Huang, Y., Chen, H., Nie, S., Comparison of immunomodulatory effects of three polysaccharide fractions from Lentinula edodes water extracts. *J. Funct. Foods*, 66, 103791, 2020.

Cheung, P.C., The nutritional and health benefits of mushrooms. *Nutr. Bull.*, *35*, 4, 292–299, 2010, https://doi.org/10.1111/j.1467-3010.2010.01859.x.

Chong, P.S., Man-Lung, F., Kah, W., Lee, W.L., Therapeutic Potential of *Hericium erinaceus* for Depressive Disorder. *Int. J. Mol. Sci.*, *21*, 1, 163, 2020, https://doi.org/10.3390/ijms21010163.

Chugh, R.M., Mittal, P., Mp, N., Arora, T., Bhattacharya, T., Chopra, H., Gautam, R.K., Fungal mushrooms: a natural compound with therapeutic applications. *Front. Pharmacol.*, 13, 925387, 2022.

Chun, S., Gopal, J., Muthu, M., Antioxidant activity of mushroom extracts/polysaccharides—Their antiviral properties and plausible antiCOVID-19 properties. *Antioxidants*, *10*, 12, 1899, 2021.

Deshpande, A.G. and Arya, A., Mushroom Biotechnology: Developing Cultivation Protocol for Four Different Mushrooms and Accessing Their Potential in Pollution Management, in: *Biology, Cultivation and Applications of Mushrooms*, A. Arya, and K. Rusevska, (Eds.), Springer, Singapore, 2022, https://doi.org/10.1007/978-981-16-6257-7_16.

de Mattos-Shipley, K.M.J., Ford, K.L., Alberti, F., Banks, A.M., Bailey, A.M., Foster, G.D., The good, the bad and the tasty: The many roles of mushrooms. *Stud. Mycol.*, 85, 125–157, 2016, ISSN 0166-0616. https://doi.org/10.1016/j.simyco.2016.11.002.

Dissanayake, A.A., Zhang, C.R., Mills, G.L., Nair, M.G., Cultivated maitake mushroom demonstrated functional food quality as determined by *in vitro* bioassays. *J. Funct. Foods*, 44, 79–85, 2018, https://doi.org/10.1016/j.jff.2018.02.031.

Dowley, A., Sweeney, T., Conway, E., Vigors, S., Ryan, M.T., Yadav, S., O'Doherty, J.V., The effects of dietary supplementation with mushroom or selenium enriched mushroom powders on the growth performance and intestinal health of post-weaned pigs. *J. Anim. Sci. Biotechnol.*, *14*, 1, 12, 2023.

Draper, M.L., Matthew Oram, The Trials of Psychedelic Therapy: LSD Psychotherapy in America. *J. Hist. Med. Allied Sci.*, *75*, 2, 221–241, 2020, https://doi.org/10.1093/jhmas/jraa002.

Duvenage, S., Rossouw, W., Villamizar-Rodríguez, G., Du Plessis, E.M., Korsten, L., Antibiotic resistance profiles of Staphylococcus spp. From white button mushrooms and handlers. *South African J. Sci.*, *117*, 11–12, 2021, https://doi.org/10.17159/SAJS.2021/8667.

Ekpe, E.K., Hinkle, C.R., Quigley, M.F., Owusu, E.H., Natural resource and biodiversity conservation in Ghana: The use of livelihoods support activities to achieve conservation objectives. *Int. J. Biodivers. Sci. Ecosyst. Serv. Manag.*, *10*, 4, 253–261, 2014, https://doi.org/10.1080/21513732.2014.971056.

El-Maradny, Y.A., El-Fakharany, E.M., Abu-Serie, M.M., Hashish, M.H., Selim, H.S., Lectins purified from medicinal and edible mushrooms: Insights into their antiviral activity against pathogenic viruses. *Int. J. Biol. Macromol.*, 179, 239–258, 2021.

El-Ramady, H., Abdalla, N., Badgar, K., Llanaj, X., Törős, G., Hajdú, P., Prokisch, J., Edible mushrooms for sustainable and healthy human food: nutritional and medicinal attributes. *Sustainability (Switzerland)*, MDPI, 14, 9, 4941, 2022.

El-Ramady, H., Abdalla, N., Fawzy, Z., Badgar, K., Llanaj, X., Törős, G., Prokisch, J., Green biotechnology of oyster mushroom (Pleurotus ostreatus L.): A sustainable strategy for myco-remediation and bio-fermentation. *Sustainability (Switzerland)*, MDPI, 14, 6, 3667, 2022, https://doi.org/10.3390/su14063667.

Firenzuoli, F., Gori, L., Lombardo, G., The medicinal mushroom Agaricus blazei Murrill: review of literature and pharmaco-toxicological problems. *Evid.-Based Complement. Alternat. Med.*, 5, 1, 3–15, 2008.

Gebreyohannes, G., Nyerere, A., Bii, C., Berhe Sbhatu, D., Determination of antimicrobial activity of extracts of indigenous wild mushrooms against pathogenic organisms. *Evid.-Based Complement. Alternat. Med., 2019*, 1, 6212673, 2019.

Goel, D.B. and Zilate, S., Potential therapeutic effects of psilocybin: A systematic review. *Cureus*, 14, 10, 2022, https://doi.org/10.7759/cureus.30214.

Gründemann, C., Reinhardt, J.K., Lindequist, U., European medicinal mushrooms: Do they have potential for modern medicine?–An update. *Phytomedicine*, 66, 153131, 2020.

González-Mariscal, J.M. and Sosa-Cortés, P.E., Insights for Modern Applications of Psilocybin Therapy from a Case Study of Traditional Mazatec Medicine. *Anthropol. Conscious.*, 33, 2, 358–384, 2022, https://doi.org/10.1111/anoc.12168.

Gopal, J. and Muthu, M., Matrix-assisted laser desorption ionization-time of flight mass spectrometry (MALDI-TOF MS) for the analysis of mushroom polysaccharides of clinical, nutritional, and medicinal significance-contemplating the achievements and eliciting future trends. *Trends Anal. Chem.*, 164, 117095, 2023, ISSN 0165-9936. https://doi.org/10.1016/j.trac.2023.117095.

Habtemariam, S., The chemistry, pharmacology and therapeutic potential of the edible mushroom Dictyophora indusiata (Vent ex. Pers.) Fischer (Synn. Phallus indusiatus). *Biomedicines*, *7*, 4, 98, 2019.

Ham, J.K., Chang, E.H., Kwon, H.J., Lee, A.S., Lim, J.G., Park, J.S., Comparison of the biological activity of Maitake (Grifola frondosa) mushrooms extracts according to cultivation methods. *Korean J. Food Sci. Technol.*, *55*, 1, 29–35, 2023, https://doi.org/10.9721/KJFST.2023.55.1.29.

Harisoesyanti, K.S., Analisis Swot Program Pemberdayaan Ekonomi Umat Pada Sentra Budi Daya Jamur Tiram. Komitmen. *J. Ilm. Manaj.*, *4*, 1, 119–136, 2023, https://doi.org/10.15575/jim.v4i1.23729.

Hrudayanath, T. and Sameer, K.S., Diversity, nutritional composition and medicinal potential of Indian mushrooms: A review. *Afr. J. Biotechnol.*, *13*, 4, 523–545, 2014, https://doi.org/10.5897/ajb2013.13446.

Huguet, C., Bourjot, M., Bellanger, J.M., Prévost, G., Urbain, A., Screening for antibacterial activity of French mushrooms against pathogenic and multi-drug resistant bacteria. *Appl. Sci.*, 12, 10, 5229, 2022, https://doi.org/10.3390/app12105229.

Hussein, L.A., Novel prebiotics and next-generation probiotics: opportunities and challenges, in: *Functional foods and nutraceuticals in metabolic and non-communicable diseases*, pp. 431–457, Academic Press, 2022, ISBN 9780128198155. https://doi.org/10.1016/B978-0-12-819815-5.00055-0.

Jeong, Y.U. and Park, Y.J., Ergosterol peroxide from the medicinal mushroom ganoderma lucidum inhibits differentiation and lipid accumulation of 3T3-L1 adipocytes. *Int. J. Mol. Sci.*, 21, 2, 2020, https://doi.org/10.3390/ijms21020460.

Jiang, H., Xiang, Y., He, X., Li, C., Lin, F., Shao, J., Li, Y., Identification and antibiotic resistance of Cronobacter spp. isolated from dried edible mushrooms. *J. Food Sci.*, *87*, 8, 3588–3598, 2022, https://doi.org/10.1111/1750-3841.16251.

Jin, X., Ruiz, B.J., Sze, D.M.Y., Chan, G.C.F., *Ganoderma lucidum* (Reishi mushroom) for cancer treatment. *Cochrane Database Syst. Rev.*, 2016, 4, Art. No.: CD007731, 2016, DOI: 10.1002/14651858.CD007731.

Kumar, A., Yadav, A.N., Mondal, R., Kour, D., Subrahmanyam, G., Shabnam, A.A., Malyan, S.K., Myco-remediation: A mechanistic understanding of contaminants alleviation from natural environment and future prospect. *Chemosphere*, *284*, 131325, 2021.

Khatian, N. and Aslam, M., Effect of Ganoderma lucidum on memory and learning in mice. *Clin. Phytosci.*, 5, 1, 2019, https://doi.org/10.1186/s40816-019-0101-7.

Kirdeeva, Y., Fedorova, O., Daks, A., Barlev, N., Shuvalov, O., How Should the Worldwide Knowledge of Traditional Cancer Healing Be Integrated with Herbs and Mushrooms into Modern Molecular Pharmacology? *Pharmaceuticals*, 15, 7, 868, 2022, https://doi.org/10.3390/ph15070868.

Kour, H., Kour, D., Kour, S., Singh, S., Hashmi, S.A.J., Yadav, A.N., Ahluwalia, A.S., Bioactive compounds from mushrooms: Emerging bioresources of food and nutraceuticals. *Food Biosci.*, 50, 102124, 2022, https://doi.org/10.1016/j.fshw.2022.07.040.

Kour, H., Kour, D., Kour, S., Singh, S., Hashmi, S.A.J., Yadav, A.N., Kumar, K., Sharma, Y.P., Ahluwalia, A.S., Bioactive compounds from mushrooms: Emerging bioresources of food and nutraceuticals. *Food Biosci.*, 50, PartB, 102124, 2022, ISSN 2212-4292, https://doi.org/10.1016/j.fbio.2022.102124.

Kumar, A., Ganoderma Lucidum: A Traditional Chinese Medicine Used For Curing Tumors. *Int. J. Pharm. Pharm. Sci.*, 13, 1–13, 2021, https://doi.org/10.22159/ijpps.2021v13i3.40614.

Kumar, H., Bhardwaj, K., Sharma, R., Nepovimova, E., Cruz-Martins, N., Dhanjal, D.S., Kuča, K., Potential usage of edible mushrooms and their residues to retrieve valuable supplies for industrial applications. *J. Fungi*, *7*, 6, 427, 2021, https://doi.org/10.3390/jof7060427.

Kurtipek, S., Ataseven, G., Kurtipek, A., İlknur Kucukosmanoglu, I., Toksoz, M.R., Resolution of Cutaneous Sarcoidosis Following Topical Application of *Ganoderma lucidum* (Reishi Mushroom). *Dermatol. Ther.*, 6, 105–109, 2016, https://doi.org/10.1007/s13555-016-0099-4.

Lo, H.C., Hsu, T.H., Lee, C.H., Extracellular polysaccharopeptides from fermented turkey tail medicinal mushroom, *Trametes versicolor* (Agaricomycetes), mitigate oxidative stress, hyperglycemia, and hyperlipidemia in rats with type 2 diabetes mellitus. *Int. J. Med. Mushrooms*, 22, 5, 417–429, 2020, https://doi.org/10.1615/IntJMedMushrooms.2020034560.

Lu, X., Brennan, M.A., Narciso, J., Guan, W., Zhang, J., Yuan, L., Brennan, C.S., Correlations between the phenolic and fibre composition of mushrooms and the glycaemic and textural characteristics of mushroom enriched extruded products. *LWT*, 118, 108730, 2020, https://doi.org/10.1016/j.lwt.2019.108730.

Lu, Y., Jia, Y., Xue, Z., Li, N., Liu, J., Chen, H., Recent Developments in *Inonotus obliquus* (Chaga mushroom) Polysaccharides: Isolation, Structural Characteristics, Biological Activities and Application. *Polymers*, 13, 9, 1441, 2021, https://doi.org/10.3390/polym13091441.

Maciejczyk, E. and Kafarski, P., Mannitol in Amanita muscaria–An osmotic blood–brain barrier disruptor enhancing its hallucinogenic action? *Med. Hypotheses*, 81, 5, 766–767, 2013, https://doi.org/10.1016/j.mehy.2013.07.048.

Martel, J., Ojcius, D.M., Chang, C.J., Lin, C.S., Lu, C.C., Ko, Y.F., Young, J.D., Anti-obesogenic and antidiabetic effects of plants and mushrooms. *Nat. Rev. Endocrinol.*, *13*, 3, 149–160, 2017.

Mendel, F., Chemistry, Nutrition, and Health-Promoting Properties of *Hericium erinaceus* (Lion's Mane) Mushroom Fruiting Bodies and Mycelia and Their Bioactive Compounds. *J. Agric. Food. Chem.*, 63, 32, 108–7123, 2015, DOI: 10.1021/acs.jafc.5b02914.

Mirończuk-Chodakowska, I., Kujawowicz, K., Witkowska, A.M., Beta-glucans from fungi: Biological and health-promoting potential in the covid-19 pandemic era. *Nutrients*, *13*, 11, 2021, https://doi.org/10.3390/nu13113960.

Mjaika, N.E., A Systematic Review of Biodiversity and Conservation of Indigenous Mushrooms (Basidiomycotina, Ascomycotina) of Central Africa Countryside: Uses, Distribution and Checklists. *Res. Ecol.*, *4*, 2, 56–66, 2022, https://doi.org/10.30564/re.v4i2.4746.

Mkhize, S.S., Cedric Simelane, M.B., Mongalo, I.N., Pooe, O.J., The effect of supplementing mushroom growing substrates on the bioactive compounds, antimicrobial activity, and antioxidant activity of Pleurotus ostreatus. *Biochem. Res. Int.*, 2022, 1, 9436614, 2022.

Moxley, A., Ebel, R., Cripps, C.L., Austin, C.G., Stein, M., Winder, M., Barriers and opportunities: Specialty cultivated mushroom production in the United States. *Sustainability*, 14, 19, 12591, 2022.

Mwangi, R.W., Macharia, J.M., Wagara, I.N., Bence, R.L., The antioxidant potential of different edible and medicinal mushrooms. *Biomed. Pharmacother.*, 147, 112621, 2022, https://doi.org/10.1016/j.biopha.2022.112621.

Naguib, A.M., Apparoo, Y., Xiong, C., Phan, C.W., Maitake Medicinal Mushroom, *Grifola frondosa* (Agaricomycetes), and Its Neurotrophic Properties: A Mini-Review. *Int. J. Med. Mushrooms*, *25*, 2, 11–22, 2023, https://doi.org/10.1615/IntJMedMushrooms.2022046849.

Nkadimeng, S.M., Nabatanzi, A., Steinmann, C.M.L., Eloff, J.N., Phytochemical, cytotoxicity, antioxidant and anti-inflammatory effects of psilocybe natalensis magic mushroom. *Plants*, *9*, 9, 1–13, 2020, https://doi.org/10.3390/plants9091127.

Nkadimeng, S.M., Steinmann, C.M.L., Eloff, J.N., Anti-inflammatory effects of four psilocybin-containing magic mushroom water extracts *in vitro* on 15-lipoxygenase activity and on lipopolysaccharide-induced cyclooxygenase-2 and inflammatory cytokines in human U937 macrophage cells. *J. Inflamm. Res.*, 14, 3729–3738, 2021, https://doi.org/10.2147/JIR.S317182.

Nowacka-Jechalke, N., Olech, M., Nowak, R., Mushroom polyphenols as chemopreventive agents, in: *Polyphenols: Prevention and treatment of human disease*, pp. 137–150, Academic Press, Elsevier, 2018, https://doi.org/10.1016/B978-0-12-813008-7.00011-4.

Novakovic, S., Djekic, I., Klaus, A., Vunduk, J., Đorđević, V., Tomovic, V., Tomasevic, I., Application of porcini mushroom (Boletus edulis) to improve the quality of frankfurters. *J. Food Process. Preserv.*, 44, 8, e14556, 2020, https://doi.org/10.1111/jfpp.14556.

Nowakowski, P., Markiewicz-Żukowska, R., Bielecka, J., Mielcarek, K., Grabia, M., Socha, K., Treasures from the forest: Evaluation of mushroom extracts as anti-cancer agents. *Biomed. Pharmacother.*, 143, 112106, 2021. https://doi.org/10.1016/j.biopha.2021.112106.

Papoutsis, K., Grasso, S., Menon, A., Brunton, N.P., Lyng, J.G., Jacquier, J.C., Bhuyan, D.J., Recovery of ergosterol and vitamin D2 from mushroom waste-Potential valorization by food and pharmaceutical industries. *Trends Food Sci. Technol.*, 99, 351–366, 2020.

Park, H.J., Current uses of mushrooms in cancer treatment and their anticancer mechanisms. *Int. J. Mol. Sci.*, *23*, 18, 10502, 2022.

Patel, S., Immunomodulatory Aspects of Medicinal Mushrooms, in: *Medicinal Mushrooms*, pp. 169–185, Springer Singapore, 2019.

Patel, S.K., Chandra, R., Dhakad, P.K., Comparative study on growth parameters and yield potential of five species of oyster mushroom. *J. Pharmacogn. Phytochem.*, *8*, 4, 152–156, 2019.

Pathak, M.P., Pathak, K., Saikia, R., Gogoi, U., Ahmad, M.Z., Patowary, P., Das, A., Immunomodulatory effect of mushrooms and their bioactive compounds in cancer: A comprehensive review. *Biomed. Pharmacother.*, 149, 112901, 2022.

Peng, H. and Shahidi, F., Bioactive Compounds and Bioactive Properties of Chaga (*Inonotus obliquus*) Mushroom: A Review. *J. Food Bioact.*, 12, 2020, https://doi.org/10.31665/jfb.2020.12245.

Pepe, M., Hesami, M., de la Cerda, K.A., Perreault, M.L., Hsiang, T., Jones, A.M.P., A journey with psychedelic mushrooms: From historical relevance to biology, cultivation, medicinal uses, biotechnology, and beyond. *Biotechnol. Adv.*, 69, 108247, 2023, ISSN 0734-9750. https://doi.org/10.1016/j.biotechadv.2023.108247.

Phillips, J.M., Ooi, S.L., Pak, S.C., Health-promoting properties of medicinal mushrooms and their bioactive compounds for the COVID-19 era—an appraisal: do the pro-health claims measure up? *Molecules*, 27, 7, 2302, 2022.

Pukalski, J. and Latowski, D., Secrets of flavonoid synthesis in mushroom cells. *Cells*, *11*, 19, 3052, 2022.

Rai, S.N., Mishra, D., Singh, P., Vamanu, E., Singh, M.P., Therapeutic applications of mushrooms and their biomolecules along with a glimpse of in silico approach in neurodegenerative diseases. *Biomed. Pharmacother.*, 137, 111377, 2021.

Rathee, S., Rathee, D., Rathee, D., Kumar, V., Rathee, P., Mushrooms as therapeutic agents. *Rev. Bras. Farmacogn.*, 22, 459–474, 2012, https://doi.org/10.1590/S0102-695X2011005000195.

Rauf, A., Jehan, N., Ahmad, Z., Mubarak, M.S., Analgesic Potential of Extracts and Derived Natural Products from Medicinal Plants, in: *Pain Relief - From Analgesics to Alternative Therapies*, IntechOpen, 2017, https://doi.org/10.5772/intechopen.68631.

Rauf, A., Joshi, P.B., Ahmad, Z., Hemeg, H.A., Olatunde, A., Naz, S., Simal-Gandara, J., Edible mushrooms as potential functional foods in amelioration of hypertension. *Phytother. Res.*, 37, 6, 2644–2660, 2023. https://doi.org/10.1002/ptr.7865.

Saltarelli, R., Palma, F., Gioacchini, A.M., Calcabrini, C., Mancini, U., De Bellis, R., Stocchi, V., Potenza, L., Phytochemical composition, antioxidant and antiproliferative activities and effects on nuclear DNA of ethanolic extract from an Italian mycelial isolate of Ganoderma lucidum. *J. Ethnopharmacol.*, 231, 464–473, 2019, ISSN 0378-8741. https://doi.org/10.1016/j.jep.2018.11.041.

Santhiya, K., Abinaya, S., Aswini, R., Nitheshlee, M., Assessment Of Button And Oyster Mushroom Nutritional Quality Using Various Drying Methods. *Carpathian J. Food Sci. Technol.*, *14*, 3, 171–181, 2022, https://doi.org/10.34302/CRPJFST/2022.14.3.15.

Sasaki, J.I., Bioactive Phytocompounds and Products Traditionally Used in Japan, in: *Modern Phytomedicine: Turning Medicinal Plants into Drugs*, pp. 79–96, Wiley-VCH Verlag GmbH & Co. KGaA, 2006.

Seo, D.J. and Choi, C., Antiviral bioactive compounds of mushrooms and their antiviral mechanisms: a review. *Viruses*, 13, 2, 350, 2021, https://doi.org/10.3390/v13020350.

Sganzerla, W.G., Todorov, S.D., da Silva, A.P.G., Research Trends in the Study of Edible Mushrooms: Nutritional Properties and Health Benefits. *Int. J. Med. Mushrooms*, *24*, 5, 1–15, 2022, https://doi.org/10.1615/IntJMedMushrooms.2022043738.

Shaffique, S., Kang, S.M., Kim, A.Y., Imran, M., Aaqil Khan, M., Lee, I.J., Current knowledge of medicinal mushrooms related to anti-oxidant properties. *Sustainability*, *13*, 14, 7948, 2021.

Shahzad, F., Anderson, D., Najafzadeh, M., The antiviral, anti-inflammatory effects of natural medicinal herbs and mushrooms and SARS-CoV-2 infection. *Nutrients*, *12*, 9, 2573, 2020.

Singh, M.P., Rai, S.N., Dubey, S.K., Pandey, A.T., Tabassum, N., Chaturvedi, V.K., Singh, N.B., Biomolecules of mushroom: A recipe of human wellness. *Crit. Rev. Biotechnol.*, 42, 6, 913–930, 2022.

Softa, M., Percoco, G., Lati, E., Bony, P., Birch Sap (Betula alba) and Chaga Mushroom (*Inonotus obliquus*) Extracts Show Anti-Oxidant, Anti-Inflammatory and DNA Protection/Repair Activity *In Vitro*. *J. Cosmet. Dermatol. Sci. Appl.*, *09*, 02, 188–205, 2019.

Sommano, S.R., Suksathan, R., Sombat, T., Seehanam, P., Sirilun, S., Ruksiriwanich, W., Leksawasdi, N., Novel perspective of medicinal mushroom cultivations: A review case for 'magic' mushrooms. *Agronomy*, 12, 12, 3185, 2022, MDPI. https://doi.org/10.3390/agronomy12123185.

Stallman, J.K. and Robinson, K., Importance of Seasonal Variation in Hawaiian Mushroom (Agaricomycetes) Basidiomata Production for Biodiversity Discovery and Conservation. *Front. Fungal Biol.*, 3, 869689, 2022, https://doi.org/10.3389/ffunb.2022.869689.

Suo, H., Hochnadel, I., Petriv, N., Franke, R., Schmidt, J., Limanska, N., Yevsa, T., Elucidating the mechanism behind and investigating the efficacy of Traditional Chinese Medicine and Traditional Tibetan Medicine in combination with standard therapeutics in hepatocellular carcinoma and cholangiocarcinoma *in vitro*. *Front. Pharmacol.*, 13, 906468, 2022, https://doi.org/10.3389/fphar.2022.906468.

Suzuki, T., Genetic sequence analysis and characterization of bioactive compounds in mushroom-forming fungi. *Biosci. Biotechnol., Biochem.*, *85*, 1, 8–12, 2021, https://doi.org/10.1093/bbb/zbaa067.

Takaku, T., Kimura, Y., Okuda, H., Isolation of an antitumor compound from Agaricus blazei Murill and its mechanism of action. *J. Nutr.*, *131*, 5, 1409–1413, 2001, https://doi.org/10.1093/jn/131.5.1409.

Tan, Y., Zeng, N.K., Xu, B., Chemical profiles and health-promoting effects of porcini mushroom (Boletus edulis): A narrative review. *Food Chem.*, 390, 133199, 2022, https://doi.org/10.1016/j.foodchem.2022.133199.

Timm, T.G., Costa, T.M., Alberton, M.D., Helm, C.V., Tavares, L.B.B., Mushroom β-glucans: application and innovation for food industry and immunotherapy. *Appl. Microbiol. Biotechnol.*, *107*, 16, 5035–5049, 2023.

Törős, G., El-Ramady, H., Prokisch, J., Velasco, F., Llanaj, X., Nguyen, D.H., Peles, F., Modulation of the gut microbiota with prebiotics and antimicrobial agents from Pleurotus ostreatus mushroom. *Foods*, 12, 10, 2010, 2023, https://doi.org/10.3390/foods12102010.

Tyls, F., Palenicek, T., Horacek, J., Neurobiology of the Effects of Psilocybin in Relation to Its Potential Therapeutic Targets, in: *Neuropathology of Drug Addictions and Substance Misuse Volume 2: Stimulants, Club and Dissociative Drugs, Hallucinogens, Steroids, Inhalants and International Aspects*, pp. 782–793, Academic Press, Elsevier, 2016, https://doi.org/10.1016/B978-0-12-800212-4.00073-X.

Unlu, A., Nayir, E., Kirca, O., Ozdogan, M., Ganoderma lucidum (reishi mushroom) and cancer. *J. Buon*, *21*, 4, 792–798, 2016.

Venturella, G., Ferraro, V., Cirlincione, F., Gargano, M.L., Medicinal mushrooms: bioactive compounds, use, and clinical trials. *Int. J. Mol. Sci.*, 22, 2, 634, 2021, https://doi.org/10.3390/ijms22020634.

Waktola, G. and Temesgen, T., Pharmacological activities of Oyster mushroom (*Pleurotus ostreatus*). *Nov. Res. Microbiol. J.*, 4, 2, 688–695, 2020, https://doi.org/10.21608/nrmj.2020.84017.

Wang, C., Liu, Y., Lan, Y., Yuan, J., Extraction of a triterpene solution and evaluation of the hypolipidemic efficacy of the Pleurotus tuber-regium (Fr.) sing sclerotium. *Foods*, *11*, 18, 2881, 2022, https://doi.org/10.3390/foods11182881.

Wong, J.H., Ng, T.B., Chan, H.H.L., Liu, Q., Man, G.C.W., Zhang, C.Z., Xia, L., Mushroom extracts and compounds with suppressive action on breast cancer: evidence from studies using cultured cancer cells, tumor-bearing animals, and clinical trials. *Appl. Microbiol. Biotechnol.*, 104, 4675–4703, 2020, https://doi.org/10.1007/s00253-020-10476-4.

Wu, Q., Dai, T., Song, J., Liu, X., Song, S., Li, L., Jacob, J.A., Effects of herbal and mushroom formulations used in Traditional Chinese Medicine on *in vitro* human cancer cell lines at the preclinical level: An empirical review of the cell killing mechanisms, in: *Process Biochemistry*, Elsevier Ltd., 2020, https://doi.org/10.1016/j.procbio.2020.04.019.

Wu, Y., Tian, W.J., Gao, S., Lian, Z.J., Wang, G.H., Lo, J.M., Chen, H.F., Secondary metabolites of petri-dish cultured *Antrodia camphorata* and their hepatoprotective activities against alcohol-induced liver injury in mice. *Chin. J. Nat. Med.*, 17, 1, 33–42, 2019, https://doi.org/10.1016/S1875-5364(19)30007-X.

Xu, M., Zhu, S., Li, Y., Xu, S., Shi, G., Ding, Z., Effect of selenium on mushroom growth and metabolism: A review. *Trends Food Sci. Technol.*, 118, 328–340, 2021, Elsevier Ltd. https://doi.org/10.1016/j.tifs.2021.10.018.

Yadav, S.K., Ir, R., Jeewon, R., Doble, M., Hyde, K.D., Kaliappan, I., Durairajan, S.S.K., A mechanistic review on medicinal mushrooms-derived bioactive compounds: Potential mycotherapy candidates for alleviating neurological disorders. *Planta Med.*, 86, 16, 1161–1175, 2020, https://doi.org/10.1055/a-1177-4834.

Yasuma, T., Toda, M., Kobori, H., Tada, N., D'alessandro-Gabazza, C.N., Gabazza, E.C., Subcritical water extracts from agaricus blazei murrill's mycelium inhibit the expression of immune checkpoint molecules and axl receptor. *J. Fungi*, 7, 8, 2021, https://doi.org/10.3390/jof7080590.

Zeb, M. and Lee, C.H., Medicinal properties and bioactive compounds from wild mushrooms native to North America. *Molecules*, 26, 2, 2021, https://doi.org/10.3390/molecules26020251.

Zhao, J., Hu, Y., Qian, C., Hussain, M., Liu, S., Zhang, A., Sun, P., The interaction between mushroom polysaccharides and gut microbiota and their effect on human health: A review. *Biology*, *12*, 1, 122, 2023, MDPI. https://doi.org/10.3390/biology12010122.

Zhao, S., Gao, Q., Rong, C., Wang, S., Zhao, Z., Liu, Y., Xu, J., Immunomodulatory effects of edible and medicinal mushrooms and their bioactive immunoregulatory products. *J. Fungi*, 6, 4, 269, 2020, https://doi.org/10.3390/jof6040269.

Zhou, L., Chen, H.P., Li, X., Liu, J.K., Ganoaustralins A and B, Unusual Aromatic Triterpenes from the Mushroom Ganoderma australe. *Pharmaceuticals*, 15, 12, 2022, https://doi.org/10.3390/ph15121520.

Zied, D.C., Alegría, F.J.G., Giménez, A.P., Enhancing the medicinal properties of agaricus subrufescens by growing practices, in: *Mushrooms: Types, Properties and Nutrition*, pp. 173–194, Nova Science Publishers, Inc., 2012.

9

Application of Mushroom in the Management of Diabetes Mellitus

Adeyomoye Olorunsola Israel[1]*, Charles Oluwaseun Adetunji[2], Oluwafemi Adebayo Oyewole[3], Oluwabukola Atinuke Popoola[4] and John Tsado Mathew[5]

[1]*Department of Physiology, University of Medical Sciences, Ondo State, Nigeria*
[2]*Applied Microbiology, Biotechnology and Nanotechnology Laboratory, Department of Microbiology, Edo State University Uzairue, Iyamho, Edo, Nigeria*
[3]*Department of Microbiology, Federal University of Technology, Minna Nigeria*
[4]*Genetics, Genomics and Bioinformatics Department, National Biotechnology Development Agency, Abuja, FCT, Nigeria*
[5]*Department of Chemistry, Ibrahim Badamasi, Babangida University Lapai, Niger, Nigeria*

Abstract

Edible mushrooms belong to a group of antidiabetic phytotherapy agents that are rich in natural compounds, e.g., fibers, alkaloids, polysaccharides, and phenolics. In addition to their antidiabetic properties, they are known to have antioxidative and antihyperlipidemic properties. Mushroom polysaccharide is a known source of prebiotics and can act via modulation of gut microbiome composition to induce insulin resistance. Furthermore, *in vivo* and *in vitro* studies have shown that mushroom polysaccharide can exhibit anti-hyperglycemic effects via a reduction in glucose absorption, improved beta cell mass, and insulin signalling. It has also been demonstrated that terpenoids derived from mushrooms function as insulin sensitizers and alpha-glucosidase activity inhibitors.

***Keywords*:** Mushroom, phytotherapy, anti-diabetic, insulin resistance, antihyperglycemia

**Corresponding author*: oadeyomoye@unimed.edu.ng

Charles Oluwaseun Adetunji, Chukwuebuka Egbuna, Oluwatosin Ademola Ijabadeniyi and Samantha C. Karunarathna (eds.) Mushroom Biotechnology for Improved Agriculture and Human Health, (209–220) © 2025 Scrivener Publishing LLC

Introduction

Millions of individuals worldwide suffer from diabetes mellitus, a metabolic, and endocrine condition [1, 2]. It is known to be a serious health concern with a great financial burden on any affected population. The disease may result in several complications if left untreated. Several synthetic anti-diabetic agents have been developed and used in the treatment of the disease. However, many of these agents are known to have serious negative or side effects. Fortunately, several natural polysaccharides have anti-diabetic properties and their use as alternatives to conventional therapy is gradually increasing especially in low-income and resource countries. Mushrooms have long been a staple of many nations' diets. They are well recognized for being high in phytonutrients, which include amino acids, dietary fiber, and polysaccharides, which are the building blocks of vital proteins [3]. Numerous health advantages and pharmacological characteristics are associated with mushrooms, such as their antidiabetic, antiviral, antibacterial, osteoprotective, nephroprotective, and hepatoprotective qualities. Mushroom extract and its bioconstituents are recognized to have a wide range of applications since both *in vitro* and *in vivo* studies have shown that wild and medicinal mushrooms possess biological activity. Mushrooms are utilized to make medications and neutraceuticals because of their phytocompositions and previously described qualities [4]. Additionally, human tissue repair and healing are significantly aided by mushrooms. Mushrooms contain biologically active chemicals that have been shown in animal and human models to demonstrate antitumor activity via enhancing innate and cell-mediated immune responses. The modulation and stimulation of key host immune responses are taught to be central in elucidating their mechanisms of antitumour activities [5]. Although mushroom has several biological activities and health benefits, this chapter discusses the antidiabetic effects of mushrooms, their biomolecules, and mechanisms of therapeutic potential.

Pathophysiology of Diabetes Mellitus

Type 2, type 1, and gestational diabetes are the most common types of diabetes in primary care settings. Type 2 diabetes is characterized by decreased insulin production, insulin resistance, and insufficient hepatic glucose production, while type 1 diabetes is an autoimmune disease characterized by the complete loss of pancreatic beta cells [6]. Identical to type 2 diabetes, gestational diabetes is identified for the first time during pregnancy. It is generally accepted that unregulated glucose control leads to hyperglycemia, which is the cause of diabetic problems. Advanced glycation end products (AGEs), the polyol pathway, the hexosamine pathway, and the activation of protein kinase C (PKC) isoforms through *de novo* synthesis of the lipid second messenger diacylglycerol (DAG) are the four molecular pathways and the main causes of tissue damage resulting from hyperglycemia. The processes causing hyperglycemic damage are causally linked to high glucose levels through hyperglycemia-induced superoxide overproduction [7]. Due to their frequent correlation with elevated free radical generation and compromised antioxidant defenses, reactive oxygen species are important in the onset, course, and severe outcomes of diabetes. Numerous mitochondrial dysfunctions and oxidative stress are associated with type 2 diabetes, according to an increasing body of studies [8]. Reduced copies and mutations of mtDNA have been connected to the pathogenesis of type 2 diabetes. An investigation into the connection between mtDNA and type 2 diabetes revealed the impact of mitochondria on glucose-stimulated insulin secretion, nuclear-encoded glucose transporters, and nuclear-encoded uncoupling proteins on beta-cell glucose toxicity. Some of the main causes of increased ROS generation in diabetes are endothelial dysfunction, oxidative damage to pancreatic cells, and altered glucose metabolism [9]. The relevance of hyperglycemia-associated oxidative stress reactions is considered within the context of the "metabolic memory" phenomena. Researchers are convinced that antioxidants should be included in the treatment plan based on a number of studies that have shown how beneficial they are in managing both types of diabetes mellitus. The development of new techniques for customized antioxidant therapy, especially focusing on ROS sources in conjunction with novel techniques for antioxidant delivery, is the main goal of contemporary therapeutic strategies for the management of diabetic mellitus [10]. Poly-ADP-ribose polymerase 1's suppression of glyceraldehyde-3-P dehydrogenase appears to be the main cause of the oxidative stress linked to diabetes, leading to a buildup of the enzyme substrate (glyceraldehyde-3-P). Elevated glyceraldehyde-3-P initiates the AGE route by encouraging the creation of diacylglycerol. This, in turn, stimulates the PKC pathway and the formation of methylglyoxal from non-enzymatic dephosphorylation of triose phosphates. These two processes are important pro-oxidative processes in diabetes [11]. Additionally, it results in an accumulation of upstream glycolytic metabolites, which overstimulates additional prooxidative pathways like the polyol and hexosamine pathways. Another theory on the cause of type 2 diabetes mellitus is that it results from a combination of environmental and genetic factors. In genetically predisposed individuals, a high-calorie diet combined with inactivity leads to the development of insulin resistance [12]. Hyperinsulinemia, which involves boosting insulin secretion, can treat this disease to a degree. However, insufficient compensation results in a condition of glucose intolerance, which, based on the available research, can progress to diabetes. Highly reactive molecules like oxygen and nitrogen species of free radicals mediate the harmful effects of hyperglycemia, which are linked to a potential rise in free fatty acids. According to Haedersdal's (2018) [13] research, reactive species are responsible for oxidative stress due to their increased synthesis or impaired clearance. These molecules also operate as signaling molecules that regulate cellular activity. These free radicals really affect macromolecules (proteins, lipids, and DNA) directly through oxidative damage or indirectly by activating specific transduction pathways that are vulnerable to stress mechanisms. Since these signaling cascades are believed to be the source of insulin resistance and decreased insulin production, the use of safe antioxidants, such as vitamin C, E, and lipoic acid, is therefore seen as a viable way to control type 2 diabetic complications [14].

Mushroom and Its Biomolecules

One of the primary sources of nutraceuticals is mushrooms. Due to their well-established health benefits, mushroom biomolecules are grabbing the attention of researchers all around the world. These include qualities that include anti-inflammatory, anti-arthritic, anti-tumor, anti-cancer, anti-HIV, anti-diabetic, anti-cholesterol, antiviral, antibacterial, antifungal, antioxidant, and anti-inflammatory [15]. Primary and secondary metabolites are found in mushrooms. Energy is produced by the primary metabolites, but the secondary metabolites offer therapeutic qualities. Fungi are considered the "best of waste" and "best of the rest" because they are good sources of nutraceuticals and because they grow on the most abundant organic waste on Earth, lignocellulosic substrate. Antioxidant molecular profiles of edible and inedible mushroom species were examined and analyzed in a study [16, 17]. Similar to edible species, the sequence of antioxidant content in non-edible wild mushrooms is phenols > flavonoids > ascorbic acid > tocopherols > carotenoids. Researchers examined the chemical makeup and biological traits of Portuguese wild mushrooms (*Lepista nuda, Lycoperdon molle, Ramaria botrytis, Cantharellus cibarius, Lycoperdon perlatum, and Tricholoma acerbum, Hypholoma fasciculare,*) in order to assess these products as sources of nutrients and nutraceuticals [18, 19]. The mushrooms under investigation contain a variety of beneficial phytochemicals, including ascorbic acid, phenols, carotenoids, tocopherols, and ascorbic acid. The fruiting bodies and extracts from six common polyporoid mushrooms (*Aphyllophorales sensulato*) found in Poland—*Laetiporus sulphureus, Fomitopsis pinicola, Daedaleopsis confragosa, Gloeophyllum sepiarium, and Piptoporus betulinus*—were evaluated for their phenolic compound content, including phenolic acids [20]. The amounts of protocatechuic acid, p-hydroxybenzoic acid, and vanillic acid were determined by HPLC. *Fomitopsis pinicola* had the highest total phenolic acid concentration (114.9 g/g dry matter), with protocatechuic acid predominating among the phenolic acids with values ranging from 17.7 to 90.0 g/g dry weight.

Mushroom and Health

Mushrooms are used in medicine and consumption for a very long time. All across the world, edible mushrooms are grown or harvested from the environment. Mushroom extracts and their secondary metabolites have garnered a lot of attention recently due to their pharmacological capabilities, which include immunomodulatory, antioxidant, antibacterial, anti-inflammatory, anti-obesity, and anticancer qualities [21].

Nutritionists and customers alike are interested in mushrooms because of their phytochemical components' potential to promote human health and reduce the risk of disease [22]. A vast quantity of scientific research on the pharmacological, phytochemical, and nutritional properties of mushrooms has been conducted in recent years. Numerous research streams indicate that mushrooms are a good source of antimicrobials and organic antioxidants [23]. Many types of mushrooms have been used as food and medicinal since ancient times. Mushrooms contain minerals, vitamins, polyphenols, and carbs. Numerous biological actions of mushrooms, such as anti-inflammatory, anti-cancer, immunomodulatory, antibacterial, hepatoprotective, and antidiabetic effects, have been demonstrated by studies. Mushrooms may therefore be used to create functional foods or preventative medicines, as they have attracted more attention recently. In addition, Chowdhury (2015) [24] discusses the treatment of several long-term ailments such as diabetes mellitus, cancer, heart disease, and neurological diseases. The molecular activity behind edible mushrooms' antioxidant qualities as well as the demand for natural, alternative antioxidant dietary sources have both grown fast. Mushrooms' chemical makeup and antioxidant capabilities have been thoroughly investigated. Edible mushrooms can be used directly to reduce oxidative stress levels by increasing antioxidant defenses through dietary supplements [25]. They have been linked to strong antioxidant capabilities whether they are wild or cultivated because of their bioactive constituents, which include polyphenols, polysaccharides, vitamins, carotenoids, and minerals. Eating edible mushrooms traditionally has been considered a delicacy because of their antioxidant and health advantages. A form of biological macromolecule known as mushroom polysaccharides is obtained from the fruiting body, mycelium, or fermentation liquid of edible mushrooms. Research on fungal polysaccharides to treat cancer, inflammatory bowel disease, metabolic diseases, and other symptoms by altering the intestinal microenvironment has increased in recent years [26]. By regulating intestinal microbiota, boosting the generation of short-chain fatty acids, enhancing the intestinal mucosal barrier, regulating lipid metabolism, and activating particular signaling pathways, mushroom polysaccharides may benefit human health. Notably, these biological functions are intimately correlated with the molecular weight, monosaccharide composition, and type of glycosidic bond of the fungal polysaccharide.

Therapeutic Potential of Mushroom in the Management of Diabetes Mellitus

Diabetes mellitus is considered a serious hazard to global health. Scientists are very interested in the pursuit of antidiabetic elements from natural sources. Due to their advantages in having a wide range of sources, structural diversity, and diverse bioactivities, mushroom polysaccharides have drawn increasing attention in the anti-diabetic field. They are thought to be an endless source of healthy active ingredients that could be used in functional foods and nutraceuticals [27]. The majority of the active polysaccharides derived from mushrooms have antioxidant action, lower oxidative stress, and delay the onset of diabetes, according to preclinical and phytochemical research [28]. In order to ascertain the mechanisms by which polysaccharides found in mushrooms effectively lower blood sugar levels and whether or not these polysaccharides could eventually be used as an alternative therapeutic option for the management and prevention of diabetes mellitus, more research is necessary [29]. Since ancient times, mushrooms have been known to have antidiabetic, antioxidant, and antihyperlipidemic benefits as a result of their high content in natural substances such as fiber, polysaccharides, phenols, and alkaloids [30]. Additionally, fungal polysaccharides function as prebiotics, influence the composition of the gut bacteria, and may thereby lessen insulin resistance. Active mushroom compounds such as comatin, glucan, Tremellastin, and Lentinan KS-2 have significant impacts on diabetes mellitus and obesity by altering cellular or biochemical processes. The anti-diabetic, hypolipidemic, and hepatoprotective properties of *Ganoderma applanatum* (MEGA, AEGA) methanol and aqueous extracts were examined in diabetic rats that were given alloxan [31]. Additionally, *in silico* testing of *Ganoderma applanatum* compounds was used to determine the binding affinities of the compounds to the enzymes. The docking study was performed using PyRx AutoDock Vina, UCSF Chimera, and Discovery Studio. *Ganoderma applanatum* MEGA and AEGA significantly decreased liver marker enzymes, lipid profile, and blood glucose after nine days of oral therapy. Myrocin C also produced the best outcomes from the docking research. Polysaccharides and terpenoids produced from several mushroom species, along with their bioactive components, have been utilized for many years in the treatment of diabetes mellitus. Research conducted *in vitro* and in animals has shown that treatment with mushroom polysaccharide has an antihyperglycemic effect by enhancing insulin signaling pathways, increasing pancreatic beta-cell mass, and reducing the efficiency of glucose absorption [32]. Through the activation of PPAR-Y, mushroom terpenoids function as insulin sensitizers and alpha-glucosidase inhibitors to reduce hyperglycemia in animal models of diabetes. The effects of *Morchella esculenta* polysaccharide (MEP) on streptozotocin (STZ)-induced type 2 diabetes (T2DM) and high-fat diet (HFD) were investigated in BALB/c mice [33]. MEP injections successfully reduced hyperglycemia and hyperlipidemia while also improving insulin sensitivity. Furthermore, in T2DM rats treated with MEP, there was a decrease in endotoxemia and pro inflammatory cytokines linked to insulin resistance, including tumor necrosis factor alpha (TNF- alpha), interleukin 1 (IL-1), and interleukin 6 (IL-6). MEP therapy also changed the colonic tight junction proteins mucin-2 protein (MUC2), occludin, claudin-1, and zonula occludens-1 (ZO-1). This decreased intestinal permeability. Furthermore, MEP therapy alters functional metabolic pathways, which affects T2DM mice's microbial communities' metagenome. All of these results point to MEP as a helpful prebiotic that helps T2DM patients' gut flora and metabolites. Due to factors including high-calorie diets and a lack of exercise, obesity has reached epidemic levels on a global scale [34]. It is well acknowledged that obesity is a medical condition that increases the risk of heart disease, cancer, and type 2 diabetes, as well as shortening life expectancy and decreasing quality of life. To tackle this epidemic, anti-obesitogenic strategies that are secure, widely accessible, and affordable are required. Antioxidants, fiber, and other phytochemicals are found in a number of plants and fungi used in as nutraceuticals. These substances also have antiobesitogenic and antidiabetic properties by modulating multiple cellular and physiological pathways [35]. Reduced appetite, altered lipid metabolism and absorption, improved insulin sensitivity, thermogenesis, and modifications to the gut flora are a few of these side effects. Diabetes is becoming more common despite the enormous efforts made to understand and cure it. In order to treat DM and its associated consequences, new forms of therapies derived from medicinal mushrooms are now being developed due to exciting new findings [36]. Medicinal mushroom fruiting bodies, cultured mycelium, and cultured broth include an increasing number of discovered active compounds with antihyperglycemic effects. Among these elements include dietary fiber and different substances, such as polysaccharides and their protein complexes. These medications' antidiabetic effects are mediated by a number of different pathways. Inhibiting glucose absorption, avoiding beta-cell damage, raising insulin release, enhancing antioxidant defenses, decreasing inflammation, altering the carbohydrate metabolism pathway, and controlling both insulin-dependent and insulin-independent signaling pathways are some of the ways that medicinal mushrooms help control blood sugar in people with diabetes [37]. Furthermore, there are still problems with the production of medicinal mushrooms, such as high levels of variability, a lack of production guidelines, and inadequate testing protocols to determine the quality of the final product. In order to guarantee the safety and biological activity of medicinal mushrooms for individuals with diabetes, prolonged use in carefully planned randomized controlled trials is necessary.

Conclusion

Scientists are very interested in the pursuit of antidiabetic agents from natural sources. Due to their advantages in broad sources, structural diversity, and numerous bioactivities, mushrooms have drawn increasing attention in the anti-diabetic field. They are thought to be an endless source of healthy active ingredients that may be used in functional foods and nutraceuticals. Antioxidant, hypolipidemic, anti-inflammatory, and gut microbiota modulators are some of the probable molecular processes that might underlie their antidiabetic effects. In the future, bridging the gap between mushroom intervention and insulin signaling pathway modulation, figuring out the structure and bioactivity of mushrooms, creating synergistic foods, and conducting well-controlled clinical trials may be very beneficial in locating valuable mushrooms for better management of diabetes.

References

1. Tao, Z., Shi, A., Zhao, J., Epidemiological Perspectives of Diabetes. *Cell Biochem. Biophys.*, *73*, 1, 181–185, 2015.
2. GBD 2021 Diabetes Collaborators, Global, regional, and national burden of diabetes from 1990 to 2021, with projections of prevalence to 2050: a systematic analysis for the Global Burden of Disease Study 2021. *Lancet (London, England)*, *402*, 10397, 203–234, 2023.
3. Grimm, D. and Wösten, H.A.B., Mushroom cultivation in the circular economy. *Appl. Microbiol. Biotechnol.*, *102*, 18, 7795–7803, 2018.
4. Bains, A., Chawla, P., Kaur, S., Najda, A., Fogarasi, M., Fogarasi, S., Bioactives from Mushroom: Health Attributes and Food Industry Applications. *Materials (Basel, Switzerland)*, *14*, 24, 7640, 2021.
5. Wasser, S.P., Medicinal Mushrooms in Human Clinical Studies. Part I. Anticancer, Oncoimmunological, and Immunomodulatory Activities: A Review. *Int. J. Med. Mushrooms*, *19*, 4, 279–317, 2017.
6. Rolo, A.P. and Palmeira, C.M., Diabetes and mitochondrial function: role of hyperglycemia and oxidative stress. *Toxicol. Appl. Pharmacol.*, *212*, 2, 167–178, 2006.
7. Lopes, J.P., Oliveira, S.M., Soares Fortunato, J., Stress oxidativo e seus efeitos na insulino-resistência e disfunção das células Beta-pancreáticas: relação com as complicações da diabetes mellitus tipo 2 [Oxidative stress and its effects on insulin resistance and pancreatic beta-cells dysfunction: relationship with type 2 diabetes mellitus complications]. *Acta Med. Portuguesa*, *21*, 3, 293–302, 2008.
8. Darenskaya, M.A., Kolesnikova, L.I., Kolesnikov, S.I., Oxidative Stress: Pathogenetic Role in Diabetes Mellitus and Its Complications and Therapeutic Approaches to Correction. *Bull. Exp. Biol. Med.*, *171*, 2, 179–189, 2021.
9. Ighodaro, O.M., Molecular pathways associated with oxidative stress in diabetes mellitus. *Biomed. Pharmacother.*, *108*, 656–662, 2018.
10. Haas, L.B., Pathophysiology of diabetes mellitus. *Nurse Pract. Forum*, *9*, 2, 42–45, 1998.
11. Dariya, B., Chalikonda, G., Srivani, G., Alam, A., Nagaraju, G.P., Pathophysiology, Etiology, Epidemiology of Type 1 Diabetes and Computational Approaches for Immune Targets and Therapy. *Crit. Rev. Immunol.*, *39*, 4, 239–265, 2019.
12. Antonioli, L., Blandizzi, C., Csóka, B., Pacher, P., Haskó, G., Adenosine signalling in diabetes mellitus–pathophysiology and therapeutic considerations. *Nat. Rev. Endocrinol.*, *11*, 4, 228–241, 2015.
13. Hædersdal, S., Lund, A., Knop, F.K., Vilsbøll, T., The Role of Glucagon in the Pathophysiology and Treatment of Type 2 Diabetes. *Mayo Clinic Proc.*, *93*, 2, 217–239, 2018.

14. Beck-Nielsen, H., Henriksen, J.E., Vaag, A., Hother-Nielsen, O., Pathophysiology of non-insulin-dependent diabetes mellitus (NIDDM). *Diabetes Res. Clin. Pract., 28 Suppl.*, S13–S25, 1995.
15. Anusiya, G., Gowthama Prabu, U., Yamini, N.V., Sivarajasekar, N., Rambabu, K., Bharath, G., Banat, F., A review of the therapeutic and biological effects of edible and wild mushrooms. *Bioengineered, 12*, 2, 11239–11268, 2021.
16. Reis, F.S., Pereira, E., Barros, L., Sousa, M.J., Martins, A., Ferreira, I.C., Biomolecule profiles in inedible wild mushrooms with antioxidant value. *Molecules (Basel, Switzerland), 16*, 6, 4328–4338, 2011.
17. Singh, M.P., Rai, S.N., Dubey, S.K., Pandey, A.T., Tabassum, N., Chaturvedi, V.K., Singh, N.B., Biomolecules of mushroom: a recipe of human wellness. *Crit. Rev. Biotechnol., 42*, 6, 913–930, 2022.
18. Venturella, G., Ferraro, V., Cirlincione, F., Gargano, M.L., Medicinal Mushrooms: Bioactive Compounds, Use, and Clinical Trials. *Int. J. Mol. Sci., 22*, 2, 634, 2021.
19. Barros, L., Venturini, B.A., Baptista, P., Estevinho, L.M., Ferreira, I.C., Chemical composition and biological properties of portuguese wild mushrooms: a comprehensive study. *J. Agric. Food. Chem., 56*, 10, 3856–3862, 2008.
20. Sulkowska-Ziaja, K., Muszynska, B., Motyl, P., Pasko, P., Ekiert, H., Phenolic compounds and antioxidant activity in some species of polyporoid mushrooms from Poland. *Int. J. Med. Mushrooms, 14*, 4, 385–393, 2012.
21. Zhang, J.J., Li, Y., Zhou, T., Xu, D.P., Zhang, P., Li, S., Li, H.B., Bioactivities and Health Benefits of Mushrooms Mainly from China. *Molecules (Basel, Switzerland), 21*, 7, 938, 2016.
22. Thu, Z.M., Myo, K.K., Aung, H.T., Clericuzio, M., Armijos, C., Vidari, G., Bioactive Phytochemical Constituents of Wild Edible Mushrooms from Southeast Asia. *Molecules (Basel, Switzerland), 25*, 8, 1972, 2020.
23. Erbiai, E.H., da Silva, L.P., Saidi, R., Lamrani, Z., Esteves da Silva, J.C.G., Maouni, A., Chemical Composition, Bioactive Compounds, and Antioxidant Activity of Two Wild Edible Mushrooms *Armillaria mellea* and *Macrolepiota procera* from Two Countries (Morocco and Portugal). *Biomolecules, 11*, 4, 575, 2021.
24. Chowdhury, M., Kubra, K., Ahmed, S., Screening of antimicrobial, antioxidant properties and bioactive compounds of some edible mushrooms cultivated in Bangladesh. *Ann. Clin. Microbiol. Antimicrob., 14*, 8, 2015.
25. Kozarski, M., Klaus, A., Jakovljevic, D., Todorovic, N., Vunduk, J., Petrović, P., Niksic, M., Vrvic, M.M., van Griensven, L., Antioxidants of Edible Mushrooms. *Molecules (Basel, Switzerland), 20*, 10, 19489–19525, 2015.
26. Zhao, J., Hu, Y., Qian, C., Hussain, M., Liu, S., Zhang, A., He, R., Sun, P., The Interaction between Mushroom Polysaccharides and Gut Microbiota and Their Effect on Human Health: A Review. *Biology, 12*, 1, 122, 2023.

27. Liu, X., Luo, D., Guan, J., Chen, J., Xu, X., Mushroom polysaccharides with potential in anti-diabetes: Biological mechanisms, extraction, and future perspectives: A review. *Front. Nutr.*, *9*, 1087826, 2022.
28. Arunachalam, K., Sreeja, P.S., Yang, X., The Antioxidant Properties of Mushroom Polysaccharides can Potentially Mitigate Oxidative Stress, Beta-Cell Dysfunction and Insulin Resistance. *Front. Pharmacol.*, *13*, 874474, 2022.
29. Das, A., Chen, C.M., Mu, S.C., Yang, S.H., Ju, Y.M., Li, S.C., Medicinal Components in Edible Mushrooms on Diabetes Mellitus Treatment. *Pharmaceutics*, *14*, 2, 436, 2022.
30. Dubey, S.K., Chaturvedi, V.K., Mishra, D., Bajpeyee, A., Tiwari, A., Singh, M.P., Role of edible mushroom as a potent therapeutics for the diabetes and obesity. *3 Biotech*, *9*, 12, 450, 2019.
31. Khursheed, R., Singh, S.K., Wadhwa, S., Gulati, M., Awasthi, A., Therapeutic potential of mushrooms in diabetes mellitus: Role of polysaccharides. *Int. J. Biol. Macromol.*, *164*, 1194–1205, 2020.
32. Hossain, M.S., Barua, A., Tanim, M.A.H., Hasan, M.S., Islam, M.J., Hossain, M.R., Emon, N.U., Hossen, S.M.M., *Ganoderma applanatum* mushroom provides new insights into the management of diabetes mellitus, hyperlipidemia, and hepatic degeneration: A comprehensive analysis. *Food Sci. Nutr.*, *9*, 8, 4364–4374, 2021.
33. Rehman, A.U., Siddiqui, N.Z., Farooqui, N.A., Alam, G., Gul, A., Ahmad, B., Asim, M., Khan, A.I., Xin, Y., Zexu, W., Song Ju, H., Xin, W., Lei, S., Wang, L., *Morchella esculenta* mushroom polysaccharide attenuates diabetes and modulates intestinal permeability and gut microbiota in a type 2 diabetic mice model. *Front. Nutr.*, *9*, 984695, 2022.
34. Wińska, K., Mączka, W., Gabryelska, K., Grabarczyk, M., Mushrooms of the Genus *Ganoderma* Used to Treat Diabetes and Insulin Resistance. *Molecules (Basel, Switzerland)*, *24*, 22, 4075, 2019.
35. Martel, J., Ojcius, D.M., Chang, C.J., Lin, C.S., Lu, C.C., Ko, Y.F., Tseng, S.F., Lai, H.C., Young, J.D., Anti-obesogenic and antidiabetic effects of plants and mushrooms. *Nat. Rev. Endocrinol.*, *13*, 3, 149–160, 2017.
36. Lo, H.C. and Wasser, S.P., Medicinal mushrooms for glycemic control in diabetes mellitus: history, current status, future perspectives, and unsolved problems (review). *Int. J. Med. Mushrooms*, *13*, 5, 401–426, 2011.
37. Aramabašić Jovanović, J., Mihailović, M., Uskoković, A., Grdović, N., Dinić, S., Vidaković, M., The Effects of Major Mushroom Bioactive Compounds on Mechanisms That Control Blood Glucose Level. *J. Fungi (Basel, Switzerland)*, *7*, 1, 58, 2021.

10

Application of Mushrooms in the Management of Cardiovascular Diseases

Enitan Emmanuella Lawal[1], Frances Iseghohi[2], Aishat Sani Rabiu[3], Amarachi Rosemary Osi[4], Oluwafemi Adebayo Oyewole[1]*, Charles Oluwaseun Adetunji[5], Oluwabukola Atinuke Popoola[6], and Adeyomoye Olorunsola Israel[7]

[1]*Department of Microbiology, Federal University of Technology, Minna, Nigeria*
[2]*Medical Biotechnology Department, National Biotechnology Development Agency, Abuja, Nigeria*
[3]*Department of Microbiology, Ibrahim Badamasi Babangida University, Lapai, Niger State, Nigeria*
[4]*Ningbo Institute of Material Technology and Engineering, Chinese Academy of Sciences, Ningbo, Zhejiang Province, P. R. China*
[5]*Department of Microbiology, Edo State University, Uzairue, Edo State, Nigeria*
[6]*Genetics, Genomics and Bioinformatics Department, National Biotechnology Development Agency, Abuja, FCT, Nigeria*
[7]*Department of Physiology, Ondo Medical University, Ondo State, Nigeria*

Abstract
A series of conditions known as cardiovascular diseases (CVDs) affect the arteries, veins, and capillaries as well as the heart (cardio). The main CVDs are heart failure, stroke, coronary artery diseases (CAD), hypertension, and arthritis, having a family history of heart disease, smoking, high blood pressure, high cholesterol, obesity, and diabetes are some of the risk factors that increase the likelihood of developing CVD. Cardiovascular disease (CVD) is estimated to have caused 17.3 million deaths worldwide in 2008, accounting for around 30% of all fatalities. It is estimated that in 2013, one million deaths worldwide and 11.3% of all deaths in Africa were caused by cardiovascular disease (CVD). This number only includes deaths that occurred in Sub-Saharan Africa. Although there are medications to treat cardiovascular diseases, there are many drawbacks, these include drugs that

**Corresponding author*: oa.oyewole@futminna.edu.ng

Charles Oluwaseun Adetunji, Chukwuebuka Egbuna, Oluwatosin Ademola Ijabadeniyi and Samantha C. Karunarathna (eds.) Mushroom Biotechnology for Improved Agriculture and Human Health, (221–238) © 2025 Scrivener Publishing LLC

can cause renal failure, muscular abnormalities, rhabdomyolysis, and haemorrhage with prolonged use, which can be fatal. There is a need for an alternative treatment, especially since research has shown that many medicinal plants, including mushrooms, have cardio-protective properties due to their distinct nutritional makeup and bioactive substances. The diverse types of medicinal mushrooms have produced about 130 different therapeutic effects, including anticancer, antiviral, cardioprotective, antioxidant, radical scavenging, and immunomodulatory properties. The cardioprotective effects of medicinal mushrooms are the main described in this chapter.

***Keywords*:** Mushroom, cardiovascular diseases, fungi, medicinal, bioactive compounds

10.1 Introduction

Cardiovascular diseases (CVDs), including atherosclerosis, angina, hypertension, and cardiac insufficiency, among others are on the rise in alarming numbers all over the world. CVDs cause the most deaths and it is predicted that by 2030, 22.2 million people will die from these conditions, primarily myocardial infarction, atherosclerosis, lung-related vascular thrombosis (embolism), aortic and cerebral aneurysms, vein aneurysms, and cerebral thrombotic stroke [1].

Heart and blood vessel illnesses known as deep vein thrombosis, pulmonary embolism, congenital heart disease, rheumatic heart disease, peripheral artery disease, and cerebrovascular disease are some examples of CVDs [2]. According to estimates, 2008 had 17.3 million deaths due to cardiovascular disease (CVD) making it the top cause of death globally and responsible for about 30% of all fatalities. Furthermore, the amount of CVD deaths on a global scale is anticipated to rise to about 23.4 million by 2030 accounting for 35% of all deaths, making it crucial for global health to prevent CVD in Asia [3].

According to the Eurostat's 2017 Statistics on Cardiovascular Disease [4], CVDs are responsible for 45% of all fatalities in Europe and according to the American Heart Association's most recent figures, CVD affects almost half of all Americans [5]. Africa, which is home to over 1 billion people, contributes significantly to the global burden of CVD. A million deaths in the sub-Saharan part of Africa alone are predicted to have been caused by CVD in 2013, representing 11.3% of all fatalities in Africa and 5.5% of all CVD-related deaths on a global scale [6].

According to Balakumar *et al.* [7], CVDs are a group of illnesses that include disorders of the peripheral vessels, hypertension, myocardial infarction, stroke, heart failure, coronary heart disease (CHD), and cardiomyopathies are a few examples. Common risk factors for CVDs include smoking, an unhealthy diet, diabetes mellitus, hyperlipidemia, elevated LDL cholesterol (LDL), suppressed HDL cholesterol (HDL), and hypertension [8]. The disparities in the CVD prevalence in populations and changes therein over time may also be explained by a number of other factors, such as socio-demographics, ethnicity, culture, dietary concerns, and behavioural problems [1].

Although there are medications to treat cardiovascular illnesses, there are many limitations, including those that can lead to renal failure, muscular abnormalities, rhabdomyolysis, and hemorrhage medication for the treatment of cardiovascular disorders with prolonged usage, which can be fatal [9]. Since the usage of conventional cardiovascular disorders has proven effective but not without side effects, making their use limited and necessitating alternative search, research has highlighted the cardioprotective properties of various medicinal plants, their roles being linked to bioactive compounds that are present in these plants [10].

The fact that rural populations in developing nations have turned to traditional herbal and plant-derived treatments in their fight against disease in the face of limited socioeconomic resources is another factor contributing to the resurgence of interest in medicinal plant products. Natural products' biological richness and structural variety outperform any synthetic drug screening library currently available, which is another factor contributing to the renewed interest in natural products [11].

Medicinal fungi and mushrooms are thought to form about 130 therapeutic activities, these functions include antitumor, immunomodulating, antioxidant, cardiovascular, antiparasitic, inflammatory, tumor-fighting, fungal, bacterial, radical scavenger, hepatoprotective, detoxifying, and anti-hypercholesterolemia properties. Long used in oriental medicine to prevent and treat a wide array of illnesses, mushrooms are now more widely utilized as supplements for diets, particularly for boosting the function of the immune system and having anticancer properties [12].

The non-photosynthetic, eukaryotic, aerobic creatures known as mushrooms, or macro-fungi, are classified into two families; Basidiomycetes and Ascomycetes, according to taxonomy classification. They generate characteristic fruiting bodies. 2000 of the more than 14,000 varieties of mushrooms, which have been categorized into 30 genera, are edible and regarded as potential therapeutic or prophylactic agents for human consumption [13]. Around the world, a variety of edible wild mushrooms are consumed, including *Amanita muscari, Agaricus arvensis, Agaricus silvaticus, Handkea utriformis, Cortinarius variicolor, Ustilago maydis, Marasmius oreades, Leccinum versipelle,* and *Suillus luteus* are few examples of the species.

Ancient civilizations used mushrooms for medical purposes, and in the modern era, Asian countries have a long history of using medicinal mushrooms. The function of mushrooms in the prevention and treatment of chronic diseases has brought about more broad recognition in the past 10 years as Westernized nations' as plant-based diets are more popular now [14]. A recent PubMed search using the keywords "mushroom and health" produced 2411 publications (on April 21, 2021); nearly 80 were reviews from 2020 to the present, demonstrating that real-world high scientists studying biology and medicine are interested in the mechanisms underlying the health benefits of mushrooms.

There are many varieties of mushrooms included in the database of traditional medicines, but those that are most helpful include *Brazilian agaricus, Flammulina velutipes, Agaricus bisporus, Lentinus edodes, Ganoderma lucidum, Trametes versicolor Coprinus comatus, Tricholoma matsutake's auricularia auricula, Grifola frondosa, Pleurotus ostreatus,* and *Cordyceps sinensis* [15].

10.2 Selected Medicinal Mushrooms

10.2.1 *Ganoderma lucidum*

The therapeutic properties of the mushroom *Ganoderma lucidum* (Figure 10.1) sometimes referred to as "lingzhi" or "reishi," are derived from its various sections, including its mycelia, spores, and fruit body [16]. Commercially, *G. Lucidum* is offered in a variety of forms, including powders, nutritional supplements, and tea [17]. For more than 2000 years, *G. lucidum* has long been a popular medication in Asian nations, primarily in China, Japan, and Korea. Later, Western societies were exposed to it [18].

Triterpenoids also possess hepatoprotective, hypocholesterolemic, anti-hypertensive, anti-histaminic, anti-tumor, and antiangiogenic properties. The unique source of the triterpene fatty acids known as ganoderic acids is *G. lucidum* [19]. The bioactive constituents of the *G. lucidum* mushrooms are medicinally used to treat a wide array of illnesses, such as cancer, leukopenia, anorexia, bronchitis, asthma, arthritis, hypertension,

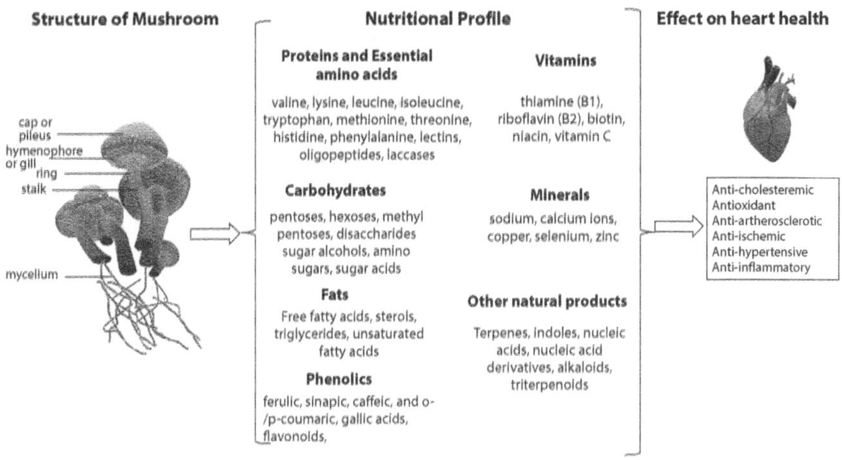

Figure 10.1 Nutritional profile of edible mushrooms [21].

nephritis, hyperlipidemia, neurasthenia, hypertension, insomnia, and atherosclerosis [20].

10.2.2 *Hericium erinaceus* (Lions Mane Mushroom)

Hericium erinaceus (Figure 10.2) is a basidiomycete of the Hericiaceae family, Russulales order, and Agaricomycetes classn [22]. *H. erinaceus* is simple to recognise when grown, because of its distinctive bas idiomes, which are made up of several solitary, generally long, dangling, fleshy spines that are initially white, then become yellowish, then brownish with age [23]. Species of the genus *Hericium* can be distinguished from one another by the presence of microscopic amyloid adorned basidiospores, appearance in single vs. multiple clumps, branching compared to unbranched hymenophore structures supporting a range of lengths of spines, and appearance in single or numerous clumps, in many parts of the world, this fungus is also referred as "Lion's Mane," "Hog's Head Fungus," "Monkey's Mushroom," "Old Man's Beard," "Pom Pom," "White Beard," and "Bearded Tooth" [24].

Hericium erinaceus is classified as a weak parasite or saprotroph since sometimes it produces fruiting bodies which may erupt from cracks or from knotholes in real hardwoods, it mainly develops on dead wood, numerous research have examined this mushroom's possible therapeutic properties or health benefits, qualities that prevent cancer, hypertension, lipidemia, and neurological diseases [25].

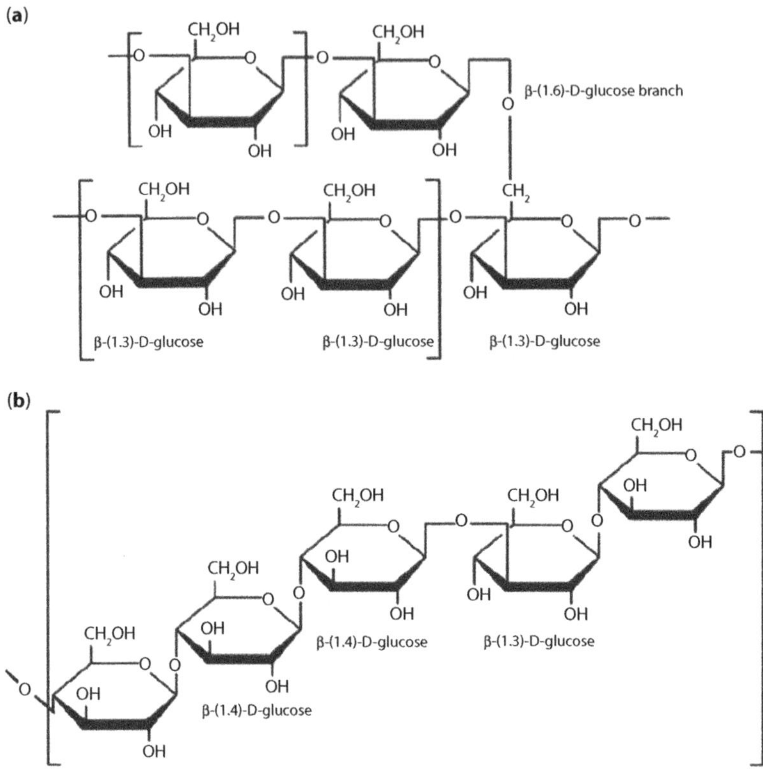

Figure 10.2 Chemical compositions of beta-glucans. (a) Side branches are coupled to the 1,3-D-glucan main chain by a 1,6 linkage. In contrast to cereal -glucans (b), which are linear and made up of D-glucose molecules connected by 1,3 and 1,4 connections, mushroom-glucans. Source; Cerletti *et al.* [43].

10.2.3 *Agaricus bisporus*

Agaricus bisporus (Figure 10.3), the most popular fungus in cultivation in the world is an *Agaricus* kind of mushroom [26]. These tasty mushrooms are currently the topic of extensive research and use in prospective medical and therapeutic applications. Lectins from *Agaricus bisporus* may be employed in medications and as food for healthy people because they have been shown to be potent immune stimulants [27]. Numerous common human ailments, such as cancer, bacterial and fungal infections, diabetes, heart disease, and skin issues, can be treated with the bioactive components isolated from *A. bisporus.* It is high in nutrients like sugars, amino acids, lipids, and minerals and may possess anti-inflammatory, anti-cancer, and antioxidant capabilities. In terms of understanding the medicinal and

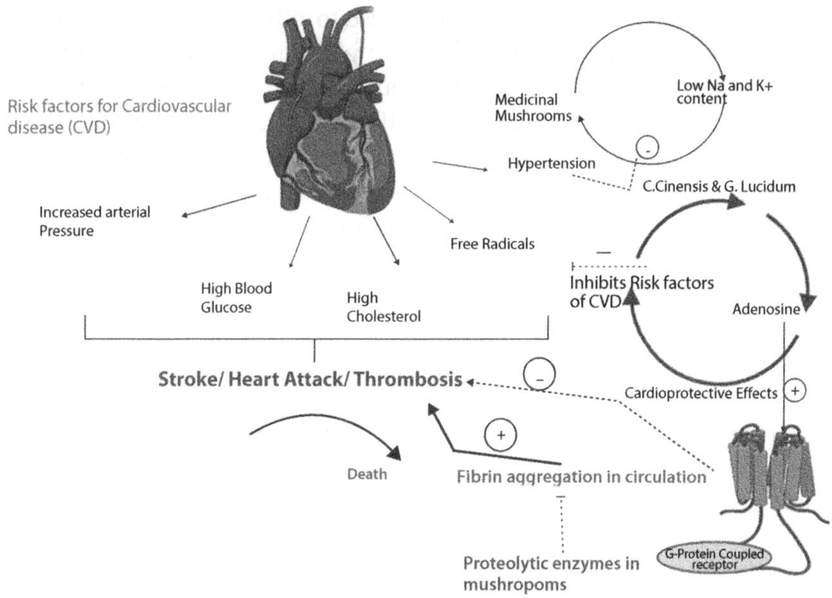

Figure 10.3 The pathway illustrating the part that medicinal mushrooms plays in heart disease [57].

culinary uses of mushrooms, *A. bisporus* has broadened the world's perspective [28]. From this fungus, tyrosinase and ergothioneine have been extracted in recent years, making them more worthwhile to consider for dietary and therapeutic applications.

10.3 Nutritional Composition of Mushrooms

Edible mushrooms have very minimal dry matter content (between 60 and 140 g/kg) due to their moisture content (between 81.8 and 94.8%) [29], but dried mushrooms have a reputation for being hygroscopic. According to Kalac [30], monosaccharides, their derivatives, and oligosaccharides with both reserve and building polysaccharides (glycans) make up around half of the dry matter in mushrooms. According to Alzand *et al.* [29], the three main types of monosaccharides, their derivatives, and oligosaccharide groups are glucose, mannitol, and trehalose.

In contrast to starch found in plants, glycogen is a reserve polysaccharide found in mushrooms. Humans cannot digest chitin, a structural polymer that accounts for 80% or more of the dry material in the cell wall of mushrooms. This limits human access to other mushroom constituents [20]. Consumable mushrooms are a plentiful source of dietary fibre, which benefits human health in a number of ways. Mushrooms and plant cell walls can both be regarded as significant sources of dietary fibre [31]. Mushrooms have more insoluble dietary fibre than soluble dietary fibre (0.32-2.20 g/100 g edible weight), ranging from 2.28 to 8.99 g/100 g edible weight, according to the kind of mushroom.

In addition to the species of mushroom, which vary greatly, wide differences in the amount of crude protein are produced by the employment of various converting factors in addition to the Kjeldahl method's determination however, to prevent overestimating the amount of crude protein a specific converting factor of 4.16 was suggested [29].

Mushrooms have long been thought of as robust sources of vitamins due to Niacin, folic acid, and traces of vitamins C, B1, D, -carotene (a precursor to vitamin A), E, and B12 which are present in elevated concentrations, along with riboflavin (vitamin B2), niacin, and folic acid and their ability to amass these vitamins, despite the fact that they are cultivated on lignocellulosic waste, is the final piece of evidence for their aptitude for biosynthesis [32]. In contrast to plants, mushrooms appear to have little carotenoids, especially those that may act as antioxidants [33].

The only naturally existing food alternative that can provide vitamin D to people who eat a vegetarian diet is mushrooms [34]. All other naturally occurring food options must originate from animal sources. Numerous wild mushroom species have large amounts of vitamin D2 (ergocalciferol), whereas farmed types have much less of it due to inadequate solar exposure [29].

A representative sample of each type of lipid component, sterols, sterol ester, phospholipids, mono-, di-, and triglycerides, as well as free fatty acids may be found in mushrooms in relatively small concentrations [35]. Total lipid content is low, often between 20 and 30 g kg1 dry matter, while concentrations above 60 g kg1 dry matter are uncommon [30]. Numerous fatty acids have been found in the lipids of mushrooms but just two stand out, accounting for at least three-quarters of the total weight of the fatty acids: linoleic acid (C18:2c,n-6) and oleic acid (C18:1c,n-9) [36]. Mushrooms have a minimal energy value of about 350 kcal kg1 fresh matter due to their low dry matter and fat contents.

There are many various mineral elements in mushrooms. Bigger differences occur more frequently in wild-growing species than in domesticated ones, perhaps with respect to the former group's more varied substrates. The average ash content of mushrooms is between 60 and 120 g kg1 dry matter, with bigger variances occurring in the former group. However, ash concentrations appear to be less variable than crude protein, lipids, and carbs (Kalac, 2013). Magnesium, calcium, copper, iron, zinc, phosphorus, magnesium, and potassium are the primary components of ash with potassium predominating and calcium and sodium at the opposite ends of the chemical elemental spectrum. However, there are certain exceptions, such as the species of the genus Morchella that have calcium concentrations ranging from 0.7 to 5.2 g kg1 DM [30, 37]. The ability of mushrooms to store nutrients from their growing media makes them particularly special which makes this property ambivalent because it can be both useful and dangerous for consumption when toxic elements (which may include heavy metals like silver, tin, cadmium, lead, mercury, chromium, and arsenic) are accumulated [29].

10.4 Bioactive Compounds in Mushrooms

The primary and secondary bioactive metabolites that give mushrooms antimicrobial, fungal, anti-inflammatory, antinociceptive, antioxidant, anti-proliferative, anti-cancer, anti-viral, hypocholesterolemic, hypoglycemia, and hypolipidemic pharmacological activities include polyketides, peptides, lectins, nucleic acids, nucleosides, fatty acids, polysaccharides, and alkaloids [38, 39].

Phellinus linteus, P. ostreatus, Hericium erinaceus, L. edodes, Ganoderma spp., G. frondosa, A. subrufescens, and *Ganoderma spp.* are among the medicinal species that have been suggested for use in a range of therapeutic procedures. They were also reported to be producers of various bioactive compound groups [40, 41]. Figure 10.1 shows the nutritional profile of edible mushrooms.

10.4.1 The Polysaccharide

Heteroglucans, the building blocks of polysaccharides, are polymers of sugars joined to one another by glycosidic or non-glycosidic bonds. Polysaccharides are macromolecules with a high potential phenotypic diversity. The α- and β- glucans are the most well-known and prevalent polysaccharides and they also have the highest capability for biological activities [42]. The polymers have type linkages in the -glucans, which mushrooms have proportionately less of type linkages that can be found in the polymers. D-glucose monomers are joined by glycosidic links at two separate sites, 1,6 and 1,3, which are in charge of the linear structure and the major branching, respectively, in the macromolecular structure of mushroom -glucans [43]. In comparison to the overall amount of -glucans, *A. auricula-judae* (47.9%), *S. variegatus* (40.6%), *S. granulatus* (49.8%), and *G. cyanescens* (54%) had the most proportion of 1,3- and 1,6-D-glucan content.

10.4.2 Terpenoids

Terpenes are hydrocarbons formed when numerous isoprene units are combined and consist of a variety of pharmacological actions, such as anti-inflammatory, antiviral, antibacterial, antimalarial, and anticancer properties [44]. Over 320 terpenoids has been collected from 22 Ganoderma species [45].

There are numerous lanostane triterpenes that could be used as cancer treatments. The majority of terpenoid compounds extracted from mushrooms are triterpenoids of the lanostane class [46]. From 29 different mushroom species, 259 terpenoids have been identified that are active biologically, are mono-, sesqui-, di-, and tri-terpenoids, and some terperniods are shown in Table 10.1.

10.4.3 Steroids

Four connected cycloalkane rings form the distinguishing structure of the chemical molecule known as a steroid [47]. They are lipophilic substances that are crucial to biological systems [48]. Many steroids with anticomplement, antibacterial, cytotoxic, antitubercular, and antiviral properties have been identified from various mushrooms [49]. Several types of mushrooms, such as *Sarcodon joedes, Ganoderma lucidum, Tomophagus cattienensis, Fomes fomentarius,* and *Hypsizigus marmoreus,* have each yielded isolated steroids [50].

Nine chemicals, including four steroids, were found in the methanolic extract of Sarcodon joedes (22E, 24R), Ergesterol peroxide (286), -6-methoxyergosta-7,22-diene3,5-diol (284), ergosta-7,22-dien-3-ol (285), and (22E,24R)3H-ergosta-5,22-dien-7-one (287).

Table 10.1 Bioactive terpenoids and their respective mushroom species [46].

Terpenoids	Mushroom Spp	Pharmacological effect
Hirsutenol A, B, & C	*Stereum hirsutum*	Antimicrobial
Pleurospiroketal A, B, & C	*Pleurotus cornucopiae*	Cytotoxic
Enokipodin E, F, G, H	*Flammulina velutipes*	Antibacterial, Antifungal, Antioxidant, and Cytotoxic
Cyathin D, E, F, G, H	*Cyathus africanus*	Cytotoxic
Erinacine A	*Hericium erinaceum*	Antibacterial
Ganorbiformin B–G	*Ganoderma orbiforme*	Cytotoxic
Lucialdehyde A–C	*G. lucidum*	Cytotoxic
Elfvingic acid A–H	*Elfvingia applanata*	Cytotoxic
Saponaceol A–C	*Tricholoma saponaceum*	Cytotoxic
Inonotic acid A	*Inonotus rickii.*	Cytotoxic
Rulepidadiol B & C	*Russula lepida*	Cytotoxic
Sarcodonin A, I, G, & M	*Sarcodon scabrosus*	Antibacterial
Trocholomalide A, B, & C	*Tricholoma spp.*	Cytotoxic

10.4.4 Phenolics

The biggest class of phytochemicals, phenolic compounds are quite known for their intriguing pharmacological properties. Flavonoids, coumarins, anthocyanins, catechins, and phenolic acid are just a few [51, 52]. There have been reports of biologically active phenolic compounds in a number of mushrooms, including *Ganoderma pfeifferi, Ganoderma fornicatum, Ganoderma colossum Albatrellus caeruleoporus, Boletus pseudocalopus, Cortinarius purpurascens, and Tricholoma oriruben* [49]. Mushroom phenolics have been discovered to be exceptional antioxidants with minimal mutagenesis potential [53].

10.4.5 Alkaloids

Alkaloid is any natural medication that has an alkaloid structure or has nitrogen in its structure, making alkaloids an important class of natural products with a significant amount of activity [44]. Only a few numbers of alkaloids from the mushroom species have been isolated [54].

10.5 Cardioprotective Effect of Mushrooms

Mushrooms have been suggested to minimise the risk of CVDs due to their wide bioactivity in conventional medicine. The therapeutic advantages of mushrooms including their anti-oxidative, anti-inflammatory, anti-atherosclerotic, anti-hypertensive, anti-obesity, hypoglycemic, hypolipidemic, and fibrinolytic properties, have been described and these advantages may be helpful in the treatment and prevention of cardiac disorders. Mushrooms, such as *G. frondosa, L. edodes,* and *P. ostreatus*, are almost perfect for diets low in calories meant to reduce CVD due to their fibre, protein, and microelement content [55].

G. lucidum, G. frondosa, H. marmoreus, L. edodes, Boletus aestivalis, Clitocybe nuda and *Pleurotus* species all contain a variety of bioactive substances, including terpenoids, peptides, isoflavones like biochanin A and formononetin, a derivative of lanosterone called fomiroid A, and lovastatin can regulate total cholesterol amount, fasting triglycerides, homocysteine, and density-low lipoproteins to help stop the onset of diabetes, oxidative stress, arterial hypertension, and other CVD [56]. The pathway illustrating the part that medicinal mushrooms play in heart disease is presented in Figure 10.3.

As a result of blood fibrin aggregation, thrombosis can also result in cardiovascular disease (CVD). Fibrinolysis, which involves using proteolytic enzymes in order to break up blood clot fibrin, is a balanced biological process [58], but thrombosis, which is caused by the accumulation of fibrin clots in arteries of the blood, leads to myocardial infarction or other CVD [59]. It is believed that mushrooms from various taxonomic and ecological groups actively produce enzymes that break down proteins, such as fibrinolytic, thrombolytic, and caseinolytic (Badalyan *et al.*, 2021). Among the other species tested (*A. bisporus, H. erinaceus, L. edodes, P. cystidiosus, P. floridanus, P. pulmonarius,* and *P. salmoneostramineus*), the *L. edodes* crude extract exhibited the maximum proteolytic activity (up to 54.28 U/mg) [60].

Triterpenes from the fungus *Ganoderma lucidum*, such as ganoderic acid C and its derivatives, have the ability to prevent manufacturing cholesterol meanwhile, others work to prevent atherosclerosis by inhibiting the angiotensin-converting enzyme or platelet aggregation [61]. According to numerous studies, lentinula mushrooms can reduce blood pressure and plasma levels of free cholesterol while also speeding up the liver's storage of lipids by removing them from circulation [62]. Additionally, rats fed with a semi-synthetic high-cholesterol diet have shown dried in comparison to the control group, *Agrocybe aegerita* considerable reduced levels of serum total cholesterol (TC), triacylglycerides, atherogenic index, hepatic TC, and total triacylglycerides [63].

Natural antioxidants in the diet, such as mushrooms, can help prevent the onset of CVD by catalase and superoxide dismutase activities are increased, free radicals are neutralised, and glutathione and malondialdehyde levels are stabilised [64].

G. lucidum utilised the diet restriction pathway and the mTOR/S6K signalling pathway, respectively, to protect model organism *Caenorhabditis elegans* from paraquat- and heavy metal-induced oxidative stress. Antioxidants are necessary for the prevention of atherosclerosis [65]. Human lymphocyte DNA has been found to be protected by *G. lucidum* from oxidative damage caused by hydrogen peroxide. Additionally, it has been discovered that *G. lucidum* modulates the production of Nrf2, which as a result influences the gene expression related to antioxidants including HO-1, GST, and NQO-1. In C2C12 myoblasts, *G. lucidum* ethanolic extract boosted Nrf2's phosphorylation, which raised HO-1. [66]. As a result, *G. lucidum*'s antioxidant ability, which is crucial for the prevention of atherosclerosis, can function on a variety of signalling pathways. *Ganoderma lucidum, Termitomyces heimii, T. mummiformis, Coriolus versicolor, G. Lentinula edodes tsugae,* and some mushrooms are high in antioxidants [57].

Mushrooms are a good dietary supplement to lower blood pressure due to their increased potassium level (182–395 mg/100 g) and low sodium content. Numerous researches have shown that mushrooms with anti-hypertensive effects include *Lentinula edodes, Pleurotus narbonensis, Ganoderma lucidum*, and *Grifola frondosa* [62].

10.6 Conclusion

Mushrooms are extensively researched for their potential medical advantages and are used as a dietary supplement in addition to common medications and treatments. The nutritional profile and active ingredient in mushrooms have a wide array of pharmacological and medicinal effects of which antiviral, antioxidant, cytotoxic, antitumor, anti-inflammatory, and anti-aging are a few of its therapeutic qualities. Although the chemicals extracted from mushrooms show great activity, not all mushrooms are nostrums, necessitating additional optimization as these compounds can be altered by altitude, temperature, humidity, as well as other environmental elements.

References

1. Wouk, J., Dekker, R.F., Queiroz, E.A., Barbosa-Dekker, A.M., β-Glucans as a panacea for a healthy heart? Their roles in preventing and treating cardiovascular diseases. *Int. J. Biol. Macromol.*, 177, 176–203, 2021.
2. Lennon, R.P., Claussen, K.A., Kuersteiner, K.A., State of the heart: an overview of the disease burden of cardiovascular disease from an epidemiologic perspective. *Prim. Care Clin. Office Pract.*, 45, 1, 1–15, 2018.
3. Ohira, T. and Iso, H., Cardiovascular disease epidemiology in Asia–an overview–. *Circ. J.*, 77, 7, 1646–1652, 2013.
4. Martinet, W., Coornaert, I., Puylaert, P., De Meyer, G.R., Macrophage death as a pharmacological target in atherosclerosis. *Front. Pharmacol.*, 10, 306, 2019.
5. Benjamin, E.J., Muntner, P., Alonso, A., Bittencourt, M.S., Callaway, C.W., Carson, A.P., American Heart Association Council on Epidemiology and Prevention Statistics Committee and Stroke Statistics Subcommittee, Heart disease and stroke statisticsn—2019 update: a report from the American Heart Association. *Circulation*, 139, 10, e56–e528, 2019.
6. Keates, A.K., Mocumbi, A.O., Ntsekhe, M., Sliwa, K., Stewart, S., Cardiovascular disease in Africa: epidemiological profile and challenges. *Nat. Rev. Cardiol.*, 14, 5, 273–293, 2017.
7. Balakumar, P., Maung-U, K., Jagadeesh, G., Prevalence and prevention of cardiovascular disease and diabetes mellitus. *Pharmacol. Res.*, 113, 600–609, 2016.
8. Wild, R.A., Carmina, E., Diamanti-Kandarakis, E., Dokras, A., Escobar-Morreale, H.F., Futterweit, W., Dumesic, D.A., Assessment of cardiovascular risk and prevention of cardiovascular disease in women with the polycystic ovary syndrome: a consensus statement by the Androgen Excess and Polycystic Ovary Syndrome (AE-PCOS) Society. *J. Clin. Endocrinol. Metab.*, 95, 5, 2038–2049, 2010.
9. Dieterle, T., Side effects and interactions of frequently used cardiovascular drugs. *Ther. Umsch. Rev. Ther.*, 72, 11-12, 701–710, 2015.
10. Adegbola, P., Aderibigbe, I., Hammed, W., Omotayo, T., Antioxidant and anti-inflammatory medicinal plants have potential role in the treatment of cardiovascular disease: a review. *Am. J. Cardiovasc. Dis.*, 7, 2, 19, 2017.

11. Mustafa, G., Arif, R., Atta, A., Sharif, S., Jamil, A., Bioactive compounds from medicinal plants and their importance in drug discovery in Pakistan. *Matrix Sci. Pharma*, 1, 1, 17–26, 2017.
12. Kück, U., Bloemendal, S., Teichert, I., Putting fungi to work: harvesting a cornucopia of drugs, toxins, and antibiotics. *PLoS Pathog.*, 10, 3, e1003950, 2014.
13. Chugh, R.M., Mittal, P., Mp, N., Arora, T., Bhattacharya, T., Chopra, H., Gautam, R.K., Fungal mushrooms: a natural compound with therapeutic applications. *Front. Pharmacol.*, 13, 925387, 2022.
14. Lam-Sidun, D., Peters, K.M., Borradaile, N.M., Mushroom-derived medicine? Preclinical studies suggest potential benefits of ergothioneine for cardiometabolic health. *Int. J. Mol. Sci.*, 22, 6, 3246, 2021.
15. Abdel-Azeem, A.M., Abdel-Azeem, M.A., Khalil, W.F., Endophytic fungi as a new source of antirheumatoid metabolites, in: *Bioactive Food as dietary interventions for arthritis and related inflammatory diseases*, pp. 355–384, Academic Press, United States, 2019.
16. El Sheikha, A.F., Nutritional profile and health benefits of Ganoderma lucidum "Lingzhi, Reishi, or Mannentake" as functional foods: Current scenario and future perspectives. *Foods*, 11, 7, 1030, 2022.
17. Kała, K., Kryczyk-Poprawa, A., Rzewinska, A., Muszyqnska, B., Fruiting bodies of selected edible mushrooms as a potential source of lovastatin. *Eur. Food Res. Technol.*, 246, 4, 713–722, 2020.
18. Badalyan, S.M., Morel, S., Barkhudaryan, A., Rapior, S., Mushrooms as Promising Therapeutic Resources: Review and Future Perspectives, in: *Mushrooms with Therapeutic Potentials: Recent Advances in Research and Development*, pp. 1–54, 2023.
19. Hansson, G., Inflammation, atherosclerosis, and coronary heart disease. *New Engl. J. Med.*, 352, 1685–1695, 2005, 10.1056/NEJMra043430.
20. Sisodiya, A.S., Kadicheeni, A.S., Das, G., Kumari, T., Archana, T.S., Kumar, D., How to tell the mushroom you are holding will kill you or heal you: A mini review, 12, 2,109–122, 2023, DOI: 10.22271/tpi.2023.v12.i2b.18725
21. Rauf, A., Joshi, P.B., Ahmad, Z., Hemeg, H.A., Olatunde, A., Naz, S., Simal-Gandara, J., Edible mushrooms as potential functional foods in amelioration of hypertension. *Phytother. Res.*, 37, 1, 2644–2660, 2023.
22. Chaiyasut, C. and Sivamaruthi, B.S., Anti-hyperglycemic property of Hericium erinaceus–A mini review. *Asian Pac. J. Trop. Biomed.*, 7, 11, 1036–1040, 2017.
23. Thongbai, B., Rapior, S., Hyde, K.D., Wittstein, K., Stadler, M., Hericium erinaceus, an amazing medicinal mushroom. *Mycol. Prog.*, 14, 1–23, 2015. Khan, M.S., Zhang, X., You, L., Fu, X., Abbasi, A.M., Structure and bioactivities of fungal polysaccharides, in: *Polysaccharides: Bioactivity and Biotechnology*, pp. 1–14, 2014.
24. He, X., Wang, X., Fang, J., Chang, Y., Ning, N., Guo, H., Zhao, Z., Structures, biological activities, and industrial applications of the polysaccharides

from Hericium erinaceus (Lion's Mane) mushroom: A review. *Int. J. Biol. Macromol.*, 97, 228–237, 2017.
25. Khan, M.A., Tania, M., Liu, R., Rahman, M.M., Hericium erinaceus: an edible mushroom with medicinal values. *J. Complement. Intgr. Med.*, 10, 1, 253–258, 2013.
26. Wa, E. and Daba, G.M., Bioactive Potential of Some Fascinating Edible Mushrooms Flammulina, Lyophyllum, Agaricus, Boletus, Letinula, Pleurotus as a Treasure of Multipurpose Therapeutic Natural Product, 6, 1, 1–10, 2022, https://doi.org/10.23880/pdraj-16000263
27. Valverde, M.E., Hernández-Pérez, T., Pareqdes-López, O., Edible mushrooms: improving human health and promoting quality life. *Int. J. Microbiol.*, 2015, 376387, 2015, https://doi.org/10.1155/2015/376387.
28. Usman, M., Murtaza, G., Ditta, A., Nutritional, medicinal, and cosmetic value of bioactive compounds in button mushroom (Agaricus bisporus): a review. *Appl. Sci.*, 11, 13, 5943, 2021.
29. Alzand, K.I., Bofaris, M.S.M., Ugis, A., Chemical composition and nutritional value of edible wild growing mushrooms: a review. *World J. Pharm. Res.*, 8, 3, 31–46, 2019.
30. Kalač, P., A review of chemical composition and nutritional value of wild-growing and cultivated mushrooms. *J. Sci. Food Agric.*, 93, 2, 209–218, 2013.
31. Kumari, K., Mushrooms as source of dietary fiber and its medicinal value: A review article. *J. Pharmacogn. Phytochem.*, 9, 2, 2075–2078, 2020.
32. Wang, Y., Liu, L., Jin, Z., Zhang, D., Microbial cell factories for green production of vitamins. *Front. Bioeng. Biotechnol.*, 9, 661562, 2021.
33. Lin, L. and Xu, J., Fungal pigments and their roles associated with human health. *J. Fungi*, 6, 4, 280, 2020.
34. Benedik, E., Sources of vitamin D for humans. *Int. J. Vitamin Nutr. Res.*, 92, 2, 118–125, 2021, https://doi.org/10.1024/0300-9831/a000733
35. Sande, D., de Oliveira, G.P., e Moura, M.A.F., de Almeida Martins, B., Lima, M.T.N.S., Takahashi, J.A., Edible mushrooms as a ubiquitous source of essential fatty acids. *Food Res. Int.*, 125, 108524, 2019.
36. Raman, J., Jang, K.Y., Oh, Y.L., Oh, M., Im, J.H., Lakshmanan, H., Sabaratnam, V., Cultivation and nutritional value of prominent Pleurotus spp.: an overview. *Mycobiology*, 49, 1, 1–14, 2021.
37. Guillamón, E., García-Lafuente, A., Lozano, M., Rostagno, M.A., Villares, A., Martínez, J.A., Edible mushrooms: role in the prevention of cardiovascular diseases. *Fitoterapia*, 81, 7, 715–723, 2010.
38. Aramabašić Jovanović, J., Mihailović, M., Uskoković, A., Grdović, N., Dinić, S., Vidaković, M., The effects of major mushroom bioactive compounds on mechanisms that control blood glucose level. *J. Fungi*, 7, 1, 58, 2021.
39. Badalyan, S.M., Barkhudaryan, A., Rapior, S., Recent progress in research on the pharmacological potential of mushrooms and prospects for their clinical

application, in: *Medicinal mushrooms: recent progress in research and development*, pp. 1–70, 2019.

40. Zhang, L., Fan, C., Liu, S., Zang, Z., Jiao, L., Chemical composition and antitumor activity of polysaccharide from *Inonotus obliquus*. *J. Med. Plants Res.*, 5, 7, 1251–1260, 2011.
41. Pereira, E., Barros, L., Martins, A., Ferreira, R., Towards chemical and nutritional inventory of Portuguese wild edible mushrooms in different habitats. *Food Chem.*, 130, 2, 394–403, 2012.
42. Venturella, G., Ferraro, V., Cirlincione, F., Gargano, M.L., Medicinal mushrooms: bioactive compounds, use, and clinical trials. *Int. J. Mol. Sci.*, 22, 2, 634, 2021.
43. Cerletti, C., Esposito, S., Iacoviello, L., Edible mushrooms and beta-glucans: Impact on human health. *Nutrients*, 13, 7, 2195, 2021.
44. Öztürk, M., Tel-Çayan, G., Muhammad, A., Terzioğlu, P., Duru, M.E., Mushrooms: a source of exciting bioactive compounds, in: *Studies in Natural Products Chemistry*, vol. 45, pp. 363–456, 2015.
45. Ruthes, A.C., Smiderle, F.R., Iacomini, M., Mushroom heteropolysaccharides: A review on their sources, structure and biological effects. *Carbohydr. Polym.*, 136, 358–375, 2016.
46. Duru, M.E. and Çayan, G.T., Biologically active terpenoids from mushroom origin: a review. *Rec. Nat. Prod.*, 9, 4, 456, 2015.
47. Meher, C.P., Sethy, S.P., Pochaiah, B., Structure and biological activities: Steroid moieties. *Res. J. Pharm. Biol. Chem. Sci.*, 4, 253–272, 2013.
48. Bhawani, S.A., Sulaiman, O., Hashim, R., Ibrahim, M.N., Thin-layer chromatographic analysis of steroids: A review. *Trop. J. Pharm. Res.*, 9, 3, 301–313, 2010.
49. Painuli, S., Semwal, P., Egbuna, C., Mushroom: Nutraceutical, mineral, proximate constituents and bioactive component, in: *Functional Foods and Nutraceuticals: Bioactive Components, Formulations and Innovations*, pp. 307–336, 2020.
50. Hien, B.T.T., Tham, L.X., Quang, D.N., Cattienoids A–C, three novel steroids from the mushroom Tomophagus cattienensis. *Fitoterapia*, 91, 125–127, 2013.
51. Tungmunnithum, D., Thongboonyou, A., Pholboon, A., Yangsabai, A., Flavonoids and other phenolic compounds from medicinal plants for pharmaceutical and medical aspects: An overview. *Medicines*, 5, 3, 93, 2018.
52. Pinacho, R., Cavero, R.Y., Astiasarán, I., Ansorena, D., Calvo, M.I., Phenolic compounds of blackthorn (Prunus spinosa L.) and influence of in vitro digestion on their antioxidant capacity. *J. Funct. Foods*, 19, 49–62, 2015.
53. Yildiz, O., Can, Z., Laghari, A.Q., Şahin, H., Malkoç, M., Wild edible mushrooms as a natural source of phenolics and antioxidants. *J. Food Biochem.*, 39, 2, 148–154, 2015.
54. Panaccione, D.G., Beaulieu, W.T., Cook, D., Bioactive alkaloids in vertically transmitted fungal endophytes. *Funct. Ecol.*, 28, 2, 299–314, 2014.

55. Khatun, S., Islam, A., Cakilcioglu, U., *et al.*, Research on mushroom as a potential source of nutraceuticals: a review on Indian perspective. *Am. J. Exp. Agric.*, 2, 1, 47–73, 2012.
56. Shibu, M.A., Agrawal, D.C., Huang, C.Y., Mushrooms: a Pandora box of cardioprotective phytochemicals, in: *Medicinal plants and fungi: recent advances in research and development*, Medicinal and aromatic plants of the world, vol. 4, D.C. Agrawal, H.S. Tsay, L.F. Shyur, Y.C. Wu, S.Y. Wang (Eds.), pp. 337–362, Springer, New York, 2017.
57. Chong, P.S., Fung, M.L., Wong, K.H., Lim, L.W., Therapeutic potential of Hericium erinaceus for depressive disorder. *Int. J. Mol. Sci.*, 21, 1, 163, 2020.
58. Hidayati, N., Fuad, H., Munandar, H., Zilda, D.S., Sulistyaningtyas, A.R., Nurrahman, N., Ethica, S.N., Potential of fibrinolytic protease enzyme from tissue of sand sea cucumber (Holothuria scabra) as thrombolysis agent. *IOP Conf. Ser.: Earth Environ. Sci.*, 743, 1, 012007, 2021, May, IOP Publishing.
59. Alkarithi, G., Duval, C., Shi, Y., Macrae, F.L., Ariëns, R.A., Thrombus structural composition in cardiovascular disease. *Arterioscler. Thromb. Vasc. Biol.*, 41, 9, 2370–2383, 2021.
60. Badalyan, S.M., Barkhudaryan, A., Rapior, S., The cardioprotective properties of Agaricomycetes mushrooms growing in the territory of Armenia. *Int. J. Med. Mushrooms*, 23, 5, 21–31, 2021, https://doi.org/10.1615/IntJMedMushrooms.2021038280
61. Rathee, S., Rathee, D., Rathee, D., Kumar, V., Rathee, P., Mushrooms as therapeutic agents. *Rev. Bras. Farmacogn.*, 22, 459–474, 2012.
62. Kundakovic, T. and Kolundzic, M., Therapeutic properties of mushrooms in managing adverse effects in the metabolic syndrome. *Curr. Top. Med. Chem.*, 13, 21, 2734–2744, 2013.
63. Ramesh, S., Majrashi, M., Almaghrabi, M., Govindarajulu, M., Fahoury, E., Fadan, M., Dhanasekaran, M., Overview of therapeutic efficacy of mushrooms, in: *Medicinal mushrooms: recent progress in research and development*, pp. 103–141, 2019.
64. Anusiya, G., Gowthama Prabu, U., Yamini, N.V., Sivarajasekar, N., Rambabu, K., Bharath, G., Banat, F., A review of the therapeutic and biological effects of edible and wild mushrooms. *Bioengineered*, 12, 2, 11239–11268, 2021.
65. Cuong, V.T., Chen, W., Shi, J., Zhang, M., Yang, H., Wang, N., *et al.*, The antioxidation and anti-aging effects of Ganoderma lucidum in Caenorhabditis elegans. *Exp. Gerontol.*, 117, 99–105, 2019.
66. Lee, Y.H., Kim, J.H., Song, C.H., Jang, K.J., Kim, C.H., Kang, J.S., *et al.*, Ethanol Extract of Ganoderma lucidum Augments Cellular Anti-oxidant Defense through Activation of Nrf2/HO-1. *J. Pharmacopuncture*, 19, 1, 59–69, 2016.

11

Application of Mushroom in the Regulation of Gut Microbiome and Maintenance of Gut Health

Adeyomoye Olorunsola Israel[1], Charles Oluwaseun Adetunji[2], Oluwafemi Adebayo Oyewole[3]*, Oluwabukola Atinuke Popoola[4] and John Tsado Mathew[5]

[1]Department of Physiology, University of Medical Sciences, Ondo State, Nigeria
[2]Applied Microbiology, Biotechnology and Nanotechnology Laboratory, Department of Microbiology, Edo State University Uzairue, Iyamho, Edo State, Nigeria
[3]Department of Microbiology, Federal University of Technology, Minna, Nigeria
[4]Genetics, Genomics and Bioinformatics Department, National Biotechnology Development Agency, Abuja, FCT, Nigeria
[5]Department of Chemistry, Ibrahim BadamasiBabangida University Lapai, Niger State, Nigeria

Abstract

Although mushrooms have been utilized for more than a millennium for therapeutic purposes, their ability to regulate gut flora and thus promote health has not yet been thoroughly understood. A substantial, mainly unexplored supply of potent novel medicinal compounds is found in mushrooms. The ability of mushroom polysaccharides to control gut microbiota by encouraging the growth of helpful bacteria and reducing pathogen levels has attracted a lot of attention recently. This opens up new applications for mushrooms in the nutraceutical and functional food industries. The majority of the health benefits of mushrooms have been linked to their ability to reverse gut microbiota dysbiosis, as mounting data has linked certain gut microbiota species exposure to health concerns. This chapter covers the uses of mushrooms and how they can function as prebiotics to encourage the growth of healthy gut microbiota and help the host's health.

**Corresponding author*: oadeyomoye@unimed.edu.ng

Charles Oluwaseun Adetunji, Chukwuebuka Egbuna, Oluwatosin Ademola Ijabadeniyi and Samantha C. Karunarathna (eds.) Mushroom Biotechnology for Improved Agriculture and Human Health, (239–250) © 2025 Scrivener Publishing LLC

Keywords: Mushroom, dysbiosis, gut microbiota, polysaccharides, health benefits

Introduction

The composition of intestinal microflora changes constantly due to influences from the immune system, the intestinal mucosa, ingested drugs, nutrition, and the microflora itself. Stressful conditions can quickly reduce microbial diversity and promote the growth of specific bacterial taxa, which can lead to dysbiosis in the gut microbiota [1]. Since stressors and natural variation work together to drive cascades of destabilizing events, the causes of gut dysbiosis frequently remain unknown. Dysbiosis can result from rapid alterations in the populations of gut microbes brought on by oxidative stress, bacteriophage induction, and bacterial toxin release [2]. Numerous diseases, including inflammatory bowel disorders and metabolic disorders including type II diabetes and obesity, are associated with intestinal dysbiosis. For a better knowledge of how the microbiota affects health and illness, it is crucial to characterize the alterations that cause intestinal dysbiosis and identify the microbial taxa that contribute to pathogenic outcomes [3]. Many multifactorial diseases, such as inflammatory, autoimmune, metabolic, neoplastic, and neurodegenerative disorders, have emerged in the recent century. Intestinal dysbiosis, or modifications in the composition and capabilities of the intestinal gut microbiome, has recently been linked to various illnesses [4]. To link dysbiosis to the pathogenesis of prevalent gut diseases, it is necessary to comprehend the mechanisms underlying the de novo production and persistence of dysbiotic microbiome topologies. Additionally, the ability to distinguish between host-microbiome relationships that are causative and secondary microbial changes that occur in tandem with disease progression is required. Mushroom fungi are extremely rare organisms that are an essential and necessary part of the environment. For many ages, mushrooms have been utilized in herbal therapy [5]. Many medicinal mushrooms, such as *Tremella, Rigidoporus, Pleurotus, Coprinus, Ganoderma, Amanita, Calocybe, Cantharellus, Cordyceps, Cortinarius, Grifola, Huitlacoche, Hydnum, Lentinus, Morchella, Agaricus, and Trametes sp.*, are vital for treating a wide range of ailments because of their many metabolic components and nutritional worth. A medicinal mushroom's morphological qualities include its size, color, consistency of the stem and cap, chemical reactions, and fungal attachment mechanism. Edible mushrooms are considered a "new generation food" because of their high nutritional content as well as their biological and functional potential [6]. Proteins and peptides found in mushrooms constitute another notable and significant group of bioactive substances. Mushroom protein concentration, which normally ranges from 19 to 35% on a dry weight basis, is greatly influenced by a number of variables, including species, substrate composition, and harvesting period. Proteins' form and structure determine how they work. The unique characteristics of fungal proteins, which often set them apart from those of animals, plants, and microbes in addition to their biological potential, are a major advantage [7]. According to recent research, lectins, fungal immunomodulatory proteins, ubiquitin-like proteins, and proteins with enzymatic activity such as laccases, ribonucleases, and other enzymes as well as ergothioneine are the most important mushroom bioactive proteins and peptides. Anti-inflammatory, anti-tumor, anti-hypertensive, immunomodulatory, antiviral, antifungal, antibacterial, anti-hypercholesterolemic or anti-hyperlipidemic, anti-tumor, and antioxidant are some of the supposed benefits of these proteins [8]. Hence, this chapter discusses the biological activities of mushrooms as neutraceuticals in the regulation of microbiota and maintenance of gut health.

Gut Microbiome in Health and Disease

Over the course of the last fifteen years, interest in the microbiome has grown. Even while gut bacteria have long been of interest to researchers studying infectious diseases, there has been a great deal of interest in the investigation of the role played by the microorganisms that inhabit the human gut [9]. For instance, multiple studies have found that the gut microbiota can change in composition during cancer, diabetes, liver disease, obesity, and even neurodegenerative diseases. New medications may be discovered in the human intestinal microbiota. One of the main characteristics of inflammatory bowel disease (IBD), ulcerative colitis, and Crohn's disease is persistent, recurrent intestinal inflammation. It is becoming more common and a concern for healthcare systems worldwide [10]. It is thought that a person's genetic predisposition triggers an aberrant and enduring immune response to gut microbes, which in turn causes IBD. IBD is caused by a complicated interaction between immune responses and genetic, environmental, or microbial factors, while the precise etiology of the condition is yet unknown. Human development and growth are influenced by gut microbes. A frequent infection with *Clostridium difficile* is just one example of how microbial composition might shed light on the temporal evolution of microbial communities and vulnerability to microbial ecology disturbances [11]. The gut flora is now known to play a major role in regulating host health. Since microbes have essentially taken over every organ in human bodies, there may be a variety of crosstalk between different organs. The intricate interactions that take place between the host and different microorganisms are being better understood due to the advent of molecular tools and techniques such as metagenomics, metabolomics, lipidomics, and metatranscriptomics [12]. It is now believed that a variety of conditions, including obesity, type 2 diabetes, hepatic steatosis, inflammatory bowel disease (IBD), and numerous malignancies, are influenced by variations in the gut microbiome. This shows that multiple pathways involved in the metabolism of lipids, glucose, and lipid products as well as immunology are impacted. The host's lifestyle and behavior may be reflected in variations in the human gut microbiome, which may also affect the blood levels of disease biomarkers [13]. Understanding the interactions between host characteristics and gut microorganisms is essential for comprehending health and illness. The microorganisms that live in the human gut were studied in new ways using developments in sequencing technology, metagenomic analysis, and bioinformatics techniques [14]. A comprehensive analysis of the gene content of the human gut microbiome revealed that its overall non-redundant genome size was 150 times greater than the human genome. Metagenomic sequencing-based *in silico* predictions are now being sought, compared to, and tested using different 'omics' technologies. A deeper knowledge of the vital interaction at mucosal interfaces will be possible due to the use of meta-omics to describe the human gut microbiome [15]. Numerous diseases are influenced by the ways in which the microbiota interacts with its host. The discovery that a large number of phages, or prokaryotic viruses, which have a high degree of host specificity and target bacteria or archaea, live in the human gut and influence microbial activity, contributes to our growing comprehension of the importance of the microbiome [16].

Additionally, it has been discovered that the gut microbiota exhibits gene exchanges more frequently than anticipated. These cutting-edge research ideas are used to uncover novel information networks vital to human health and gut homeostasis. The amount of studies into the human gut microbiota has multiplied over the last ten years [17]. Previously, the gastrointestinal system was only thought of as a digesting organ, but the development of new technology has caused scientists to reevaluate this notion. The function of the gut microbiome in metabolism, immunological defense, and behavior is gradually coming to light. Our gut microbiota changes with age, environment, stress, food, health, and drug exposure, from modifications *in utero* to those that happen quickly after birth [18]. In addition to being crucial for preserving the host's health, there is mounting evidence that the gut microbiota plays a key role in the development of numerous diseases. Age, genetics, host environment, diet, and other factors can all have an impact on the microbiota's makeup. The hypothesis that diet can impact the structure and function of the microbiome is supported by epidemiological research linking nutrition to the onset of inflammatory bowel disease [19]. Based on these findings, studies are now being carried out to find out how the gut flora and its metabolome are impacted by certain formula diets, an effective therapeutic method for Crohn's disease. Diet alters the microbiota's structure and function, in part through causing the synthesis of tiny chemicals that may have an impact on the emergence of immune-mediated and metabolic diseases [20]. It has been discovered that elements of the human gut microbiome have an impact on a variety of pathological diseases, including diabetes, cancer, and heart disease. HIV infection upsets the delicate balance in regular host-microbe interactions by altering the taxonomic make-up of the gut microbial populations as well as by interfering with regular host defense mechanisms. The numerous studies looking at the fecal microbiota have allowed the human gut microbiota to continue to show its significance in human health and disease [21]. The involvement of the microbial communities found in the small intestine is anticipated to be very relevant given the small intestine's critical role in nutritional absorption, host metabolism, and immunology.

Mushrooms and Their Derived Bioactive Molecules

The mushrooms are well-known for their distinctive flavor and esteemed biological qualities. One of the main bioactive components and a healthy component of mushrooms is its polysaccharide [22]. The structural properties and biological activity of fungal polysaccharides have been the subject of several investigations. It is believed that the many biological functions of polysaccharides are closely related to their structural or conformational properties. In a study, the chemical composition of *Pyropolyporus fomentarius* and its biological activities were investigated [23, 24]. *P. fomentarius* fruiting bodies yielded eight known triterpenoids, four known ergosterols, two rare degraded ergosterols, pyropolyncisterols A and B, and two previously discovered lupane-type pentacyclic triterpenes, 3-formyloxybetulin, and 3-formyloxybetulinic acid. Using spectroscopic studies, the chemical compositions of the fungus were identified. Nine substances demonstrated modest cytotoxicity against human cancer cell lines, whereas polyporenoic acid shown considerable cytotoxicity with IC50 values less than 10 M against SMMC-7721 and A-549. The search for bioactive natural compounds can be greatly aided by the finding of wild edible mushrooms. In a study, eight previously identified chemicals and a novel polyprenylphenol derivative called 2-geranylgeranyl-1,4-benzenediol 1-O-acetate were extracted from the edible wild mushroom *Suillus luteus*. Nuclear magnetic resonance data and high-resolution electrospray ionization mass spectrometry were utilized to determine the structure of the new molecule [25]. An indigenous medicinal fungus to Taiwan called *Antrodia camphorata* has been used to treat inflammation, cancer, and liver damage. The identification of its bioactive components, particularly terpenoids, has been focused on due to its uncommon occurrence and potent pharmacological activity. Since 1995, 162 terpenoids have been described, including triterpenoids, meroterpenoids, sesquiterpenoids, diterpenoids, and steroids. Ergostane-type triterpenoids (antcines) and meroterpenoids (anthroquinonols) are what make *A. camphorata* unique. Terpenoids have anti-cancer, anti-inflammatory, hepatoprotective, anti-diabetic, and neuroprotective properties. Folk practices from the inhabitants of Ailao Mountain in Yunnan Province, China, have included soaking *Phellinus rhabarbarinus* in wine to treat diseases like cough, gastritis, and cancer as well as to prevent infection [26]. After a comprehensive examination of the chemical constituents of *P. rhabarbarinus* fruiting bodies, 11 lanostane triterpenoids have been identified, comprising three novel ones, the phelibarins, and five ergosterols. Ganoderma is a well-known traditional Chinese medicinal mushroom whose principal ingredients include polysaccharides, triterpenoids, and meroterpenoids [27, 28]. In a study, the fruiting bodies of *Ganoderma lingzhi*, a mushroom used for centuries in East Asia as a traditional medicine and dietary supplement, yielded seven hitherto unreported sesquiterpenoids, including six gymnomitranes and a novel type of sesquiterpenoid, along with a polyketide and a known steroid. Tricholimbrins A and B, three unknown steroids, Tricholimbrins CE, and one unknown 4-chromanone derivative were found in the fruiting bodies of Tricholoma imbricatum mushrooms [29, 30]. Tricholimbrin C is a ring-rearranged steroid with an aromatic group that may have originated from ergosterol, whereas tricholimbrin A and B are two polycyclic triterpenoids that have undergone carbon degradation. It has been found that volemolide, 22-trien-7-one, demethylincisterol A3, 3,15-dihydroxyl-(22E,24R)-ergosta-5,8(14), and isocyathisterol are cytotoxic to six human cancer cell lines. The IC50 values of volemolide and Chaxin C were less than 10M, however, the 3-hydroxyl-(22E,24R)-ergosta-5,8,22-trien-7,15-dione and 3-hydroxyl-(22E,24R)-ergosta-5,8,22-trien-7, one shown a preference for cytotoxicity upon HL-60. On the other hand, A-549 was more negatively affected. Termitomyces are well-known free-growing basidiomycete fungi that are both edible and therapeutic. Studies on the health-promoting qualities of Termitomyces have been motivated by their frequent ingestion [31, 32]. The primary polysaccharide component groups in Termitomyces are responsible for many of the health benefits. Alpha-glucan homopolysaccharides are thought to be the main bioactive polysaccharides of Termitomyces, despite the fact that other heteropolysaccharides also possess biological activity. Various Termitomyces species produce polysaccharides with various molecular weights, monosaccharide compositions, and sugar connections.

Health Benefits of Edible Mushroom Associated with Maintenance of Gut Health

Since ancient times, edible mushrooms have been a mainstay of human diets. More than 100 kinds of mushrooms have been grown for their possible health advantages. Recently, there has been interest in edible mushroom polysaccharides (EMPs) because of their potential use in the treatment of cancer, IBD, and obesity [33]. There is mounting evidence that these health hazards are associated with specific types of gut flora. The increasing popularity of plant-based diets has drawn attention to the use of mushrooms and compounds derived from them in the prevention and treatment of chronic illnesses [34]. Among the anti-inflammatory components present in mushrooms are polysaccharides, fatty acids, carotenoids, mycosteroids, biometals, and phenolic and indole compounds. Through a variety of mechanisms, including the regulation of gut microbiota, the enhancement of the intestinal mucosal barrier, the alteration of lipid metabolism, and the activation of specific signaling pathways, mushroom polysaccharides have the potential to improve human health [35]. It is interesting to observe how closely the molecular weight, monosaccharide composition, and kind of glycosidic bond of the fungal polysaccharide are related to these biological processes. By promoting the growth of advantageous gut flora and boosting the production of short-chain fatty acids (SCFA), mushrooms serve as prebiotics and enhance host health [36]. Numerous studies have been conducted recently to examine the diversity and makeup of gut flora and their impact on host health. It is important to research how edible mushrooms interact with the human microbiota because this connection and the ensuing fermentation activity are what give medicinal characteristics to the mushrooms. The principal component, phenolic acids, which have already shown action in the fight against oxidative stress, are just one example of the active ingredients that might modulate processes [37]. The human gut's microbiota ferments some mushroom fungi, releasing butyric acid, propionic acid, and SCFA. Additionally, because they contain mycochemicals with antioxidant properties, mushrooms can be added to a variety of food dishes and utilized as a medicine [38, 39]. Natural foods that help fight pathogens are more popular than ever because of their clear advantages over medical treatments. Crude mushroom extracts have been the subject of numerous investigations as a functional food with antibacterial characteristics. According to beliefs that they improve immune function and promote health, mushrooms and mushroom extracts are of great interest for usage as dietary supplements [6, 40]. Certain types of mushrooms have been demonstrated to trigger the immune system to some degree, especially in *in vitro* experiments. Despite being used extensively for potential health benefits, there are surprisingly few epidemiological and experimental studies that discuss the biological activities of mushrooms after oral administration to humans or animals. Numerous studies have examined the capacity of mushroom fungi to control the activation of mononuclear cells, the phenotypic expression of cytokines, and the corresponding receptors. The anticancer properties of mushrooms have also been the subject of several investigations [41]. These investigations are crucial due to numerous components of mushrooms have the potential to exhibit considerable biological action. Higher *Ascomycetes* and *Basidiomycetes* fungi offer a variety of possible immunological and anticancer activities. They also have important health advantages and a variety of pharmacological effects. In comparison to animal and plant sources, edible mushroom proteins often have a full essential amino acid profile, satisfy nutritional requirements, and offer financial benefits. By exerting antioxidant, anticancer, angiotensin-converting enzyme (ACE), inhibitory, and antibacterial characteristics over animal proteins, mushroom proteins may promote human health [42]. Mushroom peptides, hydrolysates, and protein concentrates are utilized to enhance human health. Traditional dishes can be enhanced with edible mushrooms to boost their protein content and functional qualities. These characteristics highlight the protein found in mushrooms as a low-cost, high-quality protein that can be utilized as a meat substitute, a medication, and a treatment for malnutrition [43]. Since edible mushroom proteins are of excellent quality, are affordable, are widely accessible, and satisfy social and environmental standards, they are appropriate as a sustainable replacement for animal proteins.

Conclusion

Mushrooms are produced at an increasing rate, and their economic value is rising as well. Due to their many health advantages, the nutrient contents, functional foods, nutritional supplements, and traditional medicines made from them are becoming more popular. Mushrooms can interact with the gut microbiome and immune system, mediating both normal physiology and disease susceptibility. They can act as prebiotics altering intestinal microbiota, thereby exerting therapeutic effects. Despite the fact that mushrooms have long been used for food and medicine, study is still ongoing about their potential health advantages, particularly with regard to the regulation of intestinal microbiota and the consequent impact on intestinal disease.

References

1. Weiss, G.A. and Hennet, T., Mechanisms and consequences of intestinal dysbiosis. *Cell. Mol. Life Sci. CMLS*, *74*, 16, 2959–2977, 2017.
2. Tiffany, C.R. and Bäumler, A.J., Dysbiosis: from fiction to function. *Am. J. Physiol. Gastrointest. Liver Physiol.*, *317*, 5, G602–G608, 2019.
3. Levy, M., Kolodziejczyk, A.A., Thaiss, C.A., Elinav, E., Dysbiosis and the immune system. *Nat. Rev. Immunol.*, *17*, 4, 219–232, 2017.
4. Sitkin, S.I., Tkachenko, E.I., Vakhitov, T.Y., Metabolic dysbiosis of the gut microbiota and its biomarkers. *Eksp. Klin. Gastroenterol. = Exp. Clin. Gastroenterol.*, *12*, 12, 6–29, 2016.
5. Patel, D.K., Dutta, S.D., Ganguly, K., Cho, S.J., Lim, K.T., Mushroom-Derived Bioactive Molecules as Immunotherapeutic Agents: A Review. *Molecules (Basel, Switzerland)*, *26*, 5, 1359, 2021.
6. Sousa, A.S., Araújo-Rodrigues, H., Pintado, M.E., The Health-promoting Potential of Edible Mushroom Proteins. *Curr. Pharm. Des.*, 29, 11, 804–823, 2023.
7. Bhambri, A., Srivastava, M., Mahale, V.G., Mahale, S., Karn, S.K., Mushrooms as Potential Sources of Active Metabolites and Medicines. *Front. Microbiol.*, *13*, 837266, 2022.
8. Chen, H.P. and Liu, J.K., Secondary Metabolites from Higher Fungi. *Prog. Chem. Org. Nat. Prod.*, *106*, 1–201, 2017.
9. Cani, P.D., Human gut microbiome: hopes, threats and promises. *Gut*, *67*, 9, 1716–1725, 2018.
10. Ihekweazu, F.D. and Versalovic, J., Development of the Pediatric Gut Microbiome: Impact on Health and Disease. *Am. J. Med. Sci.*, *356*, 5, 413–423, 2018.
11. de Vos, W.M., Tilg, H., Van Hul, M., Cani, P.D., Gut microbiome and health: mechanistic insights. *Gut*, *71*, 5, 1020–1032, 2022.
12. Manos, J., The human microbiome in disease and pathology. *APMIS Acta Pathol. Microbiol. Immunol. Scand.*, *130*, 12, 690–705, 2022.
13. Manor, O., Dai, C.L., Kornilov, S.A., Smith, B., Price, N.D., Lovejoy, J.C., Gibbons, S.M., Magis, A.T., Health and disease markers correlate with gut microbiome composition across thousands of people. *Nat. Commun.*, *11*, 1, 5206, 2020.

14. Lepage, P., Leclerc, M.C., Joossens, M., Mondot, S., Blottière, H.M., Raes, J., Ehrlich, D., Doré, J., A metagenomic insight into our gut's microbiome. *Gut, 62,* 1, 146–158, 2013.
15. Cresci, G.A. and Bawden, E., Gut Microbiome: What We Do and Don't Know. *Nutr. Clin. Pract.:* official publication American Society Parenteral Enteral Nutrition, *30,* 6, 734–746, 2015.
16. Wu, G.D., The Gut Microbiome, Its Metabolome, and Their Relationship to Health and Disease, in: *Nestle Nutrition Institute workshop series*, vol. 84, pp. 103–110, 2016.
17. Milani, C., Duranti, S., Bottacini, F., Casey, E., Turroni, F., Mahony, J., Belzer, C., Delgado Palacio, S., Arboleya Montes, S., Mancabelli, L., Lugli, G.A., Rodriguez, J.M., Bode, L., de Vos, W., Gueimonde, M., Margolles, A., van Sinderen, D., Ventura, M., The First Microbial Colonizers of the Human Gut: Composition, Activities, and Health Implications of the Infant Gut Microbiota. *Microbiol. Mol. Biol. Rev. MMBR, 81,* 4, e00036–17, 2017.
18. Shi, N., Li, N., Duan, X., Niu, H., Interaction between the gut microbiome and mucosal immune system. *Mil. Med. Res., 4,* 14, 2017.
19. Liu, J., Williams, B., Frank, D., Dillon, S.M., Wilson, C.C., Landay, A.L., Inside Out: HIV, the Gut Microbiome, and the Mucosal Immune System. *J. Immunol. (Baltimore, Md.: 1950), 198,* 2, 605–614, 2017.
20. Ruigrok, R.A.A.A., Weersma, R.K., Vich Vila, A., The emerging role of the small intestinal microbiota in human health and disease. *Gut Microbes, 15,* 1, 2201155, 2023.
21. Järbrink-Sehgal, E. and Andreasson, A., The gut microbiota and mental health in adults. *Curr. Opin. Neurobiol., 62,* 102–114, 2020.
22. Ma, J.T., Ning, J.L., Wei, F.L., He, J., Liu, J.K., Chemical constituents from mushroom Geoglossum fallax and their bioactive activities. *Fitoterapia, 163,* 105326, 2022.
23. Zhang, F.L., Shi, C., Sun, L.T., Yang, H.X., He, J., Li, Z.H., Feng, T., Liu, J.K., Chemical constituents and their biological activities from the mushroom Pyropolyporus fomentarius. *Phytochemistry, 183,* 112625, 2021.
24. Yao, L., Lv, J.H., Li, J.P., An, X.Y., Cheng, G.H., Li, C.T., Li, Y., Chemical Constituents from Mushroom Suillus luteus (Agaricomycetes) and Their Bioactivities. *Int. J. Med. Mushrooms, 24,* 11, 63–71, 2022.
25. Kuang, Y., Li, B., Wang, Z., Qiao, X., Ye, M., Terpenoids from the medicinal mushroom Antrodia camphorata: chemistry and medicinal potential. *Nat. Prod. Rep., 38,* 1, 83–102, 2021.
26. Feng, T., Cai, J.L., Li, X.M., Zhou, Z.Y., Li, Z.H., Liu, J.K., Chemical Constituents and Their Bioactivities of Mushroom Phellinus rhabarbarinus. *J. Agric. Food. Chem., 64,* 9, 1945–1949, 2016.
27. Huang, X. and Nie, S., The structure of mushroom polysaccharides and their beneficial role in health. *Food Funct., 6,* 10, 3205–3217, 2015.

28. Zhao, Z.Z., Liang, X.B., Feng, W.S., Wu, Y., Zhi, Y.L., Xue, G.M., Chen, H.P., Liu, J.K., Unusual constituents from the medicinal mushroom *Ganoderma lingzhi*. *RSC Adv.*, *9*, 63, 36931–36939, 2019.
29. Yin, X., Yang, A.A., Gao, J.M., Mushroom Toxins: Chemistry and Toxicology. *J. Agric. Food. Chem.*, *67*, 18, 5053–5071, 2019.
30. Zhang, F.L., Yang, H.X., Wu, X., Li, J.Y., Wang, S.Q., He, J., Li, Z.H., Feng, T., Liu, J.K., Chemical constituents and their cytotoxicities from mushroom Tricholoma imbricatum. *Phytochemistry*, *177*, 112431, 2020.
31. Aoki, S., Aboshi, T., Shiono, Y., Kimura, K.I., Murata, T., Arai, D., Iizuka, Y., Murayama, T., Constituents of the Fruiting Body of Poisonous Mushroom Omphalotus japonicus. *Chem. Pharm. Bull.*, *68*, 5, 436–442, 2020.
32. Liu, J., Wang, Y., Wu, J., Georgiev, M.I., Xu, B., Wong, K.H., Bai, W., Tian, L., Isolation, Structural Properties, and Bioactivities of Polysaccharides from Mushrooms *Termitomyces*: A Review. *J. Agric. Food. Chem.*, *70*, 1, 21–33, 2022.
33. Zhao, J., Hu, Y., Qian, C., Hussain, M., Liu, S., Zhang, A., He, R., Sun, P., The Interaction between Mushroom Polysaccharides and Gut Microbiota and Their Effect on Human Health: A Review. *Biology*, *12*, 1, 122, 2023.
34. Zhang, H., Jiang, F., Zhang, J., Wang, W., Li, L., Yan, J., Modulatory effects of polysaccharides from plants, marine algae and edible mushrooms on gut microbiota and related health benefits: A review. *Int. J. Biol. Macromol.*, *204*, 169–192, 2022.
35. Ayimbila, F., Prayoonthien, P., Inyod, T., Haltrich, D., Keawsompong, S., Bioactive composition and modulatory effects of Hed-Tean-Rad Mushroom, *Macrocybe crassa* on gut microbiota. *3 Biotech*, *12*, 11, 314, 2022.
36. Ma, G., Du, H., Hu, Q., Yang, W., Pei, F., Xiao, H., Health benefits of edible mushroom polysaccharides and associated gut microbiota regulation. *Crit. Rev. Food Sci. Nutr.*, *62*, 24, 6646–6663, 2022.
37. Muszyńska, B., Grzywacz-Kisielewska, A., Kała, K., Gdula-Argasińska, J., Anti-inflammatory properties of edible mushrooms: A review. *Food Chem.*, *243*, 373–381, 2018.
38. Jayachandran, M., Xiao, J., Xu, B., A Critical Review on Health Promoting Benefits of Edible Mushrooms through Gut Microbiota. *Int. J. Mol. Sci.*, *18*, 9, 1934, 2017.
39. Moussa, A.Y., Fayez, S., Xiao, H., Xu, B., New insights into antimicrobial and antibiofilm effects of edible mushrooms. *Food Res. Int. (Ottawa, Ont.)*, *162*, Pt A, 111982, 2022.
40. Sganzerla, W.G., Todorov, S.D., da Silva, A.P.G., Research Trends in the Study of Edible Mushrooms: Nutritional Properties and Health Benefits. *Int. J. Med. Mushrooms*, *24*, 5, 1–18, 2022.
41. Borchers, A.T., Keen, C.L., Gershwin, M.E., Mushrooms, tumors, and immunity: an update. *Exp. Biol. Med. (Maywood, N.J.)*, *229*, 5, 393–406, 2004.

42. Mwangi, R.W., Macharia, J.M., Wagara, I.N., Bence, R.L., The antioxidant potential of different edible and medicinal mushrooms. *Biomed. Pharmacother. = Biomed. Pharmacother.*, *147*, 112621, 2022.
43. Ayimbila, F., and Keawsompong, S., Nutritional Quality and Biological Application of Mushroom Protein as a Novel Protein Alternative. *Curr. Nutr. Rep.*, 12, 2, 290–307, 2023, doi:10.1007/s13668-023-00468-x.

12

Applications of Mushrooms in the Management of Cancers

Tsado Priscilla Yetu[1*], Iseghohi Frances[2], Sani Rabiu Aishat[3], Oluwafemi Adebayo Oyewole[1], Idris Abdullahi Dabban[4], Charles Oluwaseun Adetunji[5], Oluwabukola Atinuke Popoola[6] and Adeyomoye Olorunsola Israel[7]

[1]*Department of Microbiology, Federal University of Technology, Minna, Nigeria*
[2]*Medical Biotechnology Department, National Biotechnology Development Agency, Abuja, Nigeria*
[3]*Department of Microbiology, Ibrahim Badamasi Babangida University, Lapai, Niger State, Nigeria*
[4]*Department of Biological Sciences, The Federal Polytechnic Bida, Niger State, Nigeria*
[5]*Department of Microbiology, Edo State University, Uzairue, Nigeria*
[6]*Genetics, Genomics and Bioinformatics Department, National Biotechnology Development Agency, Abuja, FCT, Nigeria*
[7]*Department of Physiology, Ondo Medical University, Ondo State, Nigeria*

Abstract
Cancer, with its numerous symptoms and high mortality rates, continues to be a global health challenge. The need for novel, tailored anticancer drugs is critical. Mushrooms have attracted attention for their possible medical importance duc to their wide species diversity. They are high in carbohydrates, proteins and polyphenolic compounds. According to recent studies, these fungi contain unique bioactive chemicals with nutraceutical that also with health-promoting effects. Tumor-fighting characteristics' main modes of operation found in some mushrooms include immunomodulation, angiogenesis inhibition, DNA repair disruption, apoptosis induction, and metastasis. Various mushroom species have been discovered to have anti-cancer properties such as *Agaricus bisporus* with anti-tumoral, anti-inflammatory, and antioxidant properties. Anti-inflammatory

**Corresponding author*: tsado.priscilla@st.futminna.edu.ng

Charles Oluwaseun Adetunji, Chukwuebuka Egbuna, Oluwatosin Ademola Ijabadeniyi and Samantha C. Karunarathna (eds.) Mushroom Biotechnology for Improved Agriculture and Human Health, (251–288) © 2025 Scrivener Publishing LLC

ergostanes found in *Antrodia cinnamomea* and cordycepin from *Cordyceps sinensis* influence cancer cells. *Coriolus versicolor* contributes immune-enhancing polysaccharides, while *Ganoderma lucidum* inhibits cancer proliferation, metastasis, and angiogenesis. *Maitake* mushrooms, with their β-glucans and D-fraction, induce apoptosis and inhibit tumor growth. *Shiitake* and *oyster mushrooms* also exhibit anticancer potential. These mushroom-derived components offer promise for innovative cancer therapies, including vaccine-based methods, immune-enhancing polysaccharides, and nanoparticles.

Keywords: Mushrooms, cancers, metabolite

12.1 Introduction

The secondary leading factor of mortality in developed countries, following cardiovascular ailments, is cancer, which poses a serious threat to the entire world [1]. Due to the distinct nature of each form of cancer, a unique treatment plan is required for each, an accurate diagnosis is crucial for efficient therapy. Surgical procedures, therapy involving medications to combat cancer, radiation treatment, and comprehensive therapy (which includes chemotherapy and precision biological treatments) represent just a selection of the numerous treatments accessible for cancer. The existing anti-cancer drugs available today lack precision in their targeting, leading to the emergence of resistance to these drugs and giving rise to various undesirable effects. These effects encompass issues such as gastrointestinal disturbances, irritation of the skin or mucous membranes, discomfort, exhaustion, suppression of bone marrow activity, and weakening of the immune system. These complications can elevate the risk of infections and exert a substantial impact on the overall well-being of individuals undergoing or recuperating from cancer therapies [2]. Reoccurrence and spreading to other parts still happen often, contingent on the specific cancer kind and the disease's progression stage, despite the increased efficacy of traditional customized cancer therapy [3]. Additionally, a lot of patients receive their diagnoses at an advanced stage, which leaves them with few other therapeutic options than restorative care [4]. Complementary and integrative medicine (CIM) is extensively acknowledged and prevalent among individuals diagnosed with cancer. According to studies, 20–50% of cancer patients supplement their standard care with CIM therapies [5]. Hence, it remains essential to develop novel, powerful anti-cancer medications that exhibit minimal toxicity. Natural substances have been emphasized in this regard as anticancer medications [6].

In recent times, the use of medicinal fungi has gained widespread popularity. These mushrooms are not only sought after for their potential in crafting remedies but also for their capacity to complement traditional chemotherapy or radiation treatments. This partnership aims to amplify the effectiveness of the therapies while concurrently diminishing undesirable repercussions. Additionally, these mushrooms offer notable nourishment beyond their therapeutic contributions. About 700 of the 14,000 different kinds of mushrooms have been shown to have therapeutic qualities. The biological functions and methods of action of mushroom chemicals have recently been made known by numerous investigations [7]. Some mushrooms and the substances that make them up have anti-cancer potential. The potential for their utilization as anti-tumor medications or as supplementary agents in cancer therapy has come to light. To determine the advantages of medicinal mushroom extracts in chemotherapy, numerous clinical trials are now being conducted [7].

12.2 Cancer

This is a cluster of conditions characterized by the unchecked proliferation and dissemination of irregular cells [8]. It evolves via the transformation of regular cells into cancerous cells in a series of stages, advancing from a pre-cancerous anomaly to a malignancy during a multi-phase advancement [9]. Cancer emerges due to a range of external elements or agents, encompassing physical triggers like ultraviolet and ionizing radiation, chemical culprits like polluted water, food, transition metals, asbestos, and aflatoxin, as well as the smoke from tobacco. Additionally, there are biological contributors such as specific viruses, bacteria, and parasites that can induce cancer. Numerous factors are acknowledged for escalating the susceptibility to cancer, which comprise dietary problems, particular ailments, insufficient physical activity, obesity, and harmful environmental substances. These factors could synergistically kickstart or amplify the process of carcinogenesis within the human body, establishing cancer as the leading cause of fatality. As per reports from the World Health Organization, the year 2020 witnessed 2.2 million instances of breast cancer, leading to 685,000 fatalities, 2.21 million diagnoses of lung cancer, culminating in 1.18 million deaths, 1.93 million occurrences of colorectal cancer, resulting in 935,000 demises, and 1.09 million reported cases of stomach cancer, ultimately causing 769,000 passings [9].

12.2.1 Types of Cancer

Cancer can be classified into two categories on the basis of tissue affected and on the basis of organ affected.

Category one, based on the affected tissue:

- Carcinomas, this type of cancer are identified by cells that envelop both the inner and outer regions of the organism, including instances like lung, breast, and colon cancer.
- Sarcomas is a type of cancer that exhibits cells situated within bones, cartilage, adipose tissue, connective tissue, muscles, and other supporting tissues.
- Lymphomas are malignancies that initiate in the lymph nodes and tissues of the immune system.
- Leukemias, this type of malignancies initiate in the marrow of bones and frequently gather within the blood circulation.
- Adenomas, these are malignancies that grow within the thyroid, the pituitary gland, the adrenal gland, and other secretory tissues.

Category two, based on the affected organ:

- Cancers of the lip and oral cavity
- Nasopharyngeal cancer
- Bladder cancer
- Non-Hodgkin lymphoma
- Cervical cancer
- Kaposi sarcoma
- Liver cancer
- Lung cancer
- Stomach cancer
- Esophageal cancer
- Colorectal cancer (Mathur *et al.*, 2015)

12.3 Mushrooms

Mushrooms are fungus. The realm of fungi is believed to stand as one of the most abundant life categories, ranking just after insects, encompassing around 5 million distinct types classified into five distinct divisions. A significant number of these types remain undiscovered. The general consensus is that mushrooms represent the reproductive structures that produce spores for larger fungi (also termed macrofungi), primarily falling within the Basidiomycota class. These, along with various Ascomycota species, find utility in customary medicine for the mitigation of ailments [10].

Mushrooms have been regarded as gourmet cuisine throughout history due to their distinct taste and mild flavor. Mushrooms have a cell wall formed of chitin, which is the same material found in lobster shells [11]. The nutritional benefits arise from a significant amount of carbs, fiber, proteins, and minimal fat content. Mushrooms possess water-soluble nutrients (such as B1, B2, B12, and C) along with fat-soluble vitamins (D and E). They also hold advantageous compounds. Numerous types of mushrooms have recently demonstrated their role as compact medicinal production units, producing a plethora of distinct components with extraordinary biological qualities. While they possess an extensive past application in Eastern medicinal practices, only modern investigations corroborate their renowned impact on enhancing overall well-being and vitality. As of late, mushrooms have emerged as an incredible reservoir of nutraceuticals, antioxidative agents, anti-neoplastic substances, prebiotic agents, immune system regulators, inflammation suppressors, heart-friendly elements, antimicrobial agents, and compounds with anti-diabetic attributes [7]. The potential to address a range of medical conditions could be supported by the presence of attributes such as immunostimulatory, antihypertensive, antiviral, antifungal, antibacterial, antioxidants, anti-inflammatory, diabetes-reducing, liver-protective, cholesterol-lowering, anti-allergy, and anti-cancer properties [12].

The significance of countering cancer is particularly crucial due to the continuously growing population of individuals affected by this disease. In the year 2020 alone, a staggering 19.3 million people received a cancer diagnosis, and about 10 million lost their lives due to this ailment. If the trend of increasing cancer incidence persists, it is estimated that there will be around 28.4 million instances of cancer by the year 2040 [13]. Despite intensive attempts to create novel anti-tumor medications and tumor-revealing systems, these diseases's incidence and mortality toll are always on the rise.

12.3.1 Major Bioactive Compounds in Medicinal Mushroom

Medicinal mushrooms have been discovered to impede the proliferation of malignant cells, due to the existence of a range of bioactively potent compounds [7]. This predominantly encompasses lipids, polysaccharides, and proteins alongside an array of low molecular weight compounds such as, alkaloids, lactones, terpenoids, sterols, phenolic agents, and lectins. These substances originate from a range of intermediates in the processes related to fundamental metabolism. Hence, metabolites from therapeutic fungi can be categorized into two groupings.

(i) Metabolites with a substantial molecular weight
(ii) Metabolites with a low molecular weight [10].

12.3.1.1 Metabolites with a Substantial Molecular Weight

The predominant metabolites with a substantial molecular weight present in medicinal mushrooms are primarily proteins, polysaccharides, and polysaccharide-protein complexes/glyco-conjugates (such as glycoproteins, glycopeptides, and proteoglycans). These components are generated via primary metabolism and play a vital role in the ongoing expansion and biomass synthesis of medicinal mushrooms. These metabolites constitute integral elements of both the mycelium and the fruiting body of the fungus.

12.3.1.1.1 Polysaccharides

The cell walls of macrofungi include an abundance of polysaccharides (carbohydrates), which are polymers that belong to structurally varied families of macromolecules. They contain a variety of chemical components with immunostimulating and anti-tumor activities, such as glucans, heteroglucans, glycans, and heteroglycans. Heteroglucans consist of side chains of monosaccharides such as galactose, glucose, arabinose, xylose, mannose, ribose, glucuronic acid, or fructose. These side chains can be combined in different ways. On the other hand, glucans are polysaccharides composed entirely of D-glucose units. Glycans, which are polysaccharides called glycans, have additional units in their core structure alongside glucose. One of the primary constituents of fungal cell walls is β-glucans. These essential β-D-glucans are linear polymers made up of D-glucose units and various monosaccharides. They exhibit a repeating pattern where D-glucose molecules are linked by glycosidic bonds in straight chains [14].

Crucial mushroom polysaccharides exhibiting anti-cancer properties.

• β-glucans

It is believed that β-glucans' anti-tumor effects are mostly achieved via triggering the host's immune system (immuno-enhancing activity/ immune-modulation activity). The majority of the time, tumor cells are not directly impacted by mushroom polysaccharides. Rather, they provide assistance to one or more of the primary biological functions, allowing the host to endure unfavorable biological pressures and enhance its ability to counteract the proliferation of cancerous cells. As outlined by [15], they are classified as biological response modifiers. Figure 12.1 shows the structure of β- glucan.

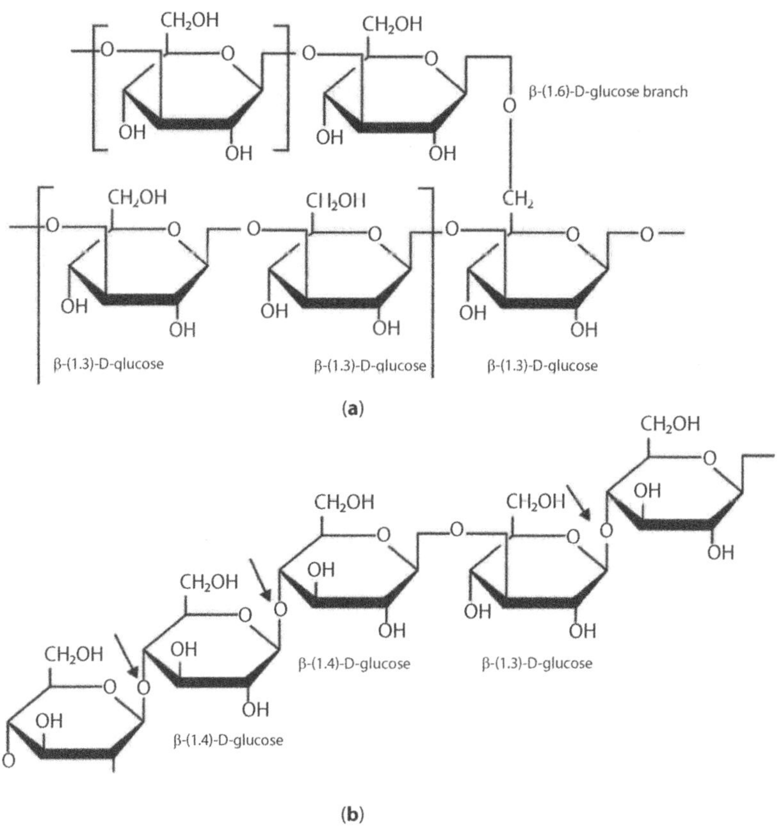

Figure 12.1 β- glucan [16].

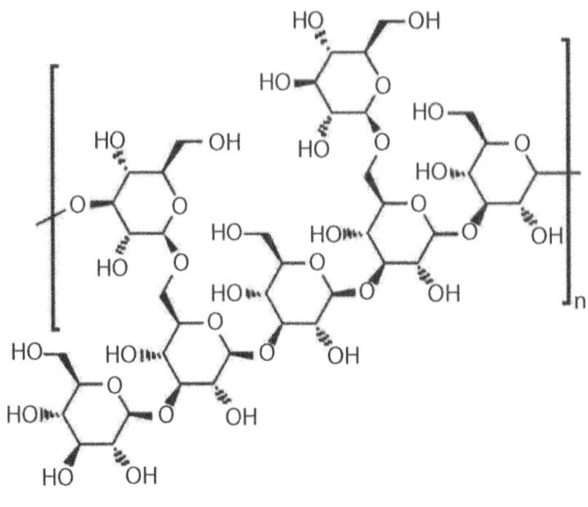

Figure 12.2 Lentinan [18].

• Lentinan
Shiitake mushroom *Lentinula edodes* produces lentinan, a polysaccharide that has anti-tumor properties. Numerous immune cells are activated by lentinan's immunostimulatory activities, which also modulate the release of cytokines and other chemical and signaling messengers that improve immune cells' capacity to engulf substances [17]. Figure 12.2 shows the structure of Lentinan.

• Schizophyllan
Numerous Asian countries have employed this complex carbohydrate, sourced from the fungus *Schizophyllum commune*, within their protocols for treating cancer. Schizophyllan's distinct gene expression kinetics for cytokines, unlike lentinan, have been established, even as they exhibit resemblances in structure and anti-tumor effects. Investigations into Schizophyllan's potential against cancer have transpired, harnessing it for tumor immunotherapy alongside radiotherapy. The efficacy of Schizophyllan against a range of carcinomas and sarcoma cell lines has been evidenced [19]. Figure 12.3 shows the structure of Schizophyllan.

• Ganoderma polysaccharides
Some of the earliest traditional remedies are mushrooms from the genus Ganoderma. *Ganoderma lucidum*, particularly, these complex carbohydrates have demonstrated an ability to enhance the natural killer cells' cytotoxic activity as well as the release of TNF and interferon from macrophages and lymphocytes, respectively [10]. Figure 12.4 shows the structure of Ganoderma polysaccharides.

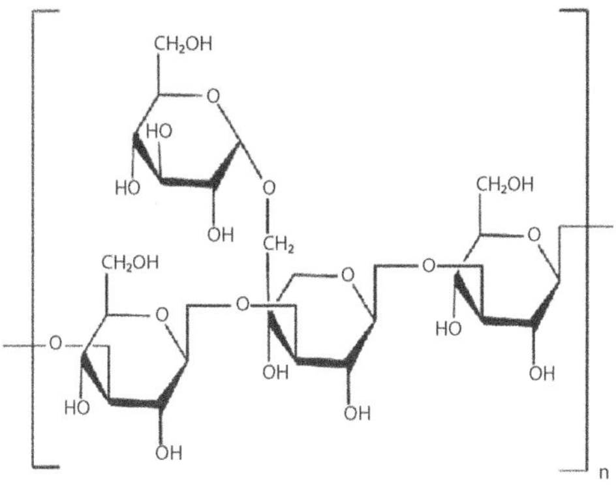

Figure 12.3 Schizophyllan [20].

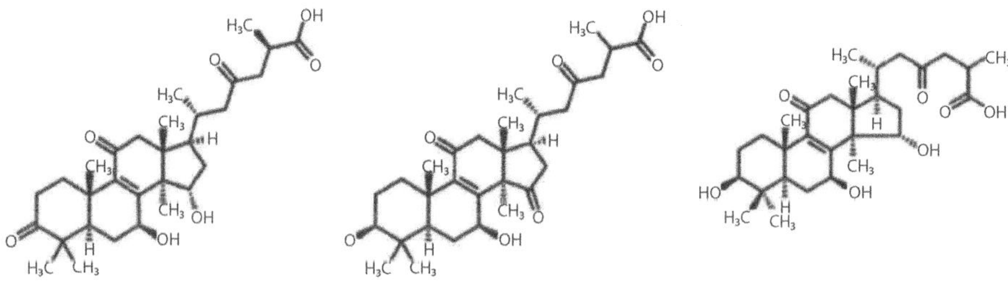

Figure 12.4 Ganoderma polysaccharides [21].

• Polysaccharide-K, (Krestin)

This particular proteoglycan, characterized by 25–38% protein residues within β-glucans, is situated within the polypore fungus *Trametes versicolor*. Its composition primarily comprises glutamic and aspartic acids which are acidic amino acids, coupled with minute quantities of arginine and lysine, which are basic amino acids, and also leucine and valine which are neutral amino acids. Krestin stands as an unparalleled protein-bound polysaccharide, having been employed as a chemo-immunotherapy agent for the treatment of various malignancies, encompassing pancreatic and pulmonary tumors [19]. Figure 12.5 shows the structure of Polysaccharide-K.

• Maitake D-Fraction Polysaccharide

A proteoglycan-linked polysaccharide (proteoglucan) stands as a bioactive compound sourced from the maitake mushroom (*Grifola frondosa*), comprising β-glucan. Maitake's D-fraction exhibits the potential to decelerate the advancement of liver, lung, and breast malignancies in individuals undergoing

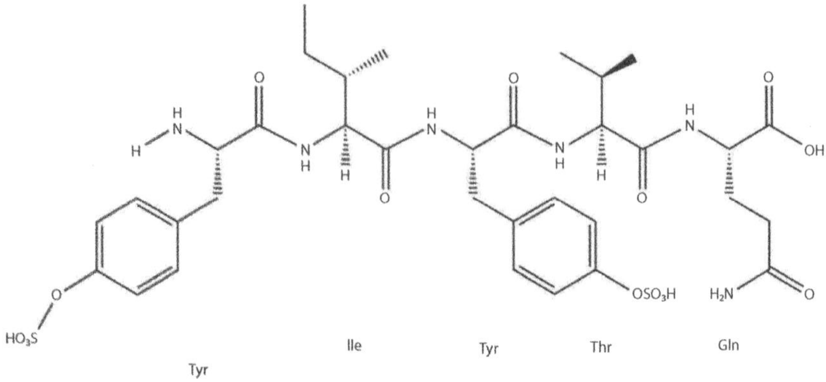

Figure 12.5 Polysaccharide-K [22].

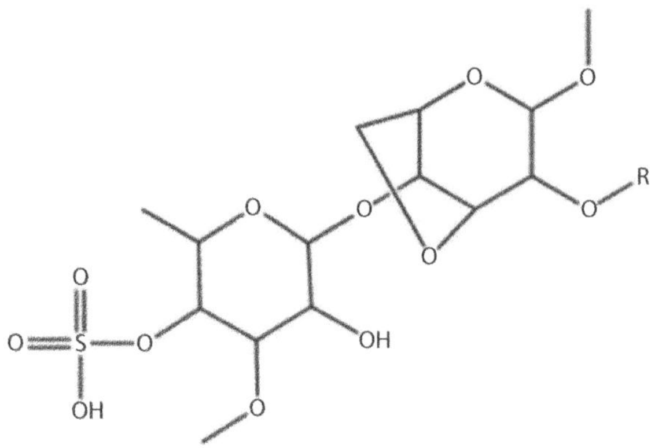

Figure 12.6 Maitake D-fraction polysaccharide [19].

concurrent immunotherapy and chemotherapy. Research indicates that the isolated fraction independently curtailed tumor marker manifestation, impeded metastasis, and enhanced the activity of natural killer (NK) cells [19]. Figure 12.6 shows the structure of Maitake D-fraction polysaccharide.

12.3.1.1.2 Bioactive Mushroom Proteins

Bioactive proteins rank second in abundance among functional components in mushrooms after the polysaccharide fraction that has received the greatest attention from researchers. Numerous proteins with important biological functions are produced by mushrooms, including fungal immunomodulatory proteins (FIPs), lectins, and ribosome-inactivating proteins (RIPs) [23].

• Lectins

Proteins called lectins have the capacity to attach to the carbohydrates on the cell surface, have the capacity to cause cell agglutination, and demonstrate various properties that hinder the growth and spread of cancer cells, leading to inhibition of proliferation and reduction in tumor formation. Numerous mushroom species, such as *Grifola frondosa Agaricus bisporus, Tricholoma mongolicum, Pleurotus ostreatus,* and *Agaricus subrufescens,* are rich in lectins [17]. Figure 12.7 shows the structure of Lectins.

• Ribosome-Inactivating Proteins

By removing some of the adenosine residues from rRNA, ribosome-inactivating proteins (RIPs) possess enzymatic capabilities that hinder the functioning of ribosomes. A fresh ribosome-inactivating protein sourced from the *Hypsizigus marmoreus* mushroom effectively curbed the growth of HepG2 hepatoma, MCF-7 breast cancer cells, and even the activity of HIV-1 reverse transcriptase [14]. Figure 12.8 shows Ribosome-inactivating proteins.

• Fungal Immunomodulatory Proteins

A fresh category of bioactive proteins possessing the capacity to pinpoint immune cells has been discovered in mushrooms, and they are known as fungal immuno-modulatory proteins (FIPs) [23]. Three of the six fungal immunomodulatory proteins (FIPs) that have been isolated and purified are FIP-gsi (*Ganoderma sinensis*), FIP-gts (*Ganoderma tsugae*), and LZ-8 (*Ganoderma lucidum*). Additionally, two more edible mushrooms are FIP-fve (*Flammulina velutipes*) and FIP-vvo (*Volvariella volvacea*) [26].

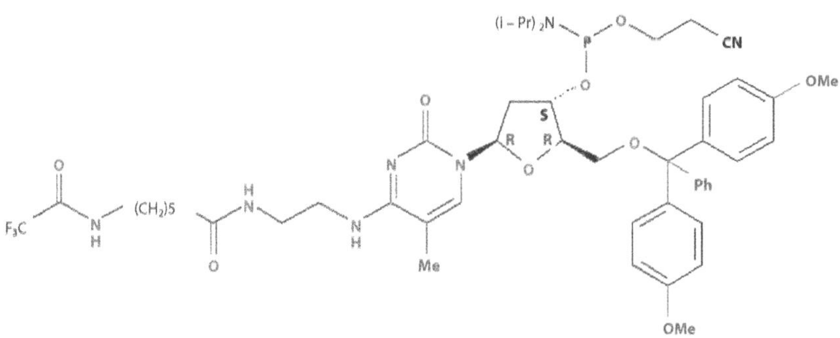

Figure 12.7 Lectins [24].

12.3.1.2 Metabolites with a Low Molecular Weight

Fungi generate an extensive array of intricate small-sized molecules with diverse chemical makeups, such as phenolic substances, polyketides, tri-terpenoids, and distinct steroids characteristic to individual mushrooms [27]. Most of these compounds are labeled as secondary metabolites as they are not involved in the main metabolic processes of fungi, which include generating energy and constructing the fundamental components of proteins, nucleic acids, and cell membranes.

Several metabolites with a low molecular weight derived from medicinal fungi possess inherent anti-cancer properties by impacting multiple cellular signaling pathways (including the NF-B pathway and the mitogen-activated protein kinase pathway [MAPK]) and impeding processes like cellular differentiation, angiogenesis, carcinogenesis, and the spread of cancer cells [10].

• Cordycepin

Cordycepin, which is a nucleoside analog known as 3'-deoxyadenosine, was initially discovered from the *Cordyceps militaris* fungus [28]. Cordycepin affects cells in two ways. Cordycepin suppresses uncontrolled cell growth and division at low dosages, On a broader scale, it hinders the binding

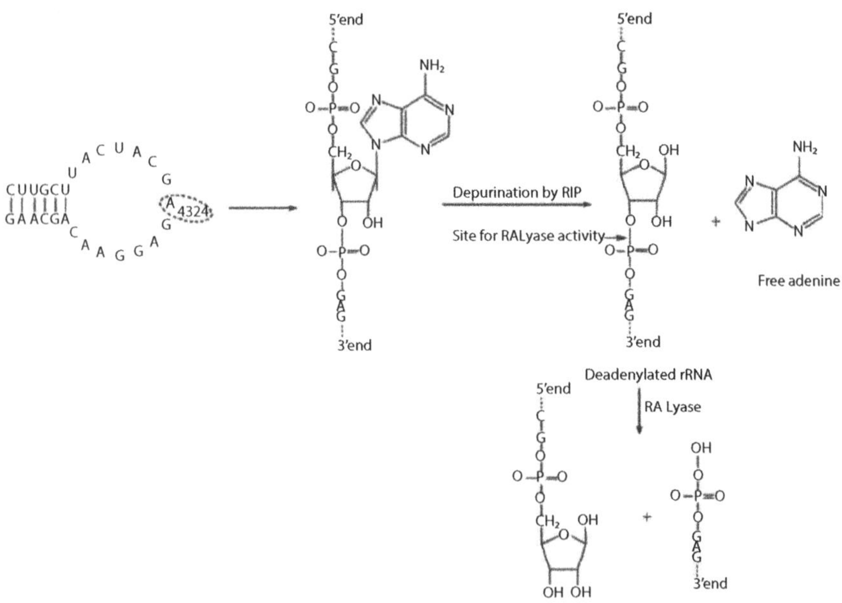

Figure 12.8 Ribosome-inactivating proteins [25].

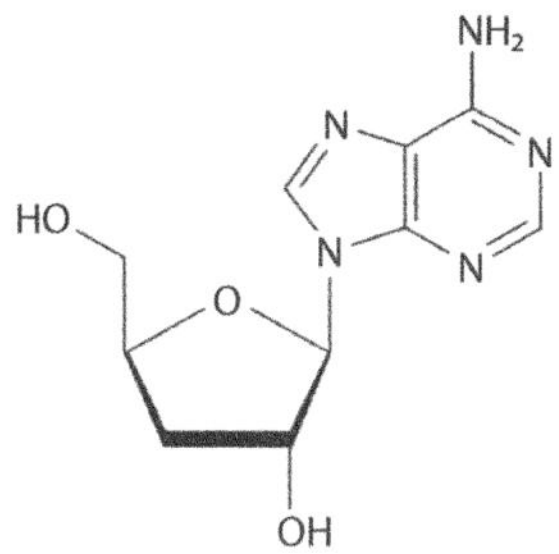

Figure 12.9 Cordycepin [30].

of cells to each other, consequently diminishing their proliferation. These outcomes are probably rooted in a common process, wherein cordycepin disrupts protein processing. At lower levels, cordycepin restrains the creation of mRNA and the formation of proteins. Meanwhile, when administered in higher quantities, cordycepin distinctly impacts the synthesis of proteins [29]. Figure 12.9 shows the structure of Cordycepin.

- Ergosterol

A compound derived from sterols, known as ergosterol peroxide (5,8-epidioxy-22E-ergosta-6,22-dien-3-ol), possesses a range of biological characteristics. These encompass strong abilities to modulate the immune system and counteract tumor development. Inadequate execution of the natural cell self-destruct process (apoptosis) encourages the expansion of tumors and the emergence of resistance against cancer drugs. Hence, agents akin to ergosterol peroxide, which induces apoptotic demise in these malignant cells, hold promise for the control of this disease [31]. Figure 12.10 shows the structure of Ergosterol.

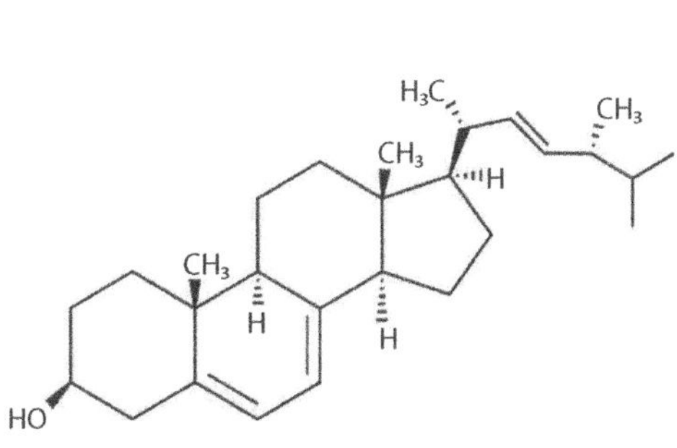

Ergosterol (**15**)

Figure 12.10 Ergosterol [32].

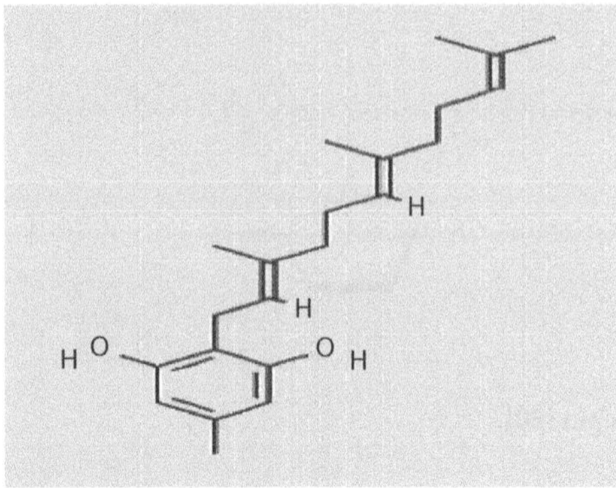

Figure 12.11 Grifolin [33].

• Grifolin

Grifolin has been recognized as a potential medication against cancer, with the ability to halt the progression of tumor formations in cancer cell cultures by inducing programmed cell death in a laboratory setting. Moreover, it was shown that the initiation of caspase-8, -9, and -3 activation along with the liberation of cytochrome c from mitochondria prompted cellular self-destruction. This process was linked with a reduction in Bcl-2 levels and a rise in Bax levels [10]. Figure 12.11 shows the structure of Grifolin.

12.3.1.3 Polyphenolic Compounds (Styrylpyrone-Class of Phenols)

Styrylpyrone-class phenols, a group of polyphenolic pigments produced by medicinal mushrooms and derived from polyketide pathways, have been linked to substantial biological effects, including anti-cancer capabilities [10].

• Hispolon

Through a mitochondrial route that is mediated by reactive oxygen species (ROS), hispolon, obtained from ethanolic extractions of the fruiting structures of *Phellinus linteus*, induces programmed cell death in human gastric cancer cells. This leads to the discharge of Cytochrome C, initiation of caspase-3, and disruption of the potential across the mitochondrial membrane. Hispolon has the capacity to suppress a few proto-oncogenes (MDM2) crucial to carcinogenesis and is a possible anti-tumor drug in bladder and breast malignancies [10].

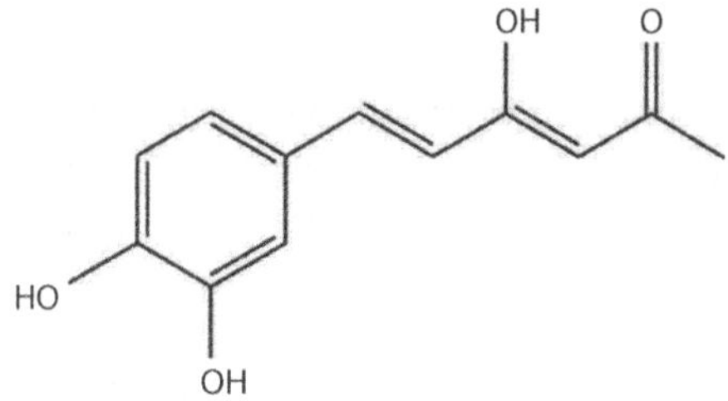

Figure 12.12 Hispolon [34].

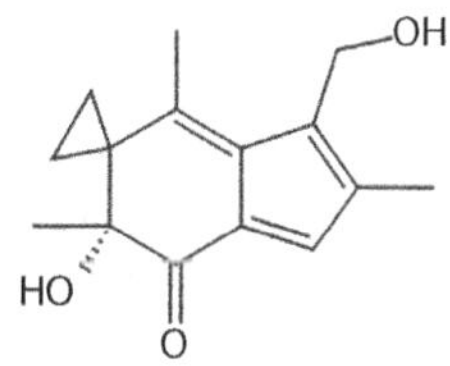

Figure 12.13 Irofulven [35].

• Irofulven

Irofulven possesses the ability to chemically bond with biological large molecules, hindering the production of DNA, and initiating programmed cell death as an agent that adds alkyl groups. Furthermore, it seems to disrupt redox equilibrium, which results in protein oxidation and mitochondrial malfunction, both of which send out a proapoptotic signal. Irofulven seems to exert a greater negative impact on cancerous cells compared to healthy cells, and the maintained balance of redox reactions provides a protective barrier for normal cells against its consequences [10]. Figure 12.12 shows the structure of Hispolon. Figure 12.13 shows the structure of Irofulven.

12.3.2 Mechanisms by Which Certain Mushrooms Exhibit Anti-Cancer Effects

The possible anti-cancer capabilities of mushrooms, especially several kinds of medicinal mushrooms, have drawn considerable interest. The strategies through which compounds from mushrooms hinder the progression of tumors have been extensively investigated in various research endeavors. Several of these bioactive chemicals found in mushrooms have demonstrated promising results in preclinical and clinical research, albeit not all mushroom species exhibit these effects.

12.3.2.1 Immune Modulation

Many compounds present in fungi have the ability to modulate the immune system, enhancing the functioning of immune cells like macrophages, natural killer cells, and T cells. This improved immune system response may aid in greater precision in tumor cell detection and targeting. The immune-stimulating properties of beta-glucans, commonly present in medicinal fungi such as Maitake (*Grifola frondosa*), Reishi (*Ganoderma lucidum*), and Shiitake (*Lentinula edodes*), are well recognized. These substances have the ability to initiate a number of immunological signaling pathways, including the complement system and toll-like receptors (TLRs), increasing immune surveillance and destroying tumor cells [36].

12.3.2.1.1 Mechanism of Immune System Modulation through Anticancer Properties of Mushrooms

According to [37], mushroom chemicals are known to suppress the innate and adaptive immune systems to combat cancer. The interaction amid host receptors for recognizing patterns (PRRs) and molecular patterns associated with invaders (PAMPs) holds immense importance for the immune system's reaction to antigen incursion. Toll-like receptors (TLRs) initiate communication routes that synchronize inherent immunity and trigger adaptive immunity against diverse pathogens, whereas PRRs start innate immunity through pathogen detection [38]. Chemical components present in mushroom cell walls, notably β-glucans, are considered significant pathogen-associated molecular patterns (PAMPs) in initiating the immune response. Receptors known as Dectin-1, which interact with β-glucans, are located on various immune cells such as dendritic cells, macrophages, neutrophils, and monocytes. The binding interaction between Dectin-1 and β-glucans triggers a process of signal transmission, leading to the activation of T cells, mitogen-activated protein kinases (MAPK), and nuclear factor kappa B (NF-kB). This ultimately results in the production of cytokines. Additionally, the pattern recognition receptor (PRR) identifies mushroom compounds by utilizing Dectin-1, toll-like receptor 2 (TLR-2), and complement receptor 3 (CR3). TLR2 binds with PAMP to initiate the adaptive immune response, while the interaction between PAMP and PRR within the innate immune system activates monocytes, dendritic cells, granulocytes, and NK cells. Consequently, immune cells become engaged, cytokines are generated, and molecules responsible for cellular adhesion are manifested [39].

Furthermore, glucans, which are pharmacologically essential mushroom components, are acid resistant and hence can travel past the stomach to the duodenum, where they interact with receptors, activating them to produce lysozyme, reactive oxygen radicals, and nitrogen oxides. These, in turn, induce the creation of cytokines, which activate phagocytes and leukocytes, resulting in local or systemic immunity. The ability of beta-glucans to activate leukocytes is affected not only by their shape, but also by their water solubility, molecular weight, and degree of substitution and branching. Their pharmacological effect is associated with interactions with particular -glucopyranose receptors on leukocytes [39].

• Influence of Mushroom Substances on the Generation of Cytokines
Studies using chemicals from mushrooms have revealed that both *in vivo* and *in vitro* treatments require different effects on the variation of genes and cytokines. Immune cells release cytokines, which are either proteins or glycoproteins, to regulate the specific and non-specific immune systems [40]. Gut defensive elements, Cells like macrophages and dendritic cells, when exposed to mushrooms or compounds found in mushrooms, become active and produce cytokines. These cytokines initiate either local or systemic immune responses. Moreover, ingestion of mushrooms or mushroom-related substances triggers the activation of these cells. Furthermore, the ingestion leads to the stimulation of intestinal epithelial cells, causing them to release a crucial cytokine called IL-7. This cytokine plays a significant role in cancer immunotherapy. When promonocytic THP-1 cells are cultured with an extract from *Agaricus blazei Murill*, a variety of cytokines are secreted. These include the IL-23 subunit of the IL-12 family, IL-1, monocyte chemoattractant protein-1 (MCP-1), granulocyte colony-stimulating factor (G-CSF), and tumor necrosis factor-alpha (TNF-alpha) [39].

• Effect of Mushroom Compounds on the Body's Immune Cells
Once identified by receptors that detect invading agents, bioactive compounds from mushrooms are introduced into cancerous cells either through injection or oral administration. This prompts immune cells to initiate a direct cellular attack on the cancer cells. In line with Wang *et al.* 2020 findings, specific chemicals derived from mushrooms have the capability to prompt the production of cytotoxic T cells and macrophages, while also inducing generalized immune responses. The extracts from *Pleurotus tuberregium* and *P. rhinoceros* have demonstrated their ability to combat tumors by fostering the maturation of lymphocytes and natural killer cells, along with enhancing the proliferation of macrophages and the T helper cell population. This leads to a shift in the CD4/CD8 ratio. These changes are accompanied by an augmentation in the weight and size of the spleen,

which is attributed to an increased presence of various immune cells such as monocytes and granulocytes, among others [39].

To eradicate cancer and enhance the weakened immune system, the intake of elements from mushrooms improves the immune system's ability to counter cancer. This process engages monocytes, macrophages, NK cells, and B cells, along with the discharge of cytokines associated with anti-tumor effects by CTLs [39].

12.3.2.2 Angiogenesis

Angiogenesis is the mechanism by which pre-existing blood vessels are utilized to create fresh ones, supporting the expansion of malignancies. This phenomenon, enabling tumors to prosper and disseminate beyond their initial location, is activated by the "angiogenic switch" [36]. The fundamental role that angiogenesis plays in cancer development is widely recognized. Endothelial cell proliferation, arrangement, migration, and infiltration all contribute to impeding this complex progression, termed the creation of novel and irregular networks of blood vessels [42]. *Coriolus versicolor* (CV) compounds have the capacity to decrease the production of pro-angiogenic cytokines such as IL-6 and IL-8 within the chronic inflammatory

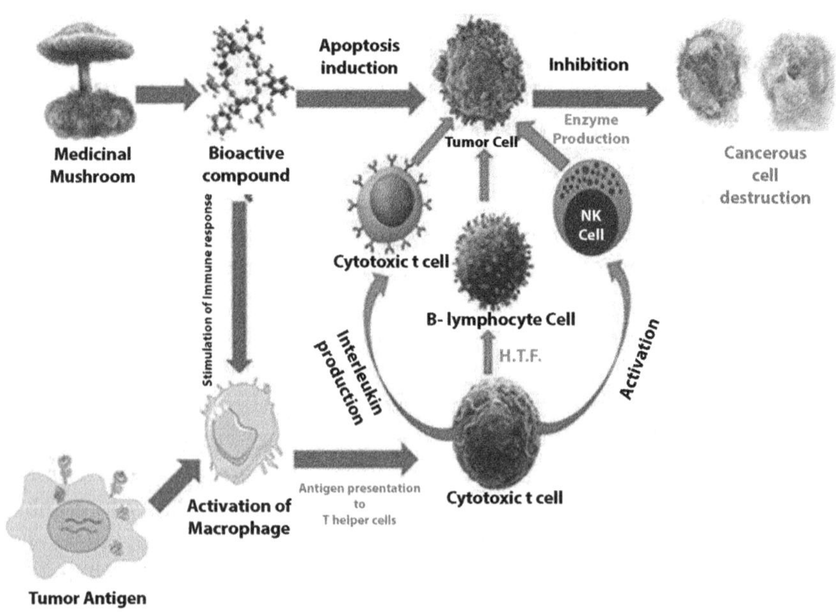

Figure 12.14 Mechanism of modulating the immune system [41].

environment, and this impact is linked to a decreased expression of TLR4 and phospho-IB. Co-culture investigations have demonstrated that the effect of CV on tumor-associated macrophages (TAMs), which are the primary origin of angiogenic factors propelling the shift towards angiogenesis, is accountable for interrupting the interaction between breast cancer cells and macrophages. Protein-bound polysaccharides may indirectly alter TAMs from M2 to M1, thereby potentially decreasing the count of matrix met [42]. Figure 12.14 shows the mechanism mushroom uses in modulating the immune system.

12.3.2.3 DNA Repair Inhibition

Some mushrooms include chemicals that impede cancer cell DNA repair pathways. This interference with DNA repair can make cancer cells more susceptible to other treatments, such as chemotherapy and radiation therapy. In a heterogeneous tumor environment, cancer stem cells or tumor-initiating cells are the most tumorigenic and treatment-resistant cells. Self-inflicted DNA DSB creation and ataxia telangiectasia mutated (ATM) activation are required for cancer cell stemness to be maintained. [43] showcased that the extract of *Ganoderma lucidum* (GLE) lessened the populations of CD44+/CD24- and ALDH1+, alongside reducing the ability of breast cancer stemness to create mammospheres. Furthermore, they observed a decrease in the expression of stemness-associated transcription factors necessary for self-renewal. GLE also inhibited tumor growth in mice when CD44+/CD24- breast cancer cells were implanted. As a result, it was hypothesized that this therapy affects breast cancer stem cells. Further research was conducted to determine whether GLE in combination with carboplatin impacts the capacity of inflammatory breast cancer cells by measuring mammosphere development. The clusters of mammary cells were appropriately disturbed, as anticipated, leading to a significant decrease in the presence of indicators linked with stem cell characteristics, like OCT4 and SOX2 [44].

12.3.2.4 Apoptosis

Apoptosis, also known as programmed cell death, is an inherent mechanism in which harmed or unwanted cells, encompassing cancerous ones, get eradicated through natural means. Specific compounds found in *Agaricus blazei Murill* (Sun Mushroom), like certain chemicals present in mushrooms, have been shown to trigger the process of apoptosis in cancerous cells [46]. Mitochondrial harm leads to the liberation of pro-apoptotic proteins, like apoptosis-inducing factor (AIF), which triggers cell death not reliant on caspase activity. In order to investigate whether Agarol induces the movement of AIF from mitochondria to the nucleus within A549 cells, a treatment involving 1 g/ml of Agarol was administered for a duration of 8 hours, subsequently measuring the expression of AIF protein within nuclear segments. Agarol, a compound derived from ergosterol found in *Agaricus blazei*, exhibits tumoricidal properties. Subsequent to the treatment, a noticeable upswing in AIF translocation to the nucleus occurred. The reduction of AIF gene expression was effectively achieved through targeted knockdown in both A549 and HSC-3 cells using AIF-specific siRNA, yielding lowered AIF protein levels within the cells. Furthermore, the introduction of siRNA greatly diminished Agarol-induced apoptosis within the cells, implying that the caspase-independent apoptosis provoked by Agarol in A549 and HSC-3 cells stems from the discharge of AIF from mitochondria to the nucleus [47]. Figure 12.15 shows the mechanism of DNA repair inhibition.

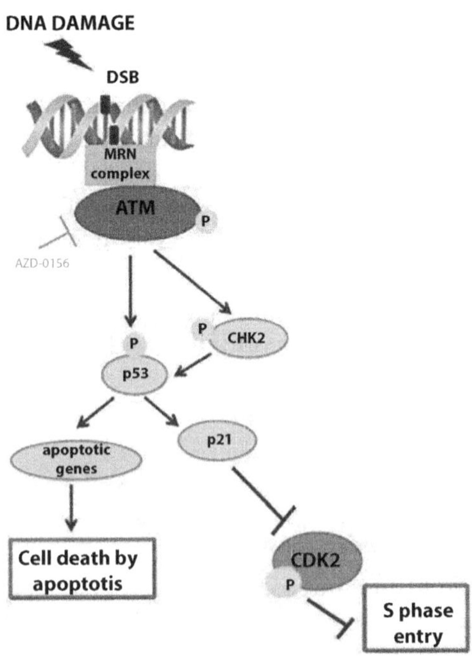

Figure 12.15 Mechanism of DNA repair inhibition [45].

12.3.2.5 *Metastasis*

The dissemination of cancer cells from the primary tumor to various parts of the organism is denoted as metastasis [49]. Mushroom chemicals can obstruct this process via mechanisms such as:

Agaricus blazei Murrill's β-d-glycans function as enhancers of the immune system, prompting the growth and triggering of white blood cells (natural killer cells, plasma cells, and T cells) and certain white blood cell types (macrophages). These stimulated cells have the capability to generate signaling molecules, mainly Interleukin-8 (IL-8) and Interleukin-6 (IL-6), along with Tumor Necrosis Factor-alpha (TNF-). TNF- initiates the activation of nuclear factor kappa B (NF-kB) within endothelial cells, which is accountable for the activation of cell adhesion molecules like E-selectin, a contributor to the spread of tumors. The low molecular weight polysaccharide from *Agaricus blazei* (LMW-ABP) is able to hinder the functions of TNF- and other signaling molecules implicated in NF-kB activation, thereby reducing NF-kB stimulation and the initiation of E-selectin [50]. Figure 12.16 shows the mechanism of apoptosis of cancer cells.

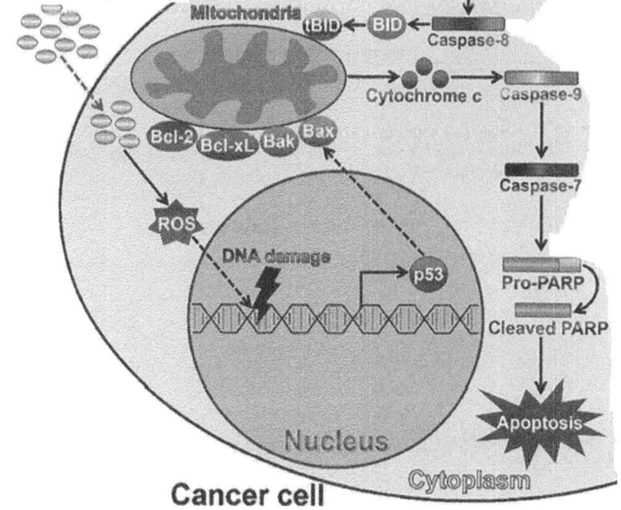

Figure 12.16 Mechanism of apoptosis [48].

12.3.3 Some Mushrooms and Their Anti-Cancer Properties

12.3.3.1 *Agaricus bisporus*

The edible basidiomycete mushroom species *Agaricus bisporus* is a member of the Agaricaceae family and is indigenous to grasslands in North America and Europe. *Agaricus bisporus* stands as one of the widely consumed mushrooms worldwide and is cultivated in over 70 nations. Its premature state is recognized by an unopened cap and can emerge in either ivory or tan shades. The tan variation is also denoted as the chestnut fungus, Italian tan fungus, Swiss tan fungus, Roman tan fungus, or champignon fungus, or push. The ivory type is also acknowledged as the typical fungus, push, ivory fungus, or champignon. When mature, with an expanded cap, it is frequently traded as portobello, portabella, or portobella, and turns brown. *Agaricus bisporus* represents a beneficial reservoir of amino acids, unsaturated fats, vitamins B and C, sterols, phenolic and indole compounds, beta-glucans, ergosterol, ergothioneine, vitamin D, and flavonoids, all of which are present in different quantities, *A. bisporus* has significant anti-inflammatory, anti-oxidant, anti-bacterial, immunomodulatory, and anti-tumor effects. Figure 12.17 shows the mechanism of metastasis. One of the bioactive elements of *A. bisporus* is conjugated

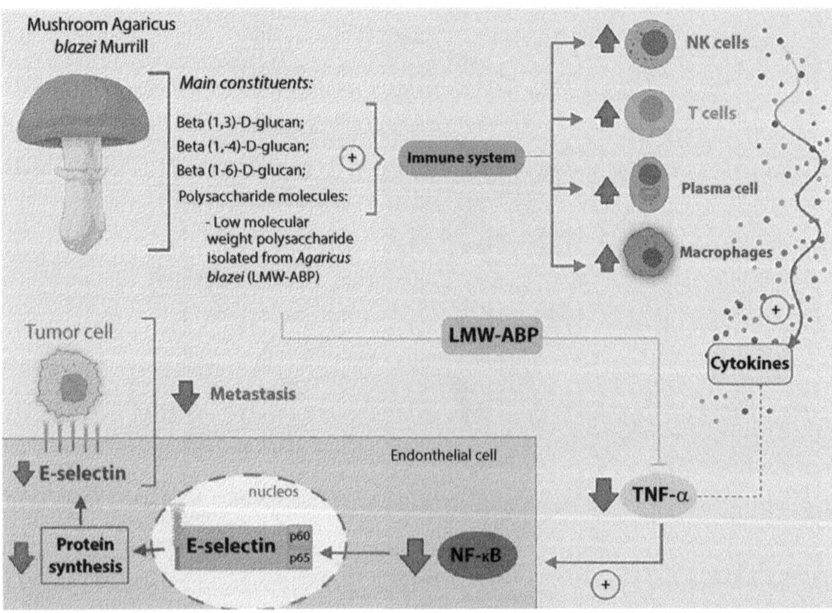

Figure 12.17 Mechanism of metastasis [50].

linoleic acid, which has a crucial role in chemoprevention and the control of carcinogenesis. In actuality, *A. bisporus* exhibits intriguing promise for preventing malignancies [51]. Conjugated linoleic acid sourced from extracts of white button mushrooms could potentially decrease the growth of MCF7 cells through the inhibition of testosterone-triggered cell proliferation, while it is unlikely to affect the growth of non-cancerous MCF-10A cells. In addition, this extract had the ability to inhibit the formation of tumors in naked mice carrying MCF-7 xenografts [52]. Numerous studies have examined the immune-modulating impacts of *A. bisporus*-derived polysaccharides, primarily focusing on β-glucans. Within this scope, there is evidence documenting the potential of *A. bisporus*-isolated linear β-(16)-d-glucan (B16) to prompt a shift in macrophage orientation towards the M1 phenotype. This phenotype is linked to anti-cancer effects, as it generates pro-inflammatory markers (IL-1, TNF-, CoX-2). In the same vein, the ability to induce the production of nitric oxide, interleukin-6, and tumor necrosis factor (TNF) was exhibited by two distinct polysaccharide portions (ABP-1 and ABP-2) acquired from *A. bisporus*, partially through the activation of the nuclear factor-kB in murine macrophages. Additionally, both fractions were effective at reducing the proliferation of human BC MCF7 cells, most likely also because of their effects on macrophages. In both animal experiments and human clinical trials, lectins found in Agaricales and other medicinal fungi demonstrated therapeutic effects against a variety of malignancies. Lectins' precise anti-cancer mechanisms, however, are still not completely understood. It is known that cytotoxicity and apoptosis are brought on by lectins' preferential binding to sugars found on the membranes of cancer cells. Lectins have the capability to impact the generation of interleukins by disturbing the functioning of the immune system. Additionally, they possess the capacity to bind with ribosomes, leading to changes in the cell's proteome and inhibiting the synthesis of proteins [52]. Moreover, the probable inhibitory effect on cell growth caused by lectins from *A. bisporus* is likely a result of the inhibition of protein entry into the cell's central region. This inhibition takes place subsequent to the movement of lectins towards the outer boundary of the nucleus, where they hinder the import of proteins that rely on specific sequences [53]. A recent discovery involved the detection and attachment of mannose by the lectin-like protein from *A. bisporus* to the cellular membranes of MCF7. This interaction triggered apoptosis when present in high concentrations (100 g/mL) and hindered cell growth at lower concentrations. What is intriguing is that due to their capacity to link with glycoproteins and glycolipids, mushroom lectins might serve as a promising reservoir for detecting anomalous glyco-markers on the cancer cell surface. For example, markers like T/Tn antigens or Lewis/Lewis X. This finding could potentially contribute to early detection, enhanced prognoses, consistent monitoring, and improved adaptability to the advancement and evolution of cancer treatments [54].

12.3.3.2 Antrodia cinnamomea

A valuable and rare edible and medicinal fungus called *A.cinnamomea* grows slowly and exclusively on decaying hollow trunks of *Cinnamomum kanehirae Hayata* (Lauraceae). Due to its scarcity and high market demand, the mushrooms' fruiting bodies are quite pricey. This situation has prompted the innovation of new cultivation methods to ensure a steady supply without excessive costs [55]. *A.Cinnamomea* is also held in high esteem for containing active compounds antioxidant, anti-cancer, anti-inflammatory, liver-protecting, anti-hypertensive, anti-hyperlipidemic, and immune-regulating properties. Glycoproteins, terpenoids, polysaccharides, benzene derivatives, lignans, ubiquinone derivatives, as well as maleic and succinic acid derivatives have all been identified as bioactive constituents within *A.cinnamomea* [56]. Ergostanes, unique to the reproductive structures, and lanostane, occurring in both reproductive structures and cultivated fungal threads, stand out as significant terpenoids in *A.cinnamomea.* Noteworthy to mention is Anticin A, which comprises an array of bioactive fungal elements. It has demonstrated remarkable effects against inflammation and tumors in studies [56]. In the context of breast cancer, Anticin-A impeded the shift from epithelial to mesenchymal traits in MCF7 cells that possess human estrogen receptors, as well as in MDA-MB-231 cells that lack estrogen receptors but have triple-negative characteristics. This achievement involved boosting the quantities of epithelial markers like E-cadherin and occludin while reducing the presence of mesenchymal markers such as N-cadherin and vimentin. At a molecular level, the functioning of Anticin-A appears to hinge on the p53 pathway. It triggers miR-200, a suppressor of ZEB1, a key player in the epithelial-mesenchymal transition (EMT) process by diminishing E-cadherin expression [57].

12.3.3.3 Cordyceps sinensis

C. sinensis (Ophiocordyceps sinensis (Berk.) along with *C. militaris* are the most renowned and extensively studied among over 680 species within the Cordyceps genus [58]. Numerous active compounds such as mannitol, ergosterol, and altered nucleosides are present in Cordyceps mushrooms.

However, cordycepin holds the primary functional element responsible for therapeutic properties, which is shared by both fungi and has been recognized for its anticancer effects on various malignancies [59]. Cordycepin, characterized by the absence of oxygen at the 30-position of its ribose component, operates as a nucleoside analog, curtailing the polyadenylation of mRNA in cancer cells [52].

12.3.3.4 *Coriolus versicolor*

In the Western world, *Coriolus versicolor*, also called Turkey tail and scientifically known as *Trametes versicolor*(L.)Lloyd, represents its present nomenclature. Belonging to the Polyporaceae family [52], this fungus finds application across Asia, North America, and Europe. Its role in traditional Chinese medicine is acknowledged, boasting a range of health benefits. In Asian countries, particularly China, this fungus has gained the reputation as a "miracle herb" for enhancing overall health, vitality, and lifespan. Since 1987 in China and 1977 in Japan, extracts from *C. versicolor* have been granted approval for combined cancer treatment alongside chemotherapy or radiotherapy [51]. Within *Coriolus versicolor*, an assortment of active compounds exists, encompassing terpenes, proteins, peptides, amino acids, purpurins, and more. Among these, PSP and PSK stand out as the most extensively investigated bioactive elements, wielding significant influence. Krestin was observed to exert direct and indirect cytotoxic effects on cancer cells *in vitro* [52], this therefore leads to the conclusion that PSP and PSK can enhance the body's innate immune response, culminating in anti-tumor advantages. However, the precise mechanism behind this action remains elusive. In particular, the inhibitory impact of *C. versicolor* extract on the proliferation of T47D, MCF7, and MDA-MB231 cells is comparable to that of the anti-cancer drug mitomycin C. The mechanism for achieving these results involves the upregulation of p53, downregulation of Bcl-2, and initiation of apoptotic cell demise [52].

12.3.3.5 *Ganoderma lucidum*

Ganoderma lucidum stands as one of the most renowned medicinal fungi worldwide. Its application extends to fostering health and extending lifespan. Among its medicinal attributes are its capability to lower blood sugar, regulate the immune system, reduce blood pressure, combat diabetes, counteract oxidation, control lipid levels, deter mutagenesis, combat aging, ward off microbes, and safeguard the liver [51]. Furthermore, it is widely employed as a complementary element in the therapy of various cancers [60].

These biological effects are believed to be orchestrated by the complex carbohydrates and secondary compounds housed within *G. lucidum*. Examples of these secondary compounds encompass ganoderic acids, ganodermic acid, ganodermic alcohols, lucidones, lucinedic acids, ergosterol, 5,6-dehydroergosterol, ergosterol peroxide (EP), and palmitic acid. These compounds manifest qualities like counteracting cancer, inhibiting metastasis, inducing cell death, and obstructing enzyme activity. Meanwhile, polysaccharides, particularly α-1,3, β-1,3, and β-1,6-D-glucans, and ganoderan, exhibit anti-angiogenic traits and bolster the immune system [60]. Consequently, these dual types of molecules play a principal role in the anti-cancer attributes of *G. lucidum*, functioning through stifling cell proliferation, restraining metastasis, impeding invasion, prompting apoptosis, and concurrently fostering immune regulation, immune stimulation, antioxidant defense, and inflammation reduction [61]. The current market houses an excess of 100 products grounded in *G. lucidum*, spanning nutraceuticals, supplements, functional foods, mycopharmaceuticals, and cosmeceuticals. Recently, an extract derived from *G. lucidum*'s sporoderm-penetrating spores (ESG) demonstrated its potential in curbing 4T1 tumor growth within living organisms, as opposed to in a lab environment. In this context, ESG substantially augmented both the population of cytotoxic T cells (Tc) and the ratio of Tc to helper T cells (Th) in the bloodstream of mice afflicted with tumors. A similar boost in Tc levels was also evident in the lymphocytes infiltrating the tumors. Notably, ESG brought about significant downregulation of programmed cell death protein-1 (PD-1) within the spleen, and cytotoxic T lymphocyte antigen-4 (CTLA-4) within the tumor. This indicates that ESG could effectively rejuvenate the T cell equilibrium by reversing the state of exhaustion through the suppression of inhibitory immune checkpoints [62].

12.3.3.6 Grifola frondosa (Maitake)

The mushroom species belonging to the Polyporales order, specifically *G. frondosa*, has a rich historical background of being used for medicinal purposes. Depending on its origin, it goes by various names: in the USA and Canada, it is called "sheep's head," "king of mushrooms," "hen-of-the-woods," and "cloud mushroom." Meanwhile, in Japan, it is referred to as "Maitake," translating to "dancing mushroom." This unique name might be attributed to its distinct appearance, characterized by petal-like growths on a thick, pale stem. Rather than solely growing on deciduous trees like beech chestnut, elm, and maple, it commonly forms clusters near the base of broadleaf trees, particularly oak trees [63]. Research has demonstrated a diverse array of health benefits associated with *G. frondosa*, encompassing, antitumoral antiviral, immunomodulation, anti-inflammatory, and antidiabetic effects. Beyond its considerable nutritional and nutraceutical value, the key components behind these positive health impacts are glucans, specifically the D-fraction. This complex glucan consists of roughly 30% protein and displays a distinct structure featuring a primary 1,6 main chain with more 1,3 branches than 1,6 branches, as well as a 1,3 main chain with 1,6 branches. Noteworthy bioactive compounds like the X-fraction, Grifolan, the MZ-fraction, and MT-glucan further supplement the D-fraction's presence within Maitake [63]. Crucially, the D-fraction within *G. frondosa* has been identified as the primary factor contributing to its anti-tumoral effects. This capability stems from its ability to influence immune responses and directly impede the growth and proliferation of various human cancer cells [63]. Extensive investigation into cancer research has centered around the Maitake D-fraction (MDF). According to *in vitro* experiments involving hormone-dependent MCF7 cells, MDF facilitates the release of cytochrome C from mitochondria, leading to cell dysfunction and apoptosis. Additionally, research on TNBC MDA-MB231 cells indicates that MDF modifies the expression of genes tied to cell growth, proliferation, and advancement. Furthermore, MDF exercises control over migratory and metastatic processes by enhancing levels of the E-cadherin protein, which fosters cell-substrate adhesion. MDF also curbs cell motility by restructuring the actin cytoskeleton and inhibiting MMP2 and MMP9, two enzymes associated with invasive behavior. Notably, *G. frondosa* polysaccharides manage cell viability by suppressing pro-survival pathways mediated by PI3K-Akt and ERK, while promoting the expression of pro-apoptotic proteins. Adding to this, the Maitake D-Fraction Pro4x has shown the ability to extend lifespan in mice with breast tumor xenografts, accomplishing this by restraining cancer invasiveness, angiogenesis, and carcinogenesis [57]. Given these findings, there is substantial potential for utilizing the medicinal properties of *G. frondosa* in the treatment of various breast tumor subtypes, including triple-negative breast cancer, hormone-dependent breast cancer, and hormone-independent breast cancer [64].

12.3.3.7 Lentinula edodes (Shitake)

Following *Agaricus bisporus*, the second most prevalent edible fungus accessible globally is *Lentinula edodes* [65]. It finds extensive employment in both Western and, more freshly, Eastern cuisine. Aside from its delightful taste, this mushroom provides advantageous effects for human health.

To illustrate, it possesses anti-oxidative qualities due to phenolic substances and ergothioneine, aids in lowering cholesterol due to ergosterol, β-glucans, and eritadenine, and contributes to reducing hypertension due to an assortment of peptides, including lenthionine [66]. Conversely, Shitake has attracted therapeutic consideration due to its immune-adjusting and antiviral attributes, along with its potent anticancer influence on various cancer types, mainly attributed to its glucan element lentinan [52]. A peptide (latcripin-7A) separated from *L. edodes* prompted cell cycle cessation and reduced mitochondrial membrane potential, causing programmed cell death in MCF7 and MDA-MB-231 breast cancer cells. Moreover, within the same cell groupings, the peptide curbed migration and increased autophagy without affecting the endurance of MCF-10A normal breast cells [67]. In the past, extracts of *L. edodes* mycelium and fruiting bodies have been exhibited to possess qualities that impede cell proliferation and encourage programmed cell death in MCF-7 cells. Furthermore, the ethyl acetate partition of *L. edodes* triggered cell cycle halt in MCF7 and MDA-MB-453 cells by elevating Bax and p21 concentrations while diminishing cdk4 and cyclin D1 concentrations [52].

12.3.3.8 Pleurotus ostreatus (Jacq.)

The fungal family Agaricaceae within the Basidiomycetes category comprises *Pleurotus ostreatus,* commonly recognized as the oyster mushroom [68]. Roughly 40 mushroom species belong to the *Pleurotus* genus, which ranks among the most extensively cultivated and ingested worldwide [69]. The ethanolic extract of *P. ostreatus* has been extensively studied for its antioxidant, antibacterial, and anticancer attributes. The phytochemicals present in *P. ostreatus* have also undergone examination. Additionally, 32 bioactive compounds identified in oyster mushrooms were evaluated for their potential to attach to the EGFR, PR, and NF-B proteins, which are excessively expressed in breast cancer [69]. The bioactive constituents present in *P. ostreatus* have been correlated with effects such as hypocholesterolemic, scavenging of free radicals, antioxidant, antiatherogenic, antitumor, and immunomodulatory impacts. These include α-glucans, β-glucans, resveratrol, lentanin, lipopolysaccharides, concanavalin A, the natural statin mevinolin, and numerous others [69]. Some of these effects have recently been associated with *P. ostreatus'* anticancer properties [69]. The methanol extract of *P. ostreatus* fruiting bodies hindered the growth of MDA-MB-231 and MCF7 cells by prompting cell cycle arrest in the G0/G1 phase, elevating the expression of the genes p21, p53, p27, and p19, and reducing the expression of genes encoding the E2f transcription factor 1,

PCNA, CDK4, CDK6, and Transcription factor DP-1 [52]. The ethanolic extract of *P. ostreatus* consistently curbed growth and proliferation in MCF7 cells while prompting oxidative stress and apoptotic cell demise. Recent in silico investigations suggest that *P. ostreatus* might be a valuable reservoir of bioactive compounds contributing to its notable antibacterial antioxidant, and anticancer effects. The six interconnected glucans present in *P. ostreatus* extracts raised the cytotoxicity of natural killer cells against breast and lung cancer cells, a phenomenon linked to the activation of nitric oxide and interferon, achieved through the overexpression of KIR2DL. The addition of Interleukin-2 boosted the compound's cytotoxic attributes [52]. Administering an oral dose of *P. ostreatus* ethanolic extract to rats with carcinogen-induced tumors led to a substantial reduction in tumor volume and an increase in body weight, with no discernible alterations in the animals' eating, drinking, or behavioral habits. Figure 12.18 shows some mushrooms with anti-cancer activities. The presence of

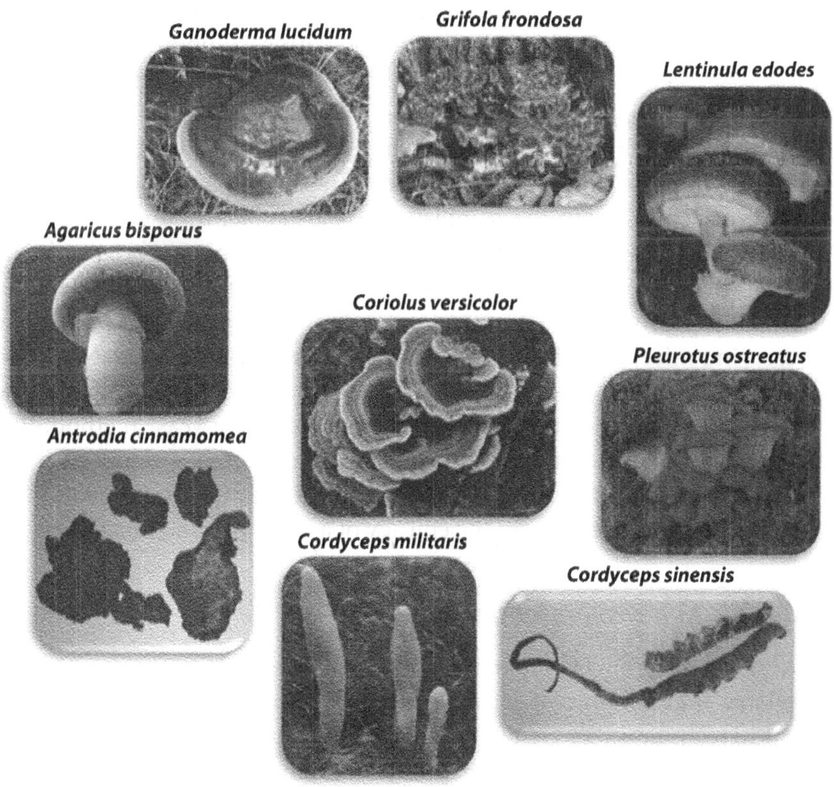

Figure 12.18 Some mushrooms with anti-cancer activities [52].

ergosterol, likely operating as an anti-estrogen blockade within receptor-mediated pathways, also brought down the expression of estrogen and progesterone receptors in ethanol extracts of P. ostreatus. This inhibition of COX2 expression triggered a cascade of prostaglandins, thereby suppressing tumor growth [70]. Within MCF7 cells, P. ostreatus polysaccharides hindered angiogenesis by downregulating the expression of VEGF.

12.3.4 Exploring Novel Mushroom-Derived Therapies for Cancer

12.3.4.1 Vaccinotherapy

In both Belgium and the United States, they developed vaccine compositions that possess preventive attributes against liver and cervical cancers associated with hepatitis B and human papillomavirus infections. There is currently no commercially available vaccine designed to treat pre-existing tumors, metastases, or relapses. Vaccines' anticancer activities can be sufficiently improved by adding immunomodulating compounds with natural and synthetic origins. Thus, based on self-descriptive glycopeptides sourced from Ehrlich's cancer, Sarcoma 37, LLC, L1210, and B16 cellular lineages, it can be inferred that doses and delivery plans for the *L. edodes* polysaccharide fraction were developed, along with plans for combining it with a vaccine. When the complement is available, these formulations enhance the ability of lymphocytes to destroy cells, boost the metabolic function of macrophages in the abdominal cavity, and increase the cell-destroying potential of blood serum in both normal and Sarcoma 37-carrying mice. Recently, it was found that an immune-regulating protein called Ling Zhi-8, sourced from the mycelia of G. lucidum, prompts the activation of dendritic cells. Through an exclusive yeast method, the genetically engineered Ling Zhi-8 protein was extensively produced. Using a preclinical model of tumors, it was proven that the previously mentioned protein significantly enhances the efficacy of a DNA vaccine for cancer [2].

12.3.4.2 Nanovectors for Drug Delivery

Because of their environmental importance, methods involving "green" chemistry and bioprocesses are now more favored than physical and chemical techniques for creating nanoparticles. In this regard, living things have a great deal of potential for producing nanoparticles. By employing *Pleurotus florida* as a reductant and a photo-irradiation process, irregularly shaped gold nanoparticles were created. It was proposed that the reduction of ions into nanoparticles was caused by flavins (flavoproteins), which are present in the mushroom extract. When subjected to sunlight, flavins soak up energy from photons and are capable of engaging in reduction-oxidation processes. The resultant nanoparticles demonstrated the ability to harm cancer cells A-549, K-562, HeLa, and MDA-MB, while showing no impact on healthy cells [2].

12.4 Conclusion

Edible and medicinal mushrooms have gained recognition for their nutritional and health advantages, including potential protection against various cancers. These mushrooms hold compounds with dual potential for cancer prevention and treatment, as well as immune system enhancement investigations employing cancerous cell cultures and animal simulations have elucidated the molecular mechanisms responsible for the cancer-fighting characteristics of mushrooms. However, the translational use of mushroom bioactives in clinical settings remains constrained, necessitating Phase-I/II studies to assess toxicity, therapeutic effects, and pharmacokinetics. Mushrooms exert their impact on tumors through diverse means, like fortifying antioxidant defenses, absorbing carcinogens, inducing tumor cell cytotoxicity, inhibiting angiogenesis, and enhancing immune responses. Combined with conventional cancer therapies, mushrooms can alleviate side effects. Their immunomodulatory attributes make them valuable in vaccinotherapy and nanovector-based drug delivery, introducing novel avenues in cancer treatment. Medicinal mushrooms, with bioactive compounds, offer promise in supplementary cancer treatment by activating immunity and controlling cancer-related pathways. Unleashing their full potential in cancer combat requires further exploration, biotechnology integration, and robust clinical investigation.

References

1. Habtemariam, S., Trametes versicolor (Synn. Coriolus versicolor) Polysaccharides in Cancer Therapy: Targets and Efficacy. *Biomedicines*, *8*, 135, 2020, doi: 10.3390/biomedicines8050135.
2. Ivanova, T.S., Krupodorova, T.A., Barshteyn, V.Y., Artamonova, A.B., Shlyakhovenko, V.A., Anticancer Substances Of Mushroom Origin. *Exp. Oncol.*, *36*, 58–66, 2014.

3. Miller, K., Leticia, N., Mariotto, A., Rowland, J., Yabroff, R., Alfono, C., Jemal, A., Kramer, J., Siegel, R., Cancer Treatment and Survivorship Statistics, 2019. *CA: Cancer Clin.*, *69*, 5, 363–385, 2019.
4. Haun, M., Estel, S., Rucker, G., Friederich, H., Villalobos, M., Thomas, M., Hartmann, M., Early Palliative Care for Adults with Advanced Cancer. *Cochrane Database Syst. Rev.*, 6, 6, 2017.
5. Ng, J.Y., Nault, H., Nazir, Z., Complementary and Integrative Medicine Mention and Recommendations: A Systematic Review and Quality Assessment of Lung Cancer Clinical Practice Guidelines. *Intgr. Med. Res.*, *10*, 1, 100452, 2021.
6. Loele, G., Chieffallo, M., Occhiuzzi, A., Deluca, M., Garofalo, A., Ragno, G., Grande, F., Anticancer Drugs: Recent Strategies to Improve Stability Profile, Pharmacokinetic and Pharmacodynamics properties. *Molecules*, *27*, 17, 5436, 2022.
7. Park, H., Review Current Uses of Mushrooms in Cancer Treatment and Their Anticancer Mechanisms. *Int. J. Mol. Sci.*, *23*, 10502, 2022, https://doi.org/10.3390/ijms231810502.
8. Mathur, G., Sumitra, N., Pramod, K.S., Cancer: An Overview. *Acad. J. Cancer Res.*, *8*, 1, 01–09, 2015.
9. Panda, S.K., Gunanidhi, S., Shasank, S.S., Luyten, W., Anticancer Activities of Mushrooms: A Neglected Source for Drug Discovery. *Pharmaceuticals*, *15*, 176, 2022, https://doi.org/10.3390/ph15020176.
10. De Silva, D.D., Sylvie, R., Françoise, F., Ali, H.B., Kevin, D.H., Medicinal mushrooms in supportive cancer therapies: an approach to anti-cancer effects and putative mechanisms of action. *J. Fungal Divers.*, *55*, 1–35, 2012, DOI 10.1007/s13225-012-0151-3.
11. Patel, S. and Goyal, A., Recent developments in mushrooms as anti-cancer therapeutics: a review. *J. Biotechnol.*, *2*, 1–15, 2012, DOI 10.1007/s13205-011-0036-2.
12. Kosanić, M., Ranković, B., Rančić, A., Stanojković, T., Evaluation of metal concentration and antioxidant, antimicrobial, and anticancer potentials of two edible mushrooms *Lactarius deliciosus* and *Macrolepiota procera*. *J. Food Drug Anal.*, *24*, 3, 477–484, 2016.
13. Sung, H., Ferlay, J., Siegel, R.L., Laversanne, M., Soerjomataram, I., Jemal, A., Bray, F., Global Cancer Statistics, GLOBOCAN estimates of incidence and mortality worldwide for 36 cancers in 185 countries, *CA: Cancer J. Clin.*, *71*, 3, 209–249, 2020.
14. Bhambri, A., Srivastava, M., Mahale, V.G., Mahale, S., Karn, S.K., Potential Sources of Active Metabolites and Medicines. *Front. Microbiol.*, 13, 837266, 2022, dio:10.3389/fmicb.2022.837266.
15. Vetvicka, V., Teplyakova, T.V., Shintyapina, A.B., Korolenko, T.A., Effects of medicinal Fungi-derived β-Glucan on Tumor progression. *J. Fungi*, *794*, 250, 2021.

16. Abuajah, C.I., Functional components and medicinal properties of Food, in: *Bioactives molecules in food*, pp. 1–34, 2017.
17. Zhang, B.-B., Guan, Y.-Y., Hu, P.-F., Chen, L., Xu, G.-R., Liu, L., Cheung, P.C.K., Production of Bioactive Metabolites by Submerged Fermentation of the Medicinal Mushroom *Antrodia Cinnamomea*: Recent Advances and Future Development. *Crit. Rev. Biotechnol.*, *39*, 541–554, 2019.
18. Bisen, P.S., Baghel, R.K., Sanodiya, B.S., Thakur, G.S., Prasad, G.B.K.S., Lentinus edodes:A Macrofungus with Pharmacological Activities. *Curr. Med. Chem.*, *17*, 2419–2430, 2010.
19. Macharia, J.M., Zhang, L., Mwangi, R.W., Rozmann, N., Kaposztas, Z., Varjas, T., Sugar, M., Alfatafta, H., Pinter, M., Bence, R., Are Chemical Compounds in Medicical Mushrooms Potent against colorectal cancer carcinogenesis and Anti-microbial growth? *Cancer Cell Int.*, *22*, 379, 2022, https://doi.org/10.1186/s12935-022-02798-2.
20. Stoica, R., Moscovici, M., Lakatos, E.S., Cioca, L., Expolysaccharides of fungal Origin: Properties and Pharmaceutical Applications. *Processes*, *11*, 2, 335, 2023.
21. Janardhanan, K.K., Ravikumar, K.S., Karuppayil, S.R., Mushroom Bioactives: Potential sources for Anti-cancer Drug Development. *Int. J. Appl. Pharm.*, *4*, 40–45, 2020.
22. Phytotechlabs, PSK-alpha, 2023, https://images.app.goo.gl/DuJbW1FWkVGiHPxi8.
23. Xu, X., Yan, H., Chen, J., Zhang, X., Bioactive proteins from mush- rooms. *J. Biotechnol. Adv.*, *29*, 667–674, 2011.
24. Butt, A.H. and Khan, Y. D. "CanLect-Pred: A Cancer Therapeutics Tool for Prediction of Target Cancerlectins Using Experiential Annotated Proteomic Sequences," in *IEEE Access*, 8, 9520–9531, 2020, doi: 10.1109/ACCESS.2019.2962002.
25. Munish, P., Kaur, I., Gupta, R.C., Ribome Inactivating Proteins from Plants Inhibiting Viruses. *Virol. Sin.*, 26, 6, 357–365, 2011, https://www.virosin.org/article/doi/10.1007/s12250-011-3223-8.
26. Li, Q., Wang, X., Chen, Y., Lin, J., Zhou, X., Cytokines expression induced by Ganoderma sinensis fungal immunomodulatory proteins (FIP-gsi) in mouse spleen cells. *J. Appl. Biochem. Biotechnol.*, *162*, 1403–1413, 2010.
27. Ferreira, I.C.F.R., Vaz, J.A., Vasconcelos, M.H., Martins, A., Compounds from wild mushrooms with antitumor potential. *Anti-Cancer Agents Med. Chem.*, *10*, 424–436, 2010.
28. Werapan, B., Nutaratat, P., Ariyaphuttarat, S., Prathumpai, W., Cordycepin production by the potential fungal strains Cordyceps militaris BCC 2819 and Cordyceps cicadae BCC 19788 in submerged culture during batch and Fed-batch fermentation. *Afr. J. Biotechnol.*, 21, 10, 483–503, 2022, DOI: 10.5897/AJB2022.17514.
29. Yoon, S., Park, S., Park, Y., The anti-cancer properties of Cordycepin and their underlying mechanisms. *Int. J. Mol. Sci.*, *19*, 10, 3027, 2018.

30. Parunyakul, K., Srisuksai, K., Charoenlappanit, S., Phaonakrop, N., Roytrakul, S., Fungfuang, W., Metabolic impacts of cordycepin on hepatic proteomic expression in streptozotocin-induced type1 diabetic mice. *PLoS One*, e0256140, 2021, DOI:10.1371/journal.pone.0256140.
31. Kang, J., Janf, J., Kumar, S., Lee, M., Nho, C.W., Dongyun, S., Jin, M., Kim, K.M., Choi, C., Seung, H.O., Ergosterol peroxide from Changa mushroom (inonotus obliquus) exhibits anti-canceractivity by down-regulation of the β-catenin pathway in colorectal cancer. *J. Ethnopharmacol.*, 173, 303–312, 2015, http://dx. doi.org/10.1016/j.jep.2015.07.030.
32. Luza, L., Gual, A., Dupont, J., Cheminform Abstract:The Partial Hydrogentation of 1, 3 –Dienes Catalyzed by Soluble Transition-Metal Nanoparticles. *ChemCATChem*, 6, 702–710, 2014, Dio:10.1002/cctc.201300673.
33. Jentsch, G.N., Zhang, X., Magolan, J., Efficient synthesis of cannabigerol, Grifolin, andpiperogalin via Alumina-promoted Allylation. *J. Nat. Prod.*, *83*, 9, 2587–2591, 2020.
34. Chang, H., Sheu, M., Yang, C., Lu, T., Chang, Y., Peng, W., Huang, S., Huang, Analgesic effects and the Mechanisms of Anti-inflammation of Hispolon in Mice. *Evid. Based Complement. Alternat. Med.*, *8*, 2011, Doi.1093/ecam/nep027.
35. Kelner, M., McMoris, T.C., Rojas, J.R., Estes, L.A., Suthipinijtham, Synergy of Irofulven in combination with various anti-metabolites, enzyme inhibitors, and miscellaneous agents in MV522 lung carcinoma cells: marked interaction with gemcitabine and 5-fluorouracil. *Invest. New Drugs*, *26*, 407–415, 2008.
36. Pathaka, M.P., Pathakb, K., Saikiab, R., Gogoib, U., Ahmadc, M.Z., Patowaryb, P., Das, A., Immunomodulatory effect of mushrooms and their bioactive compounds in cancer: A comprehensive review. *Biomed. Pharmacother.*, *149*, 112901, 2022.
37. Guggenheim, A.G., Wright, K.M., Zwickey, H., Immune Modulation From Five Major Mushrooms: Application to Integrative Oncology. *Intgr. Med.*, *13*, 1, 32–44, 2014.
38. Amarantemendes, G.P., Adjemian, S., Branco, L.M., Zanetti, L.C., Weinlich, R., Bortoluci, K.R., Pattern Recognition Receptors and the Host Cell Death Molecular Machinery. *Front. Immunol.*, *16*, 9, 2018, https://doi.org/10.3389/fimmu.2018.02379.
39. Ayeka, P.A., Review Article Potential of Mushroom Compounds as Immunomodulators in Cancer Immunotherapy: A Review. Evidence-Based Complementary Altern. Med., 2018, 1–9, 2018, https://doi.org/10.1155/2018/ 7271509.
40. Kaiser, G., Cytokines important in innate immunity, Libre texts Biology, 2022, https://bio.libretexts.org/Bookshelves/Microbiology/Microbiology_(Kaiser)/Unit_5%3A_Innate_Immunity/11.4%3A_Early_Induced_Innate_Immunity/11.3C%3A_Cytokines_Important_in_Innate_Immunity.

41. Saad, A.M., Sitohy, M.Z., Omar, B.A., El-Saadony, M.T., Sitohy, B., Mushroom; Chemistry, Bioactive Components, and Application, in: *Current Topics in Functional Food*, p. 11, 2022, DOI: http://dx.doi.org/10.5772/intechopen.104182.
42. Jędrzejewski, T., Pawlikowska, M., Sobocińska, J., Wrotek, S., COVID-19 and Cancer Diseases—The Potential of Coriolus versicolor Mushroom to Combat Global Health Challenges. *Int. J. Mol. Sci.*, *24*, 5, 4864, 2023, Doi: 10.3390/ijms24054864.
43. Chen, J., Liu, S., Su, Y., Zhang, X., ALDH1+ stem cells demonstrate more stem cell-like characteristics than CD44+/CD24–/low stem cells in different molecular subtypes of breast cancer. Transl. Cancer Res., 9, 3, 1652, 2020, DOI: 10.21037/tcr.2020.01.53.
44. Rios-Fuller, T.J., Ortiz-Soto, G., Lacourt-Ventura, M., Maldonado-Martinez, G., Cubano, L.A., Schneider, R.J., Martinez-Montemayor, M.M., Ganoderma lucidum extract (GLE) impairs breast cancer stem cells by targeting the STAT3 pathway. Oncotarget, 9, 35907–35921, 2018.
45. Suárez-Arroyo, I.J., Acevedo-Díaz, A., Ríos-Fuller, T.J., Ortiz-Soto, G., Vallejo-Calzada, R., Reyes-Chea, J., Maldonado-Martínez, G., Schneider, R.J., Martínez-Montemayor, M.M., Ganoderma lucidum enhances carboplatin chemotherapy effect by inhibiting the DNA damage response pathway and stemness. *Am. J. Cancer Res.*, *12*, 3, 1282–1294, 2022.
46. Carrassa, L. and Damia, G., DNA damage response inhibitors: Mechanisms and potential applications in cancer theraphy. *Cancer Treat. Rev.*, *60*, 139–151, 2017.
47. Tong, Z., Chu, G., Wan, C., Wang, Q., Yang, J., Meng, Z., Du, L., Yang, J., Ma, H., Multiple Metabolites Derived from Mushrooms and Their Beneficial Effect on Alzheimer's Diseases. *Nutrients*, *15*, 12, 2758, 2023, https://doi.org/10.3390/nu15122758.
48. Shimizu, T., Kawai, J., Ouchi, K., Kikuchi, H., Osima, Y., Hidemi, R., Agarol, an ergosterol derivative from Agaricus blazei, induces caspase-independent apoptosis in human cancer cells. *Int. J. Oncol.*, *48*, 4, 1670–1678, 2016, https://doi.org/10.3892/ijo.2016.3391.
49. Wang, M., Jiang, S., Zhang, Y., Li, P., Wang, K., The Multifaceted Roles of Pyroptotic Cell Death Pathways in Cancer. *Cancers*, *11*, 1313, 2019, doi: 10.3390/cancers11091313.
50. Welch, D.R. and Hurst, D.R., Defining the Hallmarks of Metastasis. *Cancer Res.*, *79*, 3011–27, 2019, doi: 10.1158/0008-5472.CAN-19-0458.
51. Bertollo, A.G., Mingoti, M.E.D., Plissari, M.E., Betti, G., Roman Junior, W.A., Luzardo, A.R., Ignácio, Z.M., Agaricus blazei Murrill mushroom: A review on the prevention and treatment of cancer. *Pharmacol. Res. Mod. Chin. Med.*, *2*, 10003, 2022.
52. Venturella, G., Ferraro, V., Cirlincione, F., Gargano, M.L., Medicinal Mushrooms: Bioactive Compounds, Use, and Clinical Trials. *Int. J. Mol. Sci.*, *22*, 634, 2021.

53. Gariboldi, B.M., Marras, E., Ferrario, N., Vivona, V., Prini, P., Vignati, F., Perletti, G., Biochar and Laccase for the Removal of Synthetic Dyes: A Review. *Int. J. Mol. Sci.*, 24, 10120, 2023, https://doi.org/10.3390/ ijms241210120.
54. Zhao, S., Gao, Q., Rong, C., Wang, S., Zhao, Z., Liu, Y., Xu, J., Immunomodulatory Effects of Edible and Medicinal Mushrooms and Their Bioactive Immunoregulatory Products. *J. Fungi*, *6*, 269, 2020.
55. Poiroux, G., Barre, A., van Damme, E.J.M., Benoist, H., Rougé, P., Plant Lectins Targeting O-Glycans at the Cell Surface as Tools for Cancer Diagnosis, Prognosis and Therapy. *Int. J. Mol. Sci.*, *18*, 1232, 2017.
56. Zhang, M., Yiran, Z., Lijuan, Z., Tian, Q., Mushroom polysaccharide lentinan for treating different types of cancers: A review of 12 years studies in China. *Prog. Mol. Biol. Transl. Sci.*, *163*, 297–329, 2019.
57. Li, H.-X., Wang, J.-J., Lu, C.-L., Gao, Y.-J., Gao, L., Yang, Z.-Q., Review of Bioactivity, Isolation, and Identification of Active Compounds from *Antrodia cinnamomea*. *J. Bioeng.*, 9, 494, 2022.
58. Wong, J.H., Ng, T.B., Chan, H.H.L., Liu, Q., Man, G.C.W., Zhang, C.Z., Guan, S., Ng, C.C.W., Fang, E.F., Wang, H., Mushroom Extracts and Compounds with Suppressive Action on Breast Cancer: Evidence from Studies Using Cultured Cancer Cells, Tumor-Bearing Animals, and Clinical Trials. *Appl. J. Microbiol. Biotechnol.*, *104*, 4675–4703, 2020.
59. Lee, D., Lee, W.-Y., Jung, K., Kwon, Y.S., Kim, D., Hwang, G.S., Kim, C.-E., Lee, S., Kang, K.S., The Inhibitory Effect of Cordycepin on the Proliferation of MCF-7 Breast Cancer Cells, and Its Mechanism: An Investigation Using Network Pharmacology-Based Analysis. *J. Biomol.*, *9*, 407, 2019.
60. Wei, C., Khan, M.A., Du, J., Cheng, J., Tania, M., Leung, E.L.-H., Fu, J., Cordycepin Inhibits Triple-Negative Breast Cancer Cell Migration and Invasion by Regulating EMT-TFs SLUG, TWIST1, SNAIL1, and ZEB1. *Front. Oncol.*, *12*, 898583, 2022.
61. Martínez-Montemayor, M.M., Ling, T., Suárez-Arroyo, I.J., Ortiz-Soto, G., Santiago-Negrón, C.L., Lacourt-Ventura, M.Y., Valentín- Acevedo, A., Lang, W.H., Rivas, F., Identification of Biologically Active *Ganoderma Lucidum* Compounds and Synthesis of Improved Derivatives That Confer Anti-Cancer Activities *in Vitro*. *Front. Pharmacol.*, *10*, 115, 2019.
62. Puri, V., Nagpal, M., Singh, I., Singh, M., Dhingra, G.A., Huanbutta, K., Dheer, D., Sharma, A., Sangnim, T.A., Comprehensive Review on Nutraceuticals: Therapy Support and Formulation Challenges. *Nutrients*, *14*, 4637, 2022.
63. Su, J., Su, L., Li, D., Shuai, O., Zhang, Y., Liang, H., Jiao, C., Xu, Z., Lai, Y., Xie, Y., Antitumor Activity of Extract From the Sporoderm-Breaking Spore of Ganoderma Lucidum: Restoration on Exhausted Cytotoxic T Cell With Gut Microbiota Remodeling. *Frontiers in Pharmacology*, 9, 1034, 2018.
64. Wu, J.Y., Siu, K.C., Geng, P., Bioactive Ingredients and Medicinal Values of *Grifola Frondosa* (Maitake). *J. Foods*, *10*, 95, 2021.
65. Alonso, E.N., Ferronato, M.J., Fermento, M.E., Gandini, N.A., Romero, A.L., Guevara, J.A., Facchinetti, M.M., Curino, A.C., Anti-tumoral and

Anti-metastatic Activity of Maitake D-Fraction in Triple-Negative Breast Cancer Cells. *J. Oncotarget*, *9*, 23396–23412, 2018.

66. Lu, H., Lou, H., Hu, J., Liu, Z., Chen, Q., Macrofungi: A Review of Cultivation Strategies, Bioactivity, and Application of Mushrooms. *J. Compr. Rev. Food Sci. Food Saf.*, *19*, 2333–2356, 2020.
67. Morales, D., Smiderle, F.R., Piris, A.J., Soler-Rivas, C., Prodanov, M., Production of a β-D-Glucan-Rich Extract from Shiitake Mushrooms (*Lentinula edodes*) by an Extraction/Microfiltration/Reverse Osmosis (Nanofiltration) Process. *Innovative Food Sci. Emerging Technol.*, *51*, 80–90, 2019.
68. Din, S.R.U., Nisar, M.A., Ramzan, M.N., Saleem, Z.M., Ghayas, H., Ahmad, B., Batool, S., Kashif, K., Guo, X., Huang, M., Zhong, M., Latcripin-7A, Derivative of Lentinula Edodes C91-3, Reduces Migration and Induces Apoptosis, Autophagy, and Cell Cycle Arrest at G1 Phase in Breast Cancer Cells. *Appl. Microbiol. Biotechnol.*, *104*, 10165–10179, 2020.
69. Okhuoya, J.A., Edible Mushrooms: As Functional Foods and Nutriceuticals. *Trop. J. Nat. Prod. Res.*, *1*, 186–187, 2017.
70. Mishra, V., Tomar, S., Yadav, P., Vishwakarma, S., Singh, M.P., Elemental Analysis, Phytochemical Screening and Evaluation of Antioxidant, Antibacterial and Anticancer Activity of Pleurotus Ostreatus through *In Vitro* and In Silico Approaches. *J. Metabolites*, 12, 821, 2022.

13

Applications of Mushrooms as Immune Boosters

Margaret Ikhiwili Oniha[1]*, Clement Shina Olusanya[2], Oluwafemi Adebayo Oyewole[2], Priscilla Yetu Tsado[2], Charles Oluwaseun Adetunji[3], Oluwabukola Atinuke Popoola[4] and Adeyomoye Olorunsola Israel[5]

[1]*Department of Biological Sciences, Covenant University, Ota, Ogun State, Nigeria*
[2]*Department of Microbiology, Federal University of Technology, Minna, Nigeria*
[3]*Department of Microbiology, Edo State University, Uzairue, Edo State, Nigeria*
[4]*Genetics, Genomics and Bioinformatics Department, National Biotechnology Development Agency, Abuja, FCT, Nigeria*
[5]*Department of Physiology, Ondo Medical University, Ondo State, Nigeria*

Abstract

Mushrooms abound with bioactive constituents that promote good health, treat diseases, and boost the immune system of humans. These organic compounds include but are not restricted to protein, vitamins, minerals, unsaturated fatty acids, bio-functional components, fungal immunomodulatory proteins (FIPs), polysaccharides, terpenes, terpenoids, sesquiterpenes, polyphenolic compounds, lactones, and steroids. Numerous notable mushrooms also exhibit therapeutic properties such as *Agaricus subrufescens, Cordyceps sinensis, Lentinula edodes,* and *Hericium erinaceus* among others. Primary immunomodulatory compounds present in numerous medicinal mushrooms include terpenes, flavonoids, lectins, terpenoids, fungal immunomodulatory proteins (FIPs), and polyoses of which the latter serves as the most common natural immunomodulators from mushrooms. Varying chemical and physical components of these immunomodulating polysaccharides are obtained with significant individual effects on the bioactivity and mechanism of action. There are three core pathways associated with fundamental molecular mechanisms of immunomodulation by mushrooms and they include signal regulation of nuclear factor kappa B (NF-κB) through the

**Corresponding author*: margaret.oniha@covenantuniversity.edu.ng

Charles Oluwaseun Adetunji, Chukwuebuka Egbuna, Oluwatosin Ademola Ijabadeniyi and Samantha C. Karunarathna (eds.) Mushroom Biotechnology for Improved Agriculture and Human Health, (289–306) © 2025 Scrivener Publishing LLC

NF-κB fundamental modulator (NEMO) composite, activation of phospholipase C-gamma (PLCγ), closely followed by calcineurin and the controlling of flagging route by which reactive oxygen species (ROS) be produced. In addition, FIPs have different functions including the inducement of antigen anchoring cells along with the discharge of cytokines including NO plus IL-12 when FIPs bind to Toll-like receptors (TLRs), promoting the propagation and distinctness of auxiliary T cells (Th0) to assemble Th1 cells together with Th2 cells, triggering of macrophages plus B cells and production of a heterogeneity of biotic determinants. Currently, research has focused on identifying antitumor components in mushroom extracts since a strong relationship exists between the human immune system and initiation of tumors, and dependence on the activation of the immune system. These compounds facilitate immune activation for cancer therapeutics, resolution of host defense-induced inflammatory reactions, and assist in the recovery of homeostasis in the recovered individuals. There are medicinal mushroom formulations that provide synergistic antitumor and immuno-modulatory functions. These functional constituents from medicinal mushrooms have been validated both from the foods and clinical assessments, hence the need for enhanced incorporation of mushrooms in foods as functional foods to achieve a robust immune system.

Keywords: Mushrooms, immune boosters, therapeutic properties, proteins, infections

13.1 Introduction

The efficient means to obviate or attenuate the severity of diverse diseases and infectious invasions is through the ingestion of foods with immune-regulating properties. The diversity of biotic components is essential for every functioning of man's immunity. Mushrooms consist of diverse salubrity-fostering biotic amalgams that serve as unsusceptible proponents that possess every ability to curb innumerable ailments as well as to prevent viral multiplication [1]. Mushrooms are macroscopic multicellular fungi with large fruiting bodies that release spores. They have lofty protein, stubby fat, and stubby energy content. In addition, mushrooms contain abundant carbohydrates, vitamins, minerals, unsaturated fatty acids, bio-functional components, fungal immunomodulatory proteins (FIPs), polysaccharides, non-primary metabolites (terpenes, terpenoids, amino acids, alkaloids, sesquiterpenes, polyphenolic compositions, lactones, steroids, nucleotide analogues, and vitamins) along with other organic substances. Many civilizations have utilized mushrooms as food and medicine since antiquity [2, 3]. All functional qualities of mushrooms are centered on their biologically active constituents, To curb chronic ailments associated with oxidative stress, these biologically active components can be excised and cleaned for use as nutriments or added to operative noshes [4]. Abundant varying edible booms abound that possess therapeutic properties. *Agaricus subrufescens* also recognised as "almond boom" is so called due to its almond flavour and it is cultivated within the United States. It produces variable biologically active substances possessing the ability to treat different maladies and had been employed as a remedial edible for cancer, diabetes mellitus, hyperlipidemia, hepatitis, and arteriosclerosis in addition to prevention. Some of its therapeutic potentials include tumour pullalation impediment, antibacterial plus antiviral actions as well as immunostimulant and antiallergy measures. Polysaccharides that include riboglucans, β-glucans, and glucomannans constitute the main part of the biologically active constituents extricated from this boom and ergosterol has been reported to possess anticancer activity. *Inonotus obliquus* known as the "Cinder Conk" or the "Chaga Fungus" is cultivated in polar atmospheric conditions on the birch evergreens globally as well as being exploited predominantly by Russian herbalists in remedy to reinstating vivacity as well as health sustenance [5]. It is a natural source of melanin due to its crusty charcoal-like coating [6]. It consists of different antioxidants constituting superoxide dismutase, melanin, and triterpenes (lupeol, inotodiol in addition to botulin) as well as polyphenols, sterols, and complex carbohydrates. It also contains betulinic acid resident in large segments in birch cortex.

Maitake, established as "signorina" or "hen of the logs," is regarded as the genuine recurrent fungus that maturates annually in the aforementioned locus, typically near oak evergreens. It is also cultivated and merchandised fresh, dried, or prepared as a supplement. Maitake is usually freshly cooked and served as a "gourmet boom" by sauciers globally, providing a flavourful savour to soups, sauces, and varying cuisines. It contains D-Fraction, a discrete proteineous-bound beta-glucan, as included in the key polyoses [7] and included in the primary constituents associated with maitake's potent immune-modulating activities. *Cordyceps sinensis* that nourishes lungs and kidneys and supports adrenal function contains all aphrodisiac elements for both women's and men's reproductive body parts [8]. β-glucans and CO-1 including cordycepin (specific to this mushroom type) are etched within *Cordyceps*. It possesses the capacities required for activating life force hence the name "Chi"-building plant. Medicinally, it is usually compared to ginseng and is commonly employed as a tonic for athletes' health recovery and performance enhancement [4, 7]. *Agaricus blazei*, commonly known as Sun boom, is characterised by a sweet almond-like flavour. It is indigenous to Brazil and it may be consumed fresh. To unbind its beneficial healing integrants, it needs treatment using hot water or an alcoholic suspension. *Agaricus* is comprised of a remarkable quantity of beta-glucans that bolster unsusceptible operation regulation. This species is connected to the white button *Agaricus bisporus* boom, generally referred to as cremini or portobella [9]. Reports have shown that the white booms present in the markets are void of numerous medicinal distinctions. In Europe and North America, among the humongous diversities, *Agaricus bisporus* (button boom) prevails in the edible mushroom markets. It belongs to the family Basidiomycetes and flourishes in European and North American pasturelands. Its rapid aging makes it develop from a miniature white smooth boom to a mammoth light brown boom [4]. Thousands of years ago, *Ganoderma lingzhi* (Reishi) had been catalogued in Asian tradition's preliminary pharmacy documentations. It is described as a potent "Shen tonic" due to its inherent tranquility on the nerve fibers and kidney method. Also, *Ganoderma lucidum*, a heterogeneity of mushrooms, assists with neuron regeneration and wakefulness treatment, hence regarded as a massive addendum for a good sleep. Reishi mushroom's mycelium consists of peptides, nucleotides, sterols, polyoses, triterpenoids, steroids, and abundant microintegrands [10, 11]. Shiitake booms (*Lentinula edodes*) subsist as one of the multitudinous prominent digestible "gourmet" mushrooms due to their wealth of rich flavour and meaty texture. They are significantly medicinal premised on the existence of a number of therapeutic composites that comprise beta-glucans with complex sugars (lentinan, emitanin, and KS-2). Extensive research had been conducted with regard to its latent medicative elements [12]. It is widely employed as a dietary addendum or a heated water extricate for improved therapeutic potency by using the whole boom, hyphal biomass, or dissociated biologically effective elements such as lentinan extricates [13]. Lion's epithet is comprehended for the neural network and cerebral matter [5, 12]. *Hericium erinaceus*, generally addressed as Lion's mane, is a mushroom thriving on trunks of dead hardwood trees such as oak. Based on its use as a nootropic and neurotrophic nutrient-rich food, it acquired its fame as a most commonly employed therapeutic species. Studies have reported its potency as a stimulator for nerve growth factors and secretion of proteins vital in the sustentation, regeneration, and viability of the neurons, all of which date back to 1990. It consists of two peculiar chemicals known as hericenones and erinacines responsible for its employment as a "nootropic" dietary supplement in the enhancement of cognitive processes. [12] also documented its usefulness in the treatment of ailments including dementia. Terpenes and terpenoids, lectins, flavonoids, fungal immunomodulatory polypeptides (FIPs), and complex saccharides explicitly D-glucans are the significant immunomodulatory compounds present in numerous medicinal mushrooms [4]. Additional functional groups such as those containing oxygen to terpenes yield terpenoids which have shown neuroprotective actions with clinical relevance. To buttress this, studies have reported that high triterpenes cum triterpenoids in *Ganoderma* sp. have demonstrated strong immunomodulating and anti-infective activities. In *Ganoderma lucidum*, terpenes and terpenoids possess the potential to alter the immune line of actions by inducing the expression of nucleic acids encoding all polypeptides within the nuclear factor (NF)-kB tracks. Heteroglycan as well as heteroglycan-peptide present in *Hericium erinaceus* mushroom is capable of altering the protections by raising the exhibition of tumour necrosis factor (TNF), interleukin (IL)-1, and IL-12 as well as promoting nitrous oxide generation [14–16].

13.2 Mushroom Composites

Studies have frequently reported that mushrooms possess bio-functional molecules with neuroprotective properties that consist of the ones with complex sugars, and those with or without side chain alterations (including polyose-peptides and complex saccharide polypeptides). Polysaccharides are obtained from diverse natural sources including foods, bacteria, and fungi among others, as shown in Figure 13.1. Polyoses constitute the

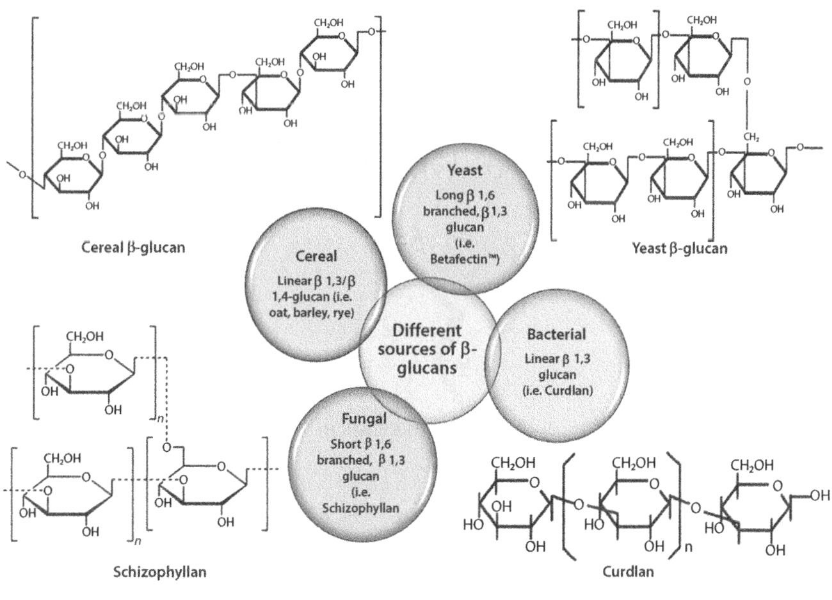

Figure 13.1 Chemical structures and branching degrees of β-glucan from diverse natural sources. Source: [39].

predominantly reported genuine antiproliferative from mushrooms. The neuroprotective complex carbohydrates are immensely varied in their dextrose elements, main sequence compound compositions, levels of furcation, configurations, subatomic weights, and supplemental substantial effects all of which exhibit remarkable effects on the biological potency and mechanism of measures of individual complex carbohydrates. Of all these polyoses, lentinan has been well identified to consist of neuroprotective and antitumour properties. It is extracted from shiitake mushrooms, along with schizophyllan, which was isolated from the *Schizophyllum commune*. These polyoses are either homoglycans (polyoses with dregs of a sole type of simple sugar molecules) or heteroglycans (complex carbohydrates with dregs of multiple dextrose molecules) and are capable of integrating alongside other atoms including oligo- or polypeptides to form peptidoglycan or polyose–polypeptide composites. More than 70% of the literature has situated the main sequence of 1,3-β-d-glucan with the insignificant numeral of minuscule divaricated sequences with 1,6-β-linkage as its characteristic of the bioactive immunomodulator polysaccharides (Figure 13.2). Primarily, larger molecular weight polyoses exhibit greater biologically active and are incapable of invading the resistant cells, but rather actively bind to cell receptors. A study reported that the peak antiproliferative function of PSK was attributed to a specified bulkiest subatomic mass section of this composite at >200 kDa [17]. Similarly, the predominant action of a polyose subset of *G. frondosa* extricate had been linked to an integrant weighing over 800 kDa [18]. However, diminished subatomic weight complex dextroses can access resistant cells from where they exert triggering effects. Both schizophyllan and lectinan comprise β-1,3-d-glucans accompanied by β1,6 branches in their structures [19]. Schizophyllan has shown efficacy

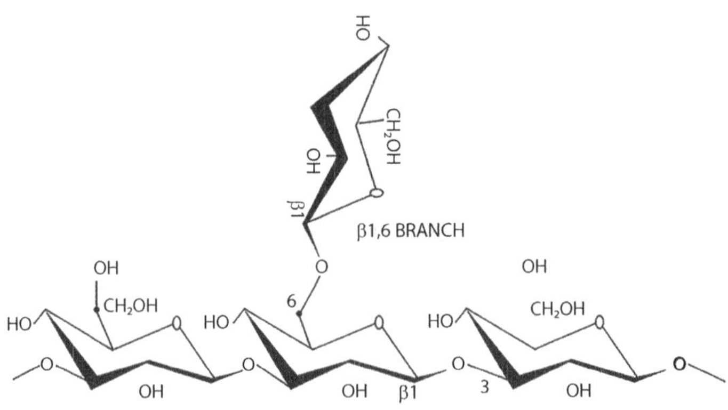

Figure 13.2 Linear structure of 1,3 glycosidic sequence of β-D-glucose monomers connected by a 1,6 glycosidic bond. Source: [40].

against head and neck cancer while lentinan exhibited neuroprotective characteristics against gastric cancer. In tumour-bearing mice, a complex dextrose-protein complex isolated from *T. giganteum* demonstrated the potential to reinstate and revamp the phagocytic operative of macrophages [20–22]. Proteins and protein-conjugate complexes which constitute the immunomodulatory substances in mushrooms have well been studied and grouped into diverse categories. These materials are categorised into two main groups consisting of lectins and FIPs. FIPs are void of a conjugate while lectins each contain a particular carbohydrate attached to a polypeptide. Findings have proven that these lectins can increase the expression of interleukins and tumour necrosis factor (TNF)-α initiate lymphocytes including stimulating the synthesis of macrophage-initiating factors in addition to other functions. Differently abundant therapeutic mushroom species consisting of these lectins include *Floccularia luteovirens, Ganoderma capense, Grifola frondosa, Pseudospermaum brinellum, Pholiota adipose, Pleurotus citrinopileatus,* and *Russuladelic.* The phytoterpenes are produced biochemically from isopentenyl pyrophosphate units from mushrooms.

The favourable neuroprotective effects of mushrooms are frequently accredited to all boom nutritive fiber preparations which basically consist of β-glucans as their constitution, configurative-anti proliferative activity interaction, and nonclinical and clinical investigations have received immense evaluations [23]. A keen focus associated with their antiproliferative prospect emerged due to configurational heterogeneity with close relation to another significant edible etymology known as flora β-glucans (β-glucans from cereals), at divergence, have no side-sequence bifurcation, nor 1→6 interconnections [21, 24–27, 29]. The core basis of mushroom β-glucans is on β-(1→3) connections in a mainstay with stubby β-(1→6) divaricates or infrequently void of divaricates (varied from extensive-divaricated chains β-(1→6) inherent in yeast). All antiproliferative and anticancer activities of mushroom β-glucans are determined by the triple helix structure [26, 30]. Generally, boom β-glucans with a subatomic mass larger than 104 Da are more effective. In addition, β-glucans, neuroprotective elements regarding varied mushrooms of unified complex dextrose (β-mannans and different complex saccharides) among others have been published [28]. [28] review situated that its complexity made it difficult to establish the complete constitution of these polyoses hence the assembly-activity connection could not be absolutely initiated. In addition, a significant number of literature have reported the immunomodulatory activities which were conducted *in vitro* or *in vivo* on animal models but did not absolutely situate the specific modus operandi of these complex saccharides in the human built.

13.3 β-Glucans and Their Nutritional Components

Nutritious food formulations of boom β-glucans are economically provided as cleansed or fractionally clarified extricates in the appearance of capsules, lozenges, and communites. All frequently published and utilized groups in nutraceutical preparations β-(1→3) (1→6)-glucans include grifolan originating from maitake β-glycan D-fraction (*G. frondosa*), lentinan (*L. edodes*), pleuran (*P. ostreatus*), as well as krestin (*T. versicolor*) [26, 30, 31]. Varying molecular weights are obtained ranging between 100 to 1000 kDa as well as lentinan with an approximate mean weight of 500 kDa and a divarication level of 0.40 (2/5) [30]. These mushroom β-glucans act primarily as performance-enhancing drugs and immune stimulants by stimulating and enhancing anticancer activities in invalids with diverse cancers and in invalids on chemotherapy or radiotherapy [32]. A large number of investigations have reported that β-glucan medicament has chronicled the introduction of programmed cell death of tumour cells [33]. Apoptosis sometimes referred to as "a guardian angel" or "cell policeman," is done through a multicomponent sequence of responses and leads to the tumour and other life-threatening biotics. Every capacity of β-glucans directed towards the trigger of apoptotic tracks or all polypeptides involved in necrobiosis creates a new sector in tumour medicament [33]. In furtherance, β-glucans trigger preventative properties against fungal, bacterial, and viral contagions such as in incurable blood-borne viral suppurations (hepatitis B, C, and D, human immune disorder virus/acquired immune disorder syndrome (HIV/AIDS), along with herpes simplex virus (HSV)) [32]. The primary molecular processes of immunomodulation by mushrooms involve three primary tracks in the composite transcriptional reaction of β-glucans [26]. These are, firstly, regulations pertaining to flagging on account of nuclear element kappa B (NF-κB) via NF-κB prerequisite modulator (NEMO) composite including analogue, via triggering of mitogen-initialized polypeptide kinase (MAPK) including the configuration of AP-1 extraction sequel to polypeptide kinase C-delta (PKCδ) trigger. NF-κB, a significant transcription feature, contributes to numerous physiological along with morbid situations encapsulating natural resistance, necrobiosis, and carcinogenesis including inflammatory procedures [33, 34]. Secondly, [35] by signalling of phospholipase C-gamma (PLCγ), and thereafter calcineurin, that moderates nuclear constituent of actuated T-cells (NFAT) while the third step involves controlling the triggering channels in which reactive oxygen strains (ROS) are produced, through moderating monocyte NADPH oxidase, which initializes the actuation of NLR family pyrin region incorporating 3 (NLRP3) polypeptide and discharge of IL-1β. Fragmentation occurs sequel to the engulfing of mushroom complex dextrose side-by-side macrophages and DCs. Disintegrated glucans (FBGs) take place through the expulsion from the biotics including to actuate natural killer (NK) living entities as well as granulocytes [36]. Actuated NK cells produce perforins and granzymes that lead to pores cum disintegration of the DNA of cancer cells, respectively [26, 36]. The body is protected from pathogens by these actuated granulocytes. Additionally, actuated macrophages with DCs can discharge cytokines plus interferons thereby actuating cytotoxic T lymphocytes (Tc) cum aider T lymphocytes (Th) also additionally conferring an antiproliferative reaction in the body's fortification in resistance to pathogens [29, 35, 36].

13.4 Antiproliferative and Other Human Health Reactions of Medicinal Mushrooms

There is a strong connection between the human immune system and the initiation of tumors. The rise in tumors on human health has received significant attention to characterise mushroom extricates/fractions/com posites possessing antitumor properties with some being dependent on the immune system's activation. Findings have presented some medicinal mushroom species that are effective for the enhancement of both immune activations for tumor medicaments and assist in resolving host defense-induced inflammatory responses as well as to situate a post-response rever sal to homeostasis in cancerous individuals. Table 13.1 shows a list of differential medicinal mushrooms and their ecological distribution across the globe. In addition, a medicinal boom constitution comprising *G. lucidum*, *L. edodes*, and *G. frondosa* exhibited anticancer besides antiproliferative potential associated with human scavenger cells [37]. The importance of useful components of medicinal booms has been cited from both clinical perspectives and foods. Since numerous remedial booms are economically groomed as edibles, an escalating trend exists of having mushrooms in addition to their integrands as addendums incorporated with edibles to originate utilitarian edibles that include the addition of novel addendum flavours or promoting operative specifics. Constituents of additional mushrooms such as *Pleurotus sajor-caju* dry powder, *A. bisporus* extracts, all freeze-dried powder from *A. aegerita*, *Suillus luteus*, as well as *Coprinopsis atramentaria* have been materialized cum employed to propagate nibbles cum cheese-related goods of immense commercial success [19, 38].

Table 13.1 Common medicinal mushrooms and their primary distributions.

Mushroom species	Common names	Family	Territorial/ecological distribution
Agrocybe aegerita	Black Poplar boom	Strophariaceae	Semitropical zone North balmy
Agaricus bisporus	Portobello boom, button mushroom, common mushroom	Agaricaceae	Poland, Germany USA, France, Ireland, Netherlands, China, United Kingdom, Italy, Spain, Canada
Amanita phalloides	Death cap	Amanitaceae	American, Asia, Europe
Boletus edulis	cep, bun bolete penny, porcini	Boletaceae	Germany, France, Swiss, China, Italy
Boletus speciosus	Red-crowned butter bolete	*Boletaceae*	America, China, Europe
Chroogomphus rutilus	copper spike	Gomphidiaceae	China
Clitocybe nebularis	Clouded conduit	Tricholomataceae	North America, North Africa, Japan, Taiwan, Europe, China
Dichomitus squalens	White-rot fungus	Polyporaceae	Trunks of conifers (e.g., larch and pine)
Flammulina velutipes	Golden indicator boom	Physalacriaceae	North America, Europe
Floccularia luteovirens (Armillaria luteovirens)	Scaly yellow boom	Agaricaceae	Meadow at 3000–4000 m above sea level
Ganoderma lucidum	Spirit plant Reishi Reitake, Ling-zhi	Ganodermataceae	Australia, Asia, Africa and America
Grifola frondosa	Hen of the logs, Maitake	Meripilaceae	China, Japan

(*Continued*)

Table 13.1 Common medicinal mushrooms and their primary distributions. (*Continued*)

Mushroom species	Common names	Family	Territorial/ecological distribution
Hericium erinaceus	Bearded tooth boom, lion's mane boom, monkey-head boom	Hericiaceae	Coniferous, Broad-leaved forest or broad-leaved mixed forest in northern balmy region (China, Western Europe, Japan, North America, and Russia)
Inonotus obliquus	Birch, clinker polypore, chaga conk	Hymenochaetaceae	Russia, China
Lentinula edodes	Black forest mushroom, golden oak boom, Shiitake	Marasmiaceae	Nepal, Northern Africa, Pacific Ocean, Japan, Papua New Guinea, the Mediterranean coast
Lignosus rhinocerotis	Tiger milk boom	Polyporaceae	China, Malaysia, Indonesia, Papua New Guinea Thailand, Philippines, Sri Lanka, Australia, and East Africa
Leucocalocybe mongolica (*Tricholoma mongolicum*)	Mongolia mushroom	Tricholomataceae	China
Marasmius oreades	Fairy ring boom	Marasmiaceae	Asia and North America
Morchella esculenta	Yellow morel, sponge morel, common morel,	Morchellaceae	China, Czechoslovakia, France, Germany, Sweden, America, India, Russia, Spain, Mexico, and Pakistan
Morchella conica	Sponge boom, black morel	Morchellaceae	Distributed under broad-leaved forest, coniferous broad-leaved mixed forest, forest edge open space and weeds
Naematelia aurantialba (*Tremella aurantialba*)	Golden tremella	*Naemateliaceae, Tremellaceae*	mutualism with *Stereum* spp., Mountain forest of quercus,

(*Continued*)

Table 13.1 Common medicinal mushrooms and their primary distributions. *(Continued)*

Mushroom species	Common names	Family	Territorial/ecological distribution
Ophiocordyceps sinensis	Himalaya viagra, caterpillar fungus	Ophiocordycipitaceae	Nepal, Southwest China
Pholiota adiposa	Chestnut boom	Strophariaceae	China
Pleurotus citrinopileatus	Tamogitake, golden oyster mushroom	Pleurotaceae	Widely cultured globally
Cerioporus squamosus (*Polyporus squamosus*)	Dryad's saddle, pheasant's back mushroom	Polyporaceae	Europe, Australia, North America, Asia
Rhodonia placenta (*Postia placenta)*	Rosy crust	Fomitopsidaceae	Globally distributed
Pseudosperma umbrinellum (*Inocybe umbrinella*)	Fibrous hat	Inocybaceae	France
Russula lepida	Rosy russula	Russulaceae	Globally distributed
Sarcodon aspratus	Black tiger paw	*Bankeraceae*	China
Schizophyllum commune	Split gill	Schizophyllaceae	Globally distributed
Stropharia rugosoannulata	Burgundy mushroom, stropharia, king stropharia, wine cap, garden giant	Strophariaceae	Asia, Europe, North America
Trametes versicolor (*Polystictus versicolor*)	Turkey tail fungus	Polyporaceae	Broad-leaf Logs, globally distributed

(Continued)

Table 13.1 Common medicinal mushrooms and their primary distributions. (*Continued*)

Mushroom species	Common names	Family	Territorial/ecological distribution
Tropicoporus linteus (Phelllinus linteus)	Mesima, black cloot fungus	*Hymenochaetaceae*	China
Volvariella volvacea	Straw mushroom	Pluteaceae	China, Southern and East Asia
Xerocomellus chrysenteron	Red cracking bolete	Boletaceae	China
Xylaria hypoxylon	Candlesnuff fungus, candlestick fungus, carbon antlers	Xylariaceae	Northern part of Europe
Xylaria nigripes	Dead moll's fingers	Xylariaceae	China

Source: [19].

All benefits encapsulate elevated polypeptides, minerals, crude fiber, or integrants along with antioxidant elements furthermore free radical root about roles. A new focus for the digestible and medicinal mushroom economies lies in the addition of mushroom components to provide diverse flavours and tastes and to enhance food nutrients.

Fungal immunomodulatory polypeptides constitute the group of polypeptides possessing highly equivalent polypeptide arrangements. These subsist as heterodimers in a dumbbell-sculptured conformation homogenous to every varying area of immune serum globulin-heavy sequences. All FIPs exhibited divergent functions which include the triggering of antigen proffer biotics including discharge of cytokines such as NO plus IL-12 when FIPs bind to Toll-like receptors (TLRs). With the activation of glycosylation of p38/MAPK plus increased assemblage of NF-κB, FIPs are able to aggrandize the propagation along with divergence of the helper T cells (Th0) for production of Th1 cells and Th2 cells, actuate macrophages as well as B cells, turn out variegated biotic determinants. In *Flammulina velutipes*, FIP-vvo can stimulate the extraction of intercell's adherence sub atomics being confined to the T cell facet by glycosylation of p38/MAPK in order to initialize Th1 cells to originate IL-2, IFN-γ for the exertion of its antiproliferative effect. FIP-vvo cannot solely actuate Th1 biotics and improve IL-2, TNF-α, and IFN-γ metamorphosis, and, in addition, impel Th2 cells to originate IL-4, B cell divergence, furthermore alteration of immune serum globulin along with the manufacture of antiserum IgE [19]. Numerous works have lent support to FIPs interaction along with TLRs can actuate other triggering routes alongside the p38/MAPK and NF-κ B as observed in FIP from *Ganoderma tsugae* (FIP-gts) with subsequent stimulation of human outlying blood leukocyte to originate IFN-γ also actuates the PI3K/Akt triggering route. FIPs primarily are available in minimal weights of the primordial brooms. Every reduced out-turn/assemblage presented a substantial structure associated with research and execution. Thus, rapid technological advancement resulted in the improvement of the origination of reunified FIPs present with other biotics including the yeast *Pichia pastoris* and the bacterium *Escherichia coli*. It has been reported that the LZ-8 nucleic acid of *G. lucidum* had shown expression in *P. pastoris* to fabricate a reunified LZ-8 polypeptide (rLZ-8). Where the reunified polypeptide is scarce of the dextrose moiety of the primordial protein, it displays equivalent biologically active for IL-2 induction as the primordial polypeptide. There has been confirmation of the expression of the FIP-vvo protein in *E. coli* [16]. Scientifically, the recombinant FIPs signified higher neuroprotective potential and influenced expansive extraction of specific cytokines compared to the extricates from the mushrooms.

Conclusion

Mushrooms possess myriad potentials ranging from the supply of nutriments to humans, improvement of the immune system up to the conferment of remedial therapies in the treatment of maladies. The key immune stimulating constituents present in booms, which are both physical and synthetic compounds, help to facilitate significant outturns including anticancer roles, particularly the medicinal strains, due to their biological activities and functionalities. These medicinal booms including but not limited to the *Ganoderma* and *Lentiluna* genera have been identified to exhibit anticancer and antiproliferative activities. Copious medicinal mushrooms are economically groomed for consumption. They are available in different preparations for the supply of nutrients to human diets. The new research facet involves the incorporation of boom constituents into foods for the enhancement of flavours and nutrients. Therefore, concerted efforts must be made to enhance functional foods containing mushrooms for an enhanced protective system.

References

1. Madhusudhanan, K., Shahina, N.K., Mathew, A., Health promoting and pharmacological compounds from mushrooms, in: *Biology, Cultivation and Applications of Mushrooms*, A. Arya, and K. Rusevska, (Eds.), pp. 149–159, Springer, Singapore, 2022, https://doi.org/10.1007/978-981-16-6257-7_5.
2. PhullAbdul-R, A.M. and Hye-Jin, P., Cordycepsmilitaris as a bio functional food source: pharmacological potential, anti-inflammatory actions and related molecular mechanisms. *Microorganisms*, 10, 2, 405, 2022, https://doi.org/10.3390/microorganisms10020405.
3. Rahman, M. A., Abdullah, N. and Aminudin, N., Interpretation of Mushroom as a Common Therapeutic Agent for Alzheimer's Disease and Cardiovascular Diseases. *Critical Reviews of Biotechnology*, 36, 6, 1131–1142, 2016. doi:10.3109/07388551.2015.1100585
4. Chugh, R.M., Mittal, P., Namratha, M.P., Arora, T., Bhattacharya, T., Chopra, H., Cavalu, S., Gautam, R.K., Fungal mushrooms: a natural compound with therapeutic applications. *Front. Pharmacol.*, 13, 1–16, 2022.
5. Kang, J.H., Jang, J.E., Mishra, S.K., Lee, H.J., Nho, C.W., Shin, D., *et al.*, Ergosterol peroxide from Chaga mushroom (*Inonotus obliquus*) exhibits anti-cancer activity by down-regulation of the β-catenin pathway in colorectal cancer. *J. Ethnopharmacol.*, 173, 303–312, 2015, doi:10.1016/j.jep.2015.07.030.
6. Balandaykin, M.E. and Zmitrovich, I.V., Review on chaga medicinal mushroom, *Inonotus obliquus* (Higher Basidiomycetes): realm of medicinal applications and approaches on estimating its resource potential. *Int. J. Med. Mushrooms*, 17, 2, 95–104, 2015.
7. Ulbricht, C., Weissner, W., Basch, E., Giese, N., Hammerness, P., Rusie-Seamon, E., *et al.*, Maitake mushroom (*Grifola frondosa*): systematic review by the natural standard research collaboration. *J. Societal Integr. Oncol.*, 7, 2, 66–72, 2009, doi:10.2310/7200.2009.0007.
8. Shashidhar, M.G., Giridhar, P., Udaya Sankar, K., Manohar, B., Bioactive principles from *Cordyceps sinensis*: a potent food supplement - a review. *J. Funct. Foods*, 5, 3, 1013–1030, 2013, doi:10.1016/j.jff.2013.04.018.
9. Blumfield, M., Abbott, K., Duve, E., Cassettari, T., Marshall, S., Fayet-Moore, F., Examining the health effects and bioactive components in *Agaricus bisporus* mushrooms: a scoping review. *J. Nutr. Biochem.*, 84, 1–17, 2020, doi:10.1016/j.jnutbio.2020.108453.
10. Cunha Zied, D. and Pardo-Giménez, A., Edible and medicinal mushrooms: technology and applications. *Funct. Foods Nutraceuticals*, 4, 12, 3756–3760, 2017.
11. Abdel-Azeem, A.M., Abdel-Azeem, M.A., Khalil, W.F., Endophytic fungi as a new source of antirheumatoid metabolites, in: *Bioactive Food as Dietary Interventions for Arthritis and Related Inflammatory Diseases*, pp.

355–384, Elsevier, Amsterdam, Netherlands, 2019, doi:10.1016/b978-0-12-813820-5.00021-0.

12. He, X., Wang, X., Fang, J., Chang, Y., Ning, N., Guo, H., *et al.*, Structures, biological activities, and industrial applications of the polysaccharides from *Hericium erinaceus* (Lion's Mane) mushroom: a review. *Int. J. Biol. Macromol.*, 97, 228–237, 2017.
13. Ganeshpurkar, A., Rai, G., Jain, A., Medicinal mushrooms: towards a new horizon. *Pharmacogn. Rev.*, 4, 8, 127–135, 2010, doi:10.4103/09737847.7090410.4103/0973-7847.70904.
14. Gao, Y.H., Zhou, S.F., Chen, G.L., Dai, X.H., Ye, J.X., A phase I/II study of a *Ganoderma lucidum* (Curt.: Fr.) P. Karst. extract (Ganopofy) in patients with advanced cancer. *Int. J. Med. Mushrooms*, 4, 207–214, 2002.
15. Lee, J.S., Cho, J.Y., Hong, E.K., Study on macrophage activation and ctructural characteristics of purified polysaccharides from the liquid culture broth of *Hericium erinaceus. Carbohydr. Polym.*, 78, 162–168, 2009.
16. El Enshasy, H.A. and Hatti-Kaul, R., Mushroom immunomodulators: unique molecules with unlimited applications. *Trends Biotechnol.*, 31, 668–677, 2013.
17. Kim, F., Sakagami, H., Tanuma, S.I., Konno, K., Stimulation of interferon-γ-induced human myelogenous leukemic cell differentiation by high molecular weight PSK subfraction. *Anticancer Res.*, 10, 55–58, 1990.
18. Adachi, Y., Ohno, N., Ohsawa, M., Oikawa, S., Yadomae, T., Change of biological activities of (1.RAR.3)-.BETA.-d-glucan from *Grifola frondosa* upon molecular weight reduction by heat treatment. *Chem. Pharm. Bull.*, 38, 477–481, 1990.
19. Zhao, S., Gao, Q., Rong, C., Wang, S., Liu, Y., Jianping Xu, J., Immunomodulatory effects of edible and medicinal mushrooms and their bioactive immunoregulatory products. *J. Fungi*, 6, 4, 1–37, 2020, https://doi.org/10.3390/jof6040269.
20. Ina, K., Kataoka, T., Ando, T., The use of lentinan for treating gastric cancer. *Anti-Cancer Agents Med. Chem.*, 13, 681–688, 2013.
21. Ngwuluka, N.C., Ochekpe, N.A., Aruoma, O.I., Functions of bioactive and intelligent natural polymers in the optimization of drug delivery, in: *Industrial applications for intelligent polymers and coatings*, pp. 165–184, Springer International Publishing, New York, USA, 2016.
22. Chakraborty, I. and Sen, I.K., Bioactive polysaccharides from natural sources: a review on the antitumor and immunomodulating activities. *Biocatalytic Agric. Biotechnol.*, 22, 101425, 2019.
23. Cerletti, C., Esposito, S., Iacoviello, L., Edible mushrooms and beta-glucans: impact on human health. *Nutrients*, 13, 2195, 2021, https://doi.org/10.3390/nu13072195.
24. Cheung, P.C.K., Mini-review on edible mushrooms as source of dietary fiber: preparation and health benefits. *Food Sci. Hum. Wellness*, 2, 162–166, 2013, https://doi.org/10.1016/j.fshw.2013.08.001.

25. Kozarski, M.S., Klaus, A.S., Niksic, M.P., *et al.*, Polysaccharides of higher fungi: biological role, structure and antioxidative activity. *Hem. Ind.*, 68, 305–320, 2014, https://doi.org/10.2298/HEMIND121114056K.
26. de Graaff, P., Govers, C., Wichers, H.J., Debets, R., Consumption of β-glucans to spice up T cell treatment of tumors: a review. *Expert Opin. Biol. Ther.*, 18, 1023–1040, 2018, https://doi.org/10.1080/14712598.2018.1523392.
27. Vunduk, J., Wan-Mohtar, W.A.A.Q.I., Mohamad, S.A., *et al.*, Polysaccharides of *Pleurotus flabellatus* strain Mynuk produced by submerged fermentation as a promising novel tool against adhesion and biofilm formation of food-borne pathogens. *LWT Food Sci. Technol.*, 112, 108221, 2019, https://doi.org/10.1016/j.lwt.2019.05.119.
28. Yin, Z., Liang, Z., Li, C., Wang, J., Ma, C., Kang, W., Immunomodulatory effects of polysaccharides from edible fungus: a review. *Food Sci. Hum. Wellness*, 10, 393–400, 2021, https://doi.org/10.1016/j.fshw.2021.04.001.
29. Kozarski, M., Klaus, A., van Griensven, L., Jakovljevic, D., Todorovic, N., Wan-Mohtard, W.A.Q.I., Vunduk, J., Mushroom β-glucan and polyphenol formulations as natural immunity boosters and balancers: nature of the application. *Food Sci. Hum. Wellness*, 12, 378–396, 2023.
30. Mironczuk-Chodakowska, I., Kujawowicz, K., Witkowska, A.M., Betaglucans from fungi: biological and health-promoting potential in the COVID-19 Pandemic Era. *Nutrients*, 13, 3960, 2021, https://doi.org/10.3390/nu13113960.
31. Wu, X., Zheng, Z., Guo, T., Wang, K., Zhang, Y., Molecular dynamics simulation of lentinan and its interaction with the innate receptor dectin-1. *Int. J. Biol. Macromol.*, 171, 527–538, 2021, https://doi.org/10.1016/j.ijbiomac.2021.01.032.
32. Gargano, M.L., Van Griensven, L.J.L.D., Isikhuemhen, O.S., Lindequist, U., Venturella, G., Wasser, S.P., Zervakis, G.I., Medicinal mushrooms: valuable biological resources of high exploitation potential. *Plant Biosyst.*, 151, 548–65, 2017, https://doi.org/10.1080/11263504.2017.1301590.
33. Wani, S.M., Gani, A., Mir, S.A., Masoodi, F.A., Khanday, F.A., β-Glucan: a dual regulator of apoptosis and cell proliferation. *Int. J. Biol. Macromol.*, 182, 1229–1237, 2021, https://doi.org/10.1016/j.ijbiomac.2021.05.065.
34. Xu, H., Zou, S., Xu, X., The β-glucan from Lentinus edodes suppresses cell proliferation and promotes apoptosis in estrogen receptor positive breast cancers. *Oncotargenetics*, 8, 86693–86709, 2017, https://doi.org/10.18632/oncotarget.21411.
35. Zhang, Y., Wang, D., Chen, Y., Liu, T., Zhang, S., Fan, H., Liu, H., Li, Y., Healthy function and high valued utilization of edible fungi. *Food Sci. Hum. Wellness*, 10, 408–420, 2021, https://doi.org/10.1016/j.fshw.2021.04.003.
36. Batbayar, S., Lee, D.H., Kim, H.W., Immunomodulation of fungal β-glucan in host defense signaling by dectin-1. *Biomol. Ther. (Seoul)*, 20, 433–445, 2012, https://doi.org/10.4062/biomolther.2012.20.5.433.

37. Mallard, B., Leach, D.N., Wohlmuth, H., Tiralongo, J., Synergistic immuno-modulatory activity in human macrophages of a medicinal mushroom formulation consisting of reishi, shiitake and maitake. *PLoS One*, 14, 11, 2019.
38. Ribeiro, A., Ruphuy, G., Lopes, J.C., Dias, M.M., Barros, L., Barreiro, F., Ferreira, I.C.F.R., Spray-drying microencapsulation of synergistic antioxidant mushroom extracts and their use as functional food ingredients. *Food Chem.*, 188, 612–618, 2015.
39. Du, B., Meenu, M., Liu, H., Xu, B., A concise review on the molecular structure and function relationship of β-Glucan. *Int. J. Mol. Sci.*, 20, 16, 4032, 2019, https://doi.org/10.3390/ijms20164032.
40. Chan, G.C., Chan, W.K., Sze, D.M., The effects of β-glucan on human immune and cancer cells. *J. Hematol. Oncol.*, 2, 25, 2009, doi:10.1186/1756-8722-2-25.

14

The Influence of Mushroom on the Taphonomic Process of Cadaver

Willy Barinem Vidona[1], Charles Oluwaseun Adetunji[2] and Charity Willy-Vidona[2*]

[1]Department of Anatomy, Edo State University Uzairue, Edo State, Nigeria
[2]Department of Microbiology, Edo State University Uzairue, Edo State, Nigeria

Abstract

The relationship between corpse taphonomic processes and mushrooms is the subject of a multidisciplinary study at the nexus of soil microbial science and forensic science. With concern about post-mortem periods and the complex dynamics of cadaver decomposition, this study explores the microbiological properties of soil fungus, namely the mushroom fruiting from burial sites, and investigates their potential as indications for forensic interpretations. The process by which cadavers decompose is essential to the cycling of nutrients and energy transformation in ecosystems. Both living and non-living factors have a complex impact on the intrinsic mechanisms of decay. Species of mushrooms, including unique genera like Penicillium, Aspergillus, Candida, and Mucor, have been identified as major contributors to the various stages of decomposition; and have been isolated from preserved bodies in archaeological investigations. The presence of certain species of mushrooms near cadaver burial sites and the relationship between particular mushrooms and decomposing remains illustrate the potential role that mushrooms may play in forensic interpretations. Notably, the quantity and timing of mushrooms, as well as their presence in the area carry significant relevance in estimating the taphonomic processes and post-mortem intervals. The research examines into the phylogeny of fungi, with a focus on mushrooms, which are the most widely known form. In the lack of a complete fossil record, molecular phylogenetic investigations, together with biochemical features and archaeological evidence, help determine evolutionary relationships. In addition, the study highlights the taphonomic process of mushrooms by describing the changes that occur in

**Corresponding author*: vidonanem@gmail.com

Charles Oluwaseun Adetunji, Chukwuebuka Egbuna, Oluwatosin Ademola Ijabadeniyi and Samantha C. Karunarathna (eds.) Mushroom Biotechnology for Improved Agriculture and Human Health, (307–316) © 2025 Scrivener Publishing LLC

the surrounding environment when corpses are released into the natural environment. This study's result emphasizes the significance of mushrooms as markers in forensic taphonomy, which helps with post-mortem interval estimation and burial site identification. The development rates of fungal spores and species are controlled by temperature and humidity conditions, despite the fact that fungi are known to be important micro-agents in cadaver decomposition also highlights the need for environmental factors in forensic investigations by this work.

***Keywords*:** Mushroom, fungi, fruiting, taphonomy, cadaver, death, decomposition, forensic investigation

14.1 Introduction

The influence of mushrooms on the taphonomic process of cadavers is a scientific inquiry at the interface of forensic science soil microbial science and examining microbial features of soil fungi and their mushroom fruiting from entombment destinations. The miniature agents that cause this mechanism can be used partly to reproduce the taphonomic past information of the body in the analysis of death scenes and post-mortem intervals. Cadaver decomposition is an essential method for cycling nutrients that plays a crucial role in the transformation of energy and its cycling process that occurs within ecosystems [1]. Intrinsic mechanism of decay is intimately affected by living factors such as mushroom and non-living factors [2]. Taphonomy, which is a process of animal decomposition and fossilization, has been influenced by the activities of microorganisms within its environment. The time and quantity of mushroom as well as its species observed within a cadaveric site have a lot of relevance in the sphere of forensic interpretations.

Without the action of cadaver-associated organisms that possessed the live host [3] and the environment milieu where the corpse falls [4], substance decomposition would involve an extremely sluggish process, leading to the formation of biochemical waste repositories [5].

Mushroom populations can colonize decomposed bodies, framing unmistakable buildup spots, at last changing over bodies into rotten corpses at the dry phase of decay [6], which frequently present apparent fungi growth [7] with specific types of Penicillium, Aspergillus, Candida, and Mucor been distinguished as the predominant genera in the swell, rot, and skeletonization stages [8]. Furthermore, most of these saprophytic genera have been disengaged from preserved dead bodies in archeological examinations [9].

There have been observations of different mushroom species in the vicinity of cadaver burial sites. Reports of sporocarp termed postputrefaction fungi abundance in the vicinity of decomposing cadavers have been published [10–13] and animal excreta [14–16]. There has been documentation of a correlation between the mushroom Hebeloma vinosophyllum and the shallowly buried dog carcasses in two distinct Japanese locations [17]. It is known that Hebeloma vinosophyllum emerges 3–8 months after soils are treated with urea or ammonium and that it continues to discharge ammonium into the soil, which allows for the sustained growth of mushroom sporophores for a duration of one to two years [18]. He noted a similar distinction between fungi that produce sporophores on animal corpses after putrefaction without the involvement of urea or ammonia and a class of fungi known as "ammonia fungi" that produce sporophores after these materials are added.

As a variety of nitrogenous substances released from cadavers into the soil present, another associated species of fungi known as ammonia fungi have been observed to be fruiting [19]. It is thought that sporophores and their fruiting sequence could be utilized to estimate the timing and taphonomic process. According to research, ammonia and other nitrogen-containing substances including nitrate, amino acids, and polypeptides are essential for these fungi to fruit [20].

Certain types of mushrooms have been found on the surfaces of decomposing corpses by experts who have done autopsies [6]. A mummified body discovered on the hard surfaces of a vacant residence has been shown to have obvious populations of Eurotium repens and Eurotium rubrum in white and yellow colors [21].

14.2 Mushroom and the Fungus Phylogeny

It is possible to argue that the most well-known kind of fungi are mushrooms. Therefore, the fruiting bodies of the fungus are called mushrooms. Similar to mold and yeast, mushrooms are classified as fungi because they are typically heterotrophic, microscopic eukaryotic organisms with cellulose in the exterior walls of their cells. Throughout the Devonian Period, a species of saprophytic fungi (Basidiomycota) were abundant, producing some of the oldest known terrestrial plantlike fossils. It is believed that the earliest known fossils of terrestrial life date from the early Silurian Period when the first simple land plants began to colonize the land, along with primitive arthropods like millipedes [22]. Yet, biochemical traits and archaeological dead body sites have proven to be helpful markers in delineating the likely evolutionary connections of fungus in the lack of a comprehensive fossil record. Newer evidence now combines molecular phylogenetic gene sequence, comparative morphology, and biochemical composition of proteins and enzymes to form a phylogenetic fungi tree for better class groupings.

These microorganisms are heterotrophic, implying they get their energy from neither light nor the decomposition of inorganic materials [23]. It is well known that fungi create a vast network of hyphae that develop into a mycelial mat. Consequently, the spread of fungi is influenced by the accessibility of carbonaceous substances that can undergo oxidation. Spores are released from the inner layer of each mushroom cap and are responsible for the propagation of mushrooms. When moved long distances by air and water currents, the spores sprout to create mycelium, which germinates on organic matter and, depending on the species, colonizes both living and dead materials. Since most organisms have mutually beneficial interactions, they typically play a critical role in recycling vital nutrients, although the species involved in the taphonomic process is observed to be the saprophytic termed microscopic mushroom attacking unhealthy host to hasten their decomposition. Despite the fact that their taphonomy is intricate, research has shown that the majority of them are found in close proximity to the sites of fungal reproduction [24]. As a result, their distribution may provide hints for additional forensic inquiry that may be beneficial. It is evident that altitude, plant type, relative humidity, amount of organic matter, and soil type all have an impact on the species of mushrooms [25] which ultimately determine the biochemical activities and actions of any species.

Usually, within the first week of deposition, molds are seen on the surfaces of decaying remains [26]. The growth of those that attach to bodies discovered in tombs happens more slowly [27].

14.3 Mushroom Taphonomic Process

The immediate environment changes in a number of ways when corpses of humans are released into the wild [28]. Due to physical suffocation and the excessive discharge of leachate, which is made up of water, carbon, phosphorus, and nitrogen that is deposited into the soil below, the plant community immediately beneath the remains is eliminated, followed by new growth [27]. At these locations, a multitude of microorganisms engage in metabolic competition with one another for cadaveric resources, altering the immediate environment's chemical and physical composition [29, 30]. Lignin and cellulose, which are complex carbon-based molecules, can be broken down by mushrooms into smaller components that are then acted upon by other microbes.

While chemical, physical, and environmental factors all play a role in the decomposition of human remains [31] invertebrates particularly bacteria and fungi, are the main drivers of this process, with mushrooms actively involved as well [29]. These organisms are crucial in the process of transforming organic materials from corpses into inorganic molecules, which are then used as a resource by plants and other living things [32].

Carbohydrates, protein, minerals, and lipids are released into the surrounding soil ecosystem when a cadaveric corpse undergoes autolysis, which results in alterations to the soft tissue. This raises the pH of the soil and increases its concentration of nitrogen [33–35], creating an environment that is beneficial to anaerobic microbes and the mushroom group [36] to promote the putrefactive phase of breakdown. A specific collection of mushroom species can be found in hard tissues that survive long after the soft tissues have died.

This process is actually a succession pattern of fruiting in which a group of fungi eventually gives way to another [27]. This succession has two phases: early and late. The more evolved fungal species include ascomycetes and saprotrophic as well as ectomycorrhizal basidiomycetes, which can form fruiting bodies within a period of one to four years following fertilization, which fruits can develop within a range of one to ten months following the process of fertilization during the initial growth phase. This split of succession is mostly caused by the use of nitrogen, as ammonia is the primary component that causes fruiting.

[37] have also reported signs of fungal colonization of the carcass within three to seven days of death.

14.4 Influence of Mushroom on Cadaver Taphonomy

It has been noted that soils containing urea or other nitrogenous substances, chemicals that release ammonia upon breakdown, ammonia fungi produce fruiting structures known as mushrooms. Typically, cadaveric taphonomic fungal fruiting is most prominent during the wet or autumnal season, and appropriate as cadaver-related fruiting occurs over burial sites usually after months and years, which signifies a late stage of decomposition [38].

On cadaveric dump in different buried sites most repeated and isolated species are Acremonium-Fusarium, Penicillium, Cladosporium, and Rhizopus [9, 39] although some are associated more with certain body parts such as mucous membranes, human hair, and clothes [40].

[41] extracted fungi from the skin of a deceased woman found in a Belgian room. This allowed them to highlight the importance of fungal growth in determining the time of death from species which include Hormodendrum sp., Mortierella sp., Penicillium chrysogenum, and Geotrichum candidum sp. The material was then incubated at the same temperature as the corpse. The woman had passed away at least eighteen days prior to the finding, as demonstrated by the fungal colonies' daily diameter measurements, which was in perfect agreement with the murderer's later confession.

On the outer layer, Aspergillus chevalieri Thom and Gliocladium clonostachys sp. were discovered on a preserved cadaver and skeletal structures found in an abandoned site [42]. These findings corroborated the theory that fungi contaminated burial locations.

Because fungi exhibit geotropism, meaning that the stalk of a mushroom develops longitudinally and the pileus transversely, they might act as markers of alternations in the crime site, such as alterations to plant branches and debris. For this reason, mushrooms have a growing influence on cadaveric taphonomy. When the substrate's position shifts, it quickly reorganizes to keep growing and functioning in a geotropic direction, and as an implication, it displays malformations that make it possible to identify alterations to the crime scene [43].

The microbial populations that participate in decomposition are present until the nutrients are exhausted, and they change in composition and amount throughout time [28].

14.5 Conclusion

As a valuable instrument in the field of forensic taphonomy, assisting in the identification of burial sites and estimation of postmortem interval, it is reported that some species of mushrooms can be viewed as very obvious indicators of cadaver decomposition, even though soil features could promote the decomposition rate. It is noted that fungi are major micro-agents of cadaver decomposition while its fruiting resulting in the mushroom species can help determine post-burial interval time useful for forensic death and burial controversies. However, the temperature and humidity conditions around the cadaver sites will influence the growth rates of the fungi spores and species for a significant revelation.

References

1. Metcalf, J.L., Xu, Z.Z., Weiss, S., Lax, S., Van Treuren, W., Hyde, E.R., Song, S.J., Amir, A., Larsen, P., Sangwan, N., Haarmann, D., Humphrey, G.C., Ackermann, G., Thompson, L.R., Lauber, C., Bibat, A., Nicholas, C., Gebert, M.J., Petrosino, J.F., Reed, S.C., Gilbert, J.A., Lynne, A.M., Bucheli, S.R., Carter, D.O., Knight, R., Microbial community assembly and metabolic function during mammalian corpse decomposition. *Science*, 351, 158–162, 2016.
2. Hyde, E.R., Haarmann, D.P., Lynne, A.M., Bucheli, S.R., Petrosino, J.F., The living dead: bacterial community structure of a cadaver at the onset and end of the bloat stage of decomposition. *PLoS One*, 8, e77733, 2013.
3. Janaway, R.C., The decay of human buried remains and their associated materials, in: *Studies in Crime: An Introduction to Forensic Archaeology*, C. Roberts, J. Hunter, A. Martin, (Eds.), pp. 58–85, Batsford, B. T. Ltd, London, 1996.
4. Tuomanen, E., Microbial inhabitants of humans - their ecology and role in health and disease. *Science*, 308, 635–635, 2005.
5. Gilbert, J.A. and Neufeld, J.D., Life in a world without microbes. *PLoS Biol.*, 12, e1002020–e1002020, 2014.
6. Hitosugi, M., Ishii, K., Yaguchi, T., Chigusa, Y., Kurosu, A., Kido, M., Nagai, T., Tokudome, S., Case report: fungi can be a useful forensic tool. *Legal Med. (Tokyo Japan)*, 8, 4, 240–242, 2006.
7. Ishii, K., Hitosugi, M., Kido, M., Yaguchi, T., Nishimura, K., Hosoya, T., Tokudome, S., An analysis of fungi detected in human cadavers. *Legal Med. (Tokyo Japan)*, 8, 3, 188–190, 2006.
8. Schwarz, P., *et al.*, Molecular identification of fungi found on decomposed human bodies in forensic autopsy cases. *Int. J. Legal Med.*, 129, 785–791, 2015.
9. Sidrim, J.J., *et al.*, Fungal microbiota dynamics as a postmortem investigation tool: focus on Aspergillus, Penicillium and Candida species. *J. Appl. Microbiol.*, 108, 1751–1756, 2010.
10. Cavka, M., *et al.*, Microbiological analysis of a mummy from the archaeological museum in Zagreb. *Coll. Antropol.*, 34, 803–805, 2010.
11. Fukiharu, T., Osaku, K., Iguchi, K., Masahiko, A., Occurrence of ammonia fungi on the forest ground after decomposition of a dog carcass. *Nat. Hist. Res.*, 6, 1, 9–14, 2000a.
12. Miller, R.N. and Hilton, O.K., New and interesting agarics from Western Australia. *Sydowia*, 39, 126–135, 1986.
13. Sagara, N., Occurrence of Laccaria proxima in the grave site of a cat. *Trans. Mycol. Soc. Jpn.*, 22, 271–275, 1981.
14. Takayama, S. and Sagara, N., The occurrence of Hebeloma vinosophyllum on soil after decomposition of the corpse of a domestic rabbit. *Trans. Mycol. Soc. Jpn.*, 22, 475–477, 1981.

15. Sagara, N., European record of the presence of a mole's nest indicated by a particular fungus. *Mammalia*, 53, 2, 301–305, 1989.
16. Sagara, N., Honda, S., Kuroyanagi, E., Takayama, S., The occurrence of Hebeloma radicosum on the dungdeposited burrows of Urotrichus talpoides (shrew mole). *Trans. Mycol. Soc. Jpn.*, 22, 441–455, 1981.
17. Sagara, N., Abe, H., Okabe, H., The persistence of moles in nesting at the same site as indicated by mushroom fruiting and nest reconstruction. *Can. J. Zool.*, 71, 1690–1693, 1993a.
18. Sagara, N., Presence of a buried mammalian carcass indicated by fungal fruiting bodies. *Nature*, 262, 816, 1976.
19. Tibbett, M. and Carter, D.O., Mushrooms and taphonomy: the fungi that mark woodland graves. *Mycologist*, 17, 20–24, 2003.
20. Sagara, N., Association of ectomycorrhizal fungi with decomposed animal wastes in forest habitats: A cleaning symbiosis? *Can. J. Bot.*, 73, Suppl. 1, S1423–S1433, 1995.
21. Wilson, A.S., Janaway, R.C., Holland, A.D., Dodson, H.I., Baran, E., Pollard, A.M., Tobin, D.J., Modelling the buried human body environment in upland climes using three contrasting field sites. *Forensic Sci. Int.*, 169, 1, 6–18, 2007.
22. Ishii, K., Hitosugi, M., Yaguchi, T., Tokudome, S., The importance of forensic mycology. *Legal Med. (Tokyo Japan)*, 9, 5, 287, 2007.
23. Dilip, K.A. and Randy, M.B., *Applied Mycology and Biotechnology. Genes Genomics*, Elsevier Science, 5, Amsterdam, 2005.
24. Boberg, J., *Litter decomposing fungi in boreal forests*, Swedish University of Agricultural Sciences, Uppsala, Sweden, 2009.
25. Galante, T.E., Horton, T.R., Swaney, D.P., 95% of basidiospores fall within 1 m of the cap: a field-and modeling-based study. *Mycologia*, 103, 6, 1175–1183, 2011.
26. Anand, T.P. and Geeta, N.P., *Shade Grown Ecofriendly Indian Coffee*, vol. 1, 2009.
27. Hawksworth, D.L. and Wiltshire, P.E.J., Forensic mycology: the use of fungi in criminal investigations. *Forensic Sci. Int.*, 206, 1-3, 1–11, 2011.
28. Dent, B.B., Forbes, S.L., Stuart, B.H., Review of human decomposition processes in soil. *Environ. Geol.*, 45, 4, 576–585, 2004.
29. Carter, D.O. and Tibbett, M., Cadaver decomposition and soil: processes, in: *Soil analysis in forensic taphonomy: chemical and biological effects of buried human remains*, M. Tibbett, and D.O. Carter, (Eds.), pp. 29–51, CRC Press Taylor and Francis Group, Boca Raton, 2008a.
30. Haelewaters, D., Hebeloma, pioneer genus in forensic mycology. *Fungi*, 6, 3, 47–48, 2013.
31. Madigan, M.T. and Martinko, J.M., *Brock: Biology of microorganisms*, 11th ed., Pearson Prentice Hall, Upper Saddle River, 2006.
32. Galloway, A., The process of decomposition: a model from the Arizona-Sonoran desert, in: *Forensic taphonomy: the postmortem fate of human*

remains, W.D. Haglund, and M.H. Sorg, (Eds.), pp. 139–150, CRC Press Taylor and Francis Group, Boca Raton, 1997.
33. Anderson, B., Meyer, J., Carter, D.O., Dynamics of ninhydrin-reactive nitrogen and pH in gravesoil during the extended postmortem interval. *J. Forensic Sci.*, 58, 5, 1348–1352, 2013.
34. Hopkins, D.W., Wiltshire, P.E.J., Turner, B.D., Microbial characteristics of soils from graves: an investigation at the interface of soil microbiology and forensic science. *Appl. Soil Ecol.*, 14, 3, 283–288, 2000.
35. Carter, D.O., Yellowlees, D., Tibbett, M., Cadaver decomposition in terrestrial ecosystems. *Die Naturwiss.*, 94, 1, 12–24, 2007.
36. Sagara, N., Ammonia fungi: a chemoecological grouping of terrestrial fungi. *Contr. Biol. Mab. Kyoto Univ.*, 24, 4, 205–90, 1975.
37. Pinar, G., Piombino-Mascali, D., Maixner, F., Zink, A., Sterflinger, K., The Capuchin Catacombs of Palermo: problems facing the conservation of an impressive burial site. *Coalition*, 25, 2–10, 2014.
38. Martínez-Ramírez, J.A., Strien, J., Walter, G., Peters, F.T., Search for fungi-specific metabolites of four model drugs in post mortem blood as potential indicators of post mortem fungal metabolism. *Forensic Sci. Int.*, 262, 173–178, 2016.
39. Samson, R.A., Hoekstra, E.S., Frisvad, J.C., *Introduction to food- and airborne fungi*, Centraalbureau voor Schimmelcultures, Utrecht, 2004.
40. van de Voorde, H. and van Dijck, P.J., Determination of the time of death by fungal growth. *Z. Rechtsmed.*, 89, 2, 75–80, 1982.
41. Tranchida, M.C., Pelizza, S.A., Elíades, L.A., The use of fungi in forensic science, a brief overview. *Can. Soc. Forensic Sci. J.*, 54, 1, 35–48, 2021.
42. Swann, L.M., Forbes, S.L., Lewis, S.W., Analytical separations of mammalian decomposition products for forensic science: a review. *Anal. Chim. Acta*, 682, 1-2, 9–22, 2010.
43. Carter, D.O. and Tibbett, M., Taphonomic mycota: fungi with forensic potential. *J. Forensic Sci.*, 48, 168–171, 2003.

15

Role of Nanobiopesticides Derived from Mushrooms: Recent Advances

John Tsado Mathew[1*], Charles Oluwaseun Adetunji[2], Abel Inobeme[3], Abdulfatai Aideye Otori[4], Elijah Yanda Shaba[5], Musah Monday[1], Yakubu Azeh[1], Jibrin Noah Akoji[6], Amos Mamman[1], Oluwafemi Adebayo Oyewole[7], Eniola K. I. T.[8] and Mohammed Bello Yerima[9]

[1]Department of Chemistry, Ibrahim Badamasi Babangida University, Lapai, Niger State, Nigeria
[2]Applied Microbiology, Biotechnology and Nanotechnology Laboratory, Department of Microbiology, Edo University, Uzairue, Auchi, Edo State, Nigeria
[3]Department of Chemistry, Edo University Uzairue, Auchi, Edo State, Nigeria
[4]Department of Chemical Engineering, Federal Polytechnic, Bida, Niger State, Nigeria
[5]Department of Chemistry, Federal University of Technology Minna, Nigeria
[6]Department of Chemistry, Baze University, Abuja, Nigeria
[7]Department of Microbiology, School of Life Sciences, Federal University of Technology, Minna, Nigeria
[8]Department of Biological Sciences, Joseph Ayo Babalola University, Ikeji Arakeji, Osun State, Nigeria
[9]Department of Microbiology, Sokoto State University, Sokoto, Sokoto State, Nigeria

Abstract

The widespread use of chemical pesticides in agriculture has raised concerns due to their adverse effects on human health and the environment. As a result, there is a growing demand for safer and sustainable alternatives. In recent years, nanobiopesticides derived from mushrooms have emerged as a promising solution for the management of agricultural pests and diseases. Mushrooms, a rich source of biologically active compounds, possess innate antimicrobial and insecticidal properties. Through the integration of nanotechnology, these beneficial

**Corresponding author*: johntsadom@gmail.com; jmathew@ibbu.edu.ng

Charles Oluwaseun Adetunji, Chukwuebuka Egbuna, Oluwatosin Ademola Ijabadeniyi and Samantha C. Karunarathna (eds.) Mushroom Biotechnology for Improved Agriculture and Human Health, (317–336)

properties can be enhanced, resulting in the development of nanobiopesticides. Nanobiopesticides are nano-sized formulations of bioactive compounds derived from mushrooms that exhibit improved efficacy, stability, and targeted delivery mechanisms. The application of nanobiopesticides in agricultural systems offers several advantages. The nano-sized particles enable better penetration and adhesion to target organisms, increasing their effectiveness against pests and pathogens. The controlled release mechanisms of nanobiopesticides allow for prolonged and sustained activity, reducing the need for frequent application. The biodegradable nature minimizes the risk of residual accumulation and environmental pollution. Additionally, the use of nanobiopesticides promotes the preservation of beneficial insects, pollinators, and natural enemies, thereby maintaining the ecological balance in agricultural ecosystems. Despite the significant potential of nanobiopesticides, certain challenges must be addressed for their successful implementation. These include standardization of manufacturing processes, formulation optimization, regulatory framework development, and cost-effectiveness. Furthermore, potential risks associated with nanomaterials, such as their impact on non-target organisms and long-term effects on soil health, require thorough investigation. Nanobiopesticides derived from mushrooms represent a promising avenue for sustainable pest and disease management in agriculture. Their unique combination of natural compounds and nanotechnology provides enhanced efficacy, targeted delivery, and reduced environmental impact.

***Keywords*:** Antimicrobial, insecticidal mushrooms, nanobiopesticides, pathogens, pests

15.1 Introduction

Agricultural pests and diseases pose significant challenges to global food production, causing substantial yield losses and compromising crop quality. Traditionally, chemical pesticides have been the primary means of pest and disease management in agriculture. However, the widespread use of chemical pesticides has raised concerns about their negative impacts on human health, beneficial organisms, and the environment. In recent years, there has been a growing interest in developing safer and more sustainable alternatives to chemical pesticides. Nanobiopesticides derived from mushrooms have emerged as a promising solution for effective pest and disease management in agriculture [43].

The management of agricultural pests and diseases is a critical aspect of ensuring global food security and sustainable agriculture. However, the use of chemical pesticides for pest control has raised concerns due to their detrimental effects on human health, beneficial organisms, and the environment. As a result, there is a growing interest in developing alternative strategies that are safer, more effective, and environmentally friendly. In recent years, nanobiopesticides derived from mushrooms have emerged as a promising solution for the management of agricultural pests and diseases [52].

Mushrooms are a diverse group of fungi known for their medicinal and nutritional properties. They have been used in traditional medicine for centuries due to their antimicrobial and insecticidal properties. Mushrooms contain a wide range of bioactive compounds, including polysaccharides, proteins, phenols, and terpenes, which contribute to their pest and disease control potential. With the advancements in nanotechnology, researchers have been able to harness the inherent properties of mushrooms and develop nanobiopesticides that offer enhanced efficacy and targeted delivery mechanisms [18].

Nanobiopesticides are nano-sized formulations of bioactive compounds derived from mushrooms. The integration of nanotechnology into these formulations allows for improved stability, controlled release, and increased bioavailability of the active ingredients. Nanoparticles, such as liposomes, nanoemulsions, and nanocapsules, are commonly employed to encapsulate and deliver bioactive compounds. The small size and high surface area-to-volume ratio of nanoparticles enables better penetration and adhesion to target organisms, enhancing their effectiveness against pathogens and pests [31, 40].

One of the key advantages of nanobiopesticides derived from mushrooms is their reduced environmental impact compared to chemical pesticides. These formulations are often biodegradable, minimizing the risk of residual accumulation in soil, water, and food products. Additionally, nanobiopesticides can specifically target pests and diseases, reducing the need for broad-spectrum applications and minimizing the potential harm to non-target organisms, including pollinators, natural enemies, as well as beneficial insects [15].

Moreover, nanobiopesticides offer controlled and sustained release mechanisms, allowing for prolonged activity and reducing the frequency of application. This not only improves the efficiency of pest and disease management but also reduces the overall chemical load in agricultural systems. The controlled release mechanisms also contribute to minimizing the development of pesticide resistance in target organisms, which is a significant concern associated with conventional chemical pesticides [19].

The development and application of nanobiopesticides from mushrooms do present some challenges. Standardization of manufacturing processes and formulation optimization are essential to ensure consistent quality and efficacy. The regulatory framework for nanobiopesticides needs

Table 15.1 Main aspects of nanobiopesticides derived from mushrooms in the management of agricultural pests and diseases.

Advantages	Challenges	Opportunities	Implications	References
High efficacy against pests and diseases	Limited knowledge of mushroom species	Exploration of new mushroom species	Expansion of nanobiopesticide options	[15]
Reduced toxicity towards non-target organisms	Delivery system optimization	Development of targeted delivery systems	Enhanced precision and reduced environmental impact	[33]
Bioactive properties of mushroom compounds	Field validation and commercialization	Field trials and real-world evaluation	Widespread adoption and commercial availability	[16]
Eco-friendly and sustainable	Regulatory frameworks and safety assessments	Regulation development and implementation	Ensuring safe and responsible use	[28]

to be developed to address safety, labeling, and registration requirements. The cost-effectiveness of nanobiopesticides also needs to be considered to ensure accessibility for farmers, particularly in resource-limited settings. Furthermore, it is crucial to thoroughly assess the potential risks associated with nanobiopesticides, such as their impact on non-target organisms and long-term effects on soil health. Research efforts are needed to understand the ecological implications and the fate of nanoparticles in agricultural ecosystems (Table 15.1) [1].

15.2 Environmental and Health Concerns with Chemical Pesticides

Chemical pesticides have been widely used in agriculture to protect crops from diseases and pests. However, their use raises significant environmental and health concerns. Understanding these concerns is crucial for developing sustainable alternatives and promoting safer practices in pest management. There are some main environmental and health concerns associated with chemical pesticides:

Ecological Impact: Chemical pesticides can have detrimental effects on non-target organisms, including beneficial insects, fish, birds, and amphibians. Pesticides may disrupt ecosystems by harming pollinators, natural enemies of pests, and other organisms critical for maintaining ecological balance. This can lead to a decline in biodiversity and disrupt the functioning of ecosystems [44].

Water Contamination: Pesticides can contaminate water bodies by runoff from agricultural fields. This contamination can have harmful effects on aquatic organisms and their habitats. Pesticides can also leach into groundwater, posing risks to drinking water sources. Long-lasting pesticides may persist in the environment, causing prolonged contamination [25].

Soil Degradation: Chemical pesticides can have adverse effects on soil health and quality. They may damage beneficial soil organisms, such as microorganisms as well as earthworms, which play essential roles in nutrient cycling and soil fertility. Prolonged pesticide use can lead to soil degradation, reduced biodiversity, and loss of soil productivity [13].

Pesticide Resistance: Repeated and excessive use of chemical pesticides can contribute to the development of pesticide resistance in target pests and diseases. This resistance reduces the effectiveness of pesticides, requiring higher doses or alternative chemicals, leading to a vicious cycle of increasing chemical use. Pesticide resistance can result in economic losses and a greater reliance on chemical interventions [48].

Human Health Risks: Exposure to chemical pesticides can pose menace to human health. Farmers, farmworkers, and individuals living near agricultural areas are particularly vulnerable. Pesticides can enter the body through inhalation, ingestion, or skin contact. Acute pesticide poisoning can cause symptoms ranging from mild irritation to severe illness or even death. Chronic exposure to low levels of pesticides has been related to several health concerns, such as neurological disorders, cancer, reproductive problems, and endocrine disruption [21].

Residue in Food: Pesticide residues can remain on crops even after application, leading to potential exposure through the consumption of contaminated food products. Regulatory agencies set maximum residue limits (MRLs) to ensure food safety. However, the long-term effects of cumulative exposure to multiple pesticide residues and the potential interactions between different chemicals are still areas of concern [49].

Environmental Persistence: Some chemical pesticides have long persistence in the environment, meaning they remain active for extended periods. This persistence can lead to bioaccumulation in organisms and impact food chains. Pesticides can also travel long distances through air currents, leading to contamination in areas far from the initial application site [41].

15.3 Mushrooms as a Source of Bioactive Compounds

Mushrooms, with their diverse range of species and unique properties, have long been recognized as a valuable source of bioactive compounds. These bioactive compounds are responsible for the various medicinal, nutritional, and functional properties exhibited by mushrooms. The bioactive compounds found in mushrooms and their potential applications are:

Polysaccharides: Mushrooms are rich in polysaccharides, particularly β-glucans. These complex carbohydrates have shown various biological activities, including immunomodulatory, anti-inflammatory, antioxidant, and anticancer effects. β-glucans from mushrooms have been studied for their potential to boost the immune system, promote gut health, and combat oxidative stress [20].

Proteins and Peptides: Mushrooms contain proteins and peptides with diverse biological activities. These include antimicrobial, antitumor, and immunomodulatory properties. Some mushroom proteins have shown potential as natural alternatives to synthetic antimicrobial agents, contributing to the development of antimicrobial strategies in various applications [32].

Phenolic Compounds: Phenolic compounds, such as phenolic acids, flavonoids, and tannins, are abundant in mushrooms. These compounds possess antioxidant properties, scavenging free radicals and protecting cells against oxidative damage. Additionally, phenolic compounds have been associated with potential anti-inflammatory, antimicrobial, and anticancer effects [3, 38].

Terpenes: Mushrooms are known to produce a wide range of terpenes, including triterpenes and sesquiterpenes. These compounds exhibit diverse biological activities, such as antimicrobial, anti-inflammatory, and anticancer effects. Some mushroom-derived terpenes have shown promising anticancer activity by inhibiting tumor growth and inducing apoptosis in cancer cells [4, 18].

Lectins: Lectins are proteins found in mushrooms that bind specifically to certain carbohydrate molecules. These compounds have been studied for their potential to modulate immune responses and inhibit the growth of cancer cells. Some lectins derived from mushrooms possess antiviral activity and have been explored as potential therapeutic agents against viral infections [36].

Ergosterol and Sterols: Mushrooms are a natural source of ergosterol, a precursor of vitamin D2. Upon exposure to UV light, ergosterol is converted into vitamin D2, which is important for bone health and immune function. Mushrooms enriched with vitamin D2 have gained popularity as a plant-based source of this essential vitamin [27].

The bioactive compounds in mushrooms offer a range of potential applications:

The bioactive compounds found in mushrooms have been utilized in traditional medicine for their therapeutic properties. Mushrooms have been studied for their potential in treating various health conditions, including cancer, cardiovascular diseases, diabetes, and immune-related disorders. Mushroom extracts or bioactive compounds are incorporated into functional foods and nutraceutical products due to their potential health benefits. Mushroom-based dietary supplements, powders, and extracts are available in the market as natural health-promoting products. The antioxidant and anti-inflammatory properties of mushroom extracts make them valuable ingredients in cosmetic and skincare formulations. Mushroom-derived compounds are used in products aimed at reducing signs of aging, improving skin hydration, and protecting against environmental damage. Some bioactive compounds from mushrooms have applications in biotechnology and bioremediation. For instance, enzymes derived from certain mushroom species have been utilized in industrial processes, such as food processing and textile production. Mushrooms also have the ability to degrade and detoxify environmental pollutants, contributing to bioremediation efforts [10, 11, 24].

15.4 Antimicrobial and Insecticidal Properties of Mushrooms

Mushrooms possess notable antimicrobial and insecticidal properties, which make them potential candidates for developing natural pesticides and antimicrobial agents. The bioactive compounds present in mushrooms contribute to these properties. Let us explore the antimicrobial and insecticidal potential of mushrooms:

15.4.1 Antimicrobial Properties

Antibacterial Activity: Several mushrooms have demonstrated antibacterial activity against a wide range of pathogenic bacteria. For example, extracts from species such as Agaricus bisporus, Pleurotus ostreatus, Ganoderma lucidum, and Lentinula edodes have shown inhibitory effects against bacteria like Salmonella typhimurium, Staphylococcus aureus, Pseudomonas aeruginosa, as well as Escherichia coli [11, 20].
Antifungal Activity: Mushrooms are known to possess antifungal properties, which can be attributed to various bioactive compounds. They have shown effectiveness against common fungal pathogens, including Candida species, Aspergillus species, and dermatophytes. For instance, extracts from Ganoderma spp., Trametes spp., and Grifola frondosa have exhibited antifungal activity against these pathogens [2, 12].
Antiviral Activity: Certain mushrooms have demonstrated antiviral activity against a range of viruses. Extracts from species like Lentinula edodes, Grifola frondosa, and Pleurotus ostreatus have exhibited inhibitory effects against viruses such as herpes simplex virus, human immunodeficiency virus (HIV), as well as influenza virus [14, 46].

15.4.2 Antiparasitic Activity

Some mushrooms have shown potential in combating parasites. For example, extracts from Ganoderma lucidum and Trametes versicolor have exhibited activity against parasites such as Leishmania spp., Plasmodium spp., and Trypanosoma spp [51].

Insecticidal Properties: Insect Growth Regulation: Certain mushrooms and their bioactive compounds have demonstrated insect growth-regulating properties. They can disrupt insect development, affect molting, and inhibit reproduction. These effects have been observed in various insect pests, including agricultural pests like aphids, caterpillars, and beetles [17].
Insecticidal Activity: Some mushrooms contain bioactive compounds that exhibit insecticidal properties, causing direct mortality or behavioral effects on insects. For instance, extracts from species like Metarhizium anisopliae, Beauveria bassiana, and Cordyceps spp. have been utilized as natural insecticides for controlling pests such as mosquitoes, flies, and termites [53].
Repellent Effects: Certain mushroom extracts have shown repellent effects against insects. They can deter pests from settling on plants or entering specific areas. These repellent properties have been studied in relation to mosquitoes, ticks, and other pests [34].

The antimicrobial and insecticidal properties of mushrooms have gained attention as potential alternatives to chemical pesticides and antimicrobial agents. Harnessing these properties can contribute to the development of sustainable and eco-friendly strategies for pest and disease management in agriculture as well as natural antimicrobial products for various applications. However, further research is needed to explore the mechanisms of action, optimize extraction methods, and assess the safety and efficacy of mushroom-derived compounds for practical applications [30].

15.5 Nanotechnology and its Applications in Agriculture

Nanotechnology, which includes the control as well as manipulation of resources at the nanoscale level (typically less than 100 nanometers), has shown great potential in revolutionizing various fields, including agriculture [26]. The application of nanotechnology in agriculture, often referred to as nanagriculture or nanotechnology-enabled agriculture, offers innovative solutions to improve crop production, enhance plant protection, and promote sustainable farming practices. Some main applications of nanotechnology in agriculture:

Nanosensors and Precision Farming: Nanosensors are used to monitor various parameters in agriculture, such as soil moisture, nutrient levels, and pH. These tiny sensors can provide real-time data and help farmers make informed decisions about irrigation, fertilization, and crop management. Precision farming techniques based on nanosensors enable targeted application of resources, optimizing inputs and reducing environmental impact [45].

Nanofertilizers: Nanotechnology has the potential to enhance the efficiency of nutrient delivery to plants. Nanoscale nutrient formulations can improve nutrient uptake, reduce fertilizer loss through leaching and volatilization, and provide controlled-release mechanisms. Nanofertilizers can also be designed to target specific plant tissues or deliver nutrients in a time-released manner, promoting optimal plant growth and minimizing environmental pollution [39].

Nanopesticides and Crop Protection: Nanotechnology offers innovative approaches to pest and disease management. Nanopesticides can deliver targeted and controlled release of active ingredients, enhancing their efficacy while minimizing environmental impact. Nanoencapsulation of pesticides improves their stability, reduces off-target effects, and enables efficient delivery systems, such as nanoemulsions or nanocapsules. Nanoparticles can also be engineered to have antimicrobial properties, inhibiting the growth of pathogens and reducing the need for chemical pesticides [20].

Smart Delivery Systems: Nanotechnology enables the development of smart delivery systems for agricultural inputs. Nanocarriers, such as liposomes, nanogels, and nanoparticles, can encapsulate and protect active ingredients, facilitating their controlled release. These delivery systems can improve the bioavailability and stability of agrochemicals, reduce their degradation, and provide targeted delivery to specific plant tissues or pests [47].

Nanobiosensors for Disease Detection: Nanoscale biosensors are being developed to detect and monitor plant diseases rapidly and accurately. These biosensors can detect the presence of pathogens or disease-related biomarkers in plants or the environment, enabling early disease diagnosis and prompt intervention. Nanobiosensors offer high sensitivity, specificity, and speed, facilitating disease management and preventing yield losses [9, 23].

Nanomaterials for Soil Remediation: Nanotechnology provides innovative solutions for soil remediation and environmental cleanup. Nanomaterials, such as nanoparticles and nanocomposites, can remove pollutants, heavy metals, and contaminants from soil, water, and air. These nanomaterials have the potential to enhance soil fertility, improve water quality, and mitigate the impact of pollution on agricultural ecosystems [6–8, 42].

Post-Harvest Preservation: Nanotechnology offers solutions for post-harvest preservation and food quality maintenance. Nanocoatings, for example, can be applied to fruits and vegetables to delay ripening, reduce spoilage, and extend shelf life. Nanosensors integrated into packaging materials can monitor food freshness, detect spoilage, and provide information on food safety [37].

15.6 Mechanisms of Action of Nanobiopesticides

Nanobiopesticides, which are nanotechnology-based formulations of biopesticides, offer innovative approaches to pest management in agriculture. These formulations utilize nanoparticles to enhance the efficacy, stability, and targeted delivery of biopesticide active ingredients. The mechanisms of action of nanobiopesticides involve various processes that contribute to their effectiveness in controlling pests and diseases. There are some major mechanisms of action of nanobiopesticides:

Enhanced Penetration and Adhesion: Nanobiopesticides can improve the penetration and adhesion of active ingredients to target pests or plant surfaces. The nanoparticles present in the formulation can facilitate the entry of biopesticide molecules into pests' bodies or plant tissues, enhancing their overall efficacy. The nanoscale size and surface properties of nanoparticles allow better interaction with pests and improve the adhesion of biopesticides to plant surfaces, leading to increased pest control [22].
Controlled Release: Nanotechnology enables the development of controlled-release systems for biopesticides. Nanocarriers, such as nanoparticles or nanocapsules, can encapsulate biopesticide active ingredients, protecting them from degradation and providing controlled release over time. This controlled release ensures a sustained presence of biopesticides, extending their effectiveness and reducing the need for frequent applications [35].
Targeted Delivery: Nanobiopesticides can be designed for targeted delivery to specific pests or plant tissues. By functionalizing nanoparticles with ligands or antibodies, biopesticides can be selectively delivered to pests' specific receptors or plant cells, increasing their specificity and reducing off-target effects. This targeted delivery approach minimizes the exposure of non-target organisms to biopesticides, enhancing their safety and environmental sustainability [12, 35].
Synergistic Effects: Nanobiopesticides can exhibit synergistic effects when combined with other compounds or strategies. Nanoparticles can enhance the efficacy of biopesticides by improving their solubility, stability, and bioavailability. Additionally, nanoparticles themselves can possess inherent

antimicrobial or insecticidal properties, contributing to the overall efficacy of nanobiopesticides [1].

Adjuvant Effects: Nanoparticles in nanobiopesticides can act as adjuvants, enhancing the activity of biopesticide formulations. The presence of nanoparticles can enhance the stability and dispersibility of biopesticide active ingredients, improving their performance and persistence. Nanoparticles can also enhance the penetration of biopesticides through the cuticle or cell walls of pests, facilitating their entry into target tissues [22].

Disruption of Biological Processes: Nanobiopesticides can disrupt vital biological processes in pests, leading to their control or suppression. For instance, nanoparticles can interfere with the physiology, metabolism, or reproductive processes of pests, inhibiting their growth, development, or reproduction. These disruptions can be achieved through various mechanisms, such as enzyme inhibition, cell membrane disruption, or interference with signaling pathways [29].

Reduced Resistance Development: The use of nanobiopesticides can help mitigate the development of resistance in pests. By improving the effectiveness and targeted delivery of biopesticides, nanotechnology reduces the selection pressure on pests, making it more challenging for them to develop resistance. This can prolong the lifespan and efficacy of biopesticides and contribute to sustainable pest management practices [1].

15.7 Benefits and Advantages of Nanobiopesticides

Nanobiopesticides, which are nanotechnology-based formulations of biopesticides, offer several benefits and advantages over conventional chemical pesticides and even traditional biopesticides. These advantages make nanobiopesticides a promising alternative for pest management in agriculture. Some main benefits of nanobiopesticides are:

Increased Efficacy: Nanobiopesticides exhibit enhanced efficacy compared to conventional biopesticides. The use of nanotechnology allows for improved solubility, stability, and bioavailability of active ingredients. Nanoparticles can enhance the penetration and adhesion of biopesticides, improving their overall effectiveness in controlling pests and diseases. This increased efficacy leads to better pest control outcomes and reduces the reliance on chemical pesticides [22].

Reduced Environmental Impact: Nanobiopesticides offer a more environmentally friendly approach to pest management. They have lower toxicity and environmental persistence compared to chemical pesticides, reducing the risk of environmental pollution and adverse effects on non-target organisms. Nanobiopesticides can also be designed to have reduced off-target effects, minimizing their impact on beneficial insects, pollinators, and other organisms in the ecosystem [15].

Reduced Pesticide Resistance: The use of nanobiopesticides can help mitigate the development of resistance in pests. Nanotechnology allows for controlled release and sustained presence of biopesticides, making it more challenging for pests to develop resistance. This prolongs the effectiveness of biopesticides, reduces the need for frequent applications, and promotes sustainable pest management practices [50].

Improved Safety: Nanobiopesticides are generally safer for human health compared to chemical pesticides. They have lower toxicity and pose fewer health risks to farmers, consumers, and workers handling the products. Nanobiopesticides also have reduced risks of pesticide residues in agricultural products, contributing to food safety and reducing potential health hazards [1].

Compatibility with Integrated Pest Management (IPM): Nanobiopesticides are compatible with integrated pest management approaches. They can be integrated into IPM strategies that combine multiple pest control methods, such as cultural practices, biological control agents, and pest monitoring. Nanobiopesticides provide an additional tool in the toolbox of IPM, offering effective and sustainable pest control options [15].

Potential for Reduced Application Frequency and Dosage: The enhanced efficacy and controlled-release properties of nanobiopesticides can reduce the frequency and dosage of applications required. This results in cost savings for farmers, reduced labor, and lower overall pesticide use, contributing to more sustainable and economically viable pest management practices [47].

15.8 Conclusion and Future Perspectives

15.8.1 Conclusion

The utilization of nanobiopesticides derived from mushrooms has demonstrated great potential in the management of agricultural pests and diseases. Through their unique properties and mechanisms of action, these nanobiopesticides offer several advantages over conventional chemical pesticides. Nobiopesticides derived from mushrooms have shown high efficacy in controlling a wide range of pests and diseases. Their nanoscale size allows for enhanced penetration and targeting, ensuring better coverage and efficacy on the target organisms. Moreover, these nanobiopesticides have demonstrated lower toxicity towards non-target organisms, reducing the risk of environmental contamination and harm to beneficial insects. Mushrooms contain bioactive compounds with inherent antimicrobial, antifungal, and insecticidal properties. By harnessing these properties through nanotechnology, scientists have been able to develop highly potent nanobiopesticides. These formulations exhibit long-lasting effects, providing extended protection against pests and diseases. Furthermore, the encapsulation of mushroom-derived bioactive compounds in nanoparticles enhances their stability, increasing their shelf life and reducing the need for frequent applications. Another notable advantage of nanobiopesticides is their eco-friendly nature. They are biodegradable and pose minimal risks to human health and the environment. This makes them a sustainable alternative to conventional chemical pesticides, which often leave behind harmful residues and contribute to environmental pollution. Furthermore, the use of nanobiopesticides derived from mushrooms promotes the utilization of natural resources and supports the development of a bio-based economy. Mushrooms are readily available and can be cultivated sustainably, making them a renewable source for nanobiopesticide production. Overall, nanobiopesticides derived from mushrooms provide an encouraging solution for the sustainable management of agricultural pests and diseases. Further development and research in this arena will undoubtedly uncover new mushroom species and innovative nanotechnology applications, expanding the range of available nanobiopesticides and contributing to the advancement of environmentally friendly and effective pest management strategies in agriculture.

15.8.2 Future Perspectives

Looking ahead, the future perspectives of nanobiopesticides derived from mushrooms in the management of agricultural pests and diseases are promising. Continued research and development in this field hold significant potential for advancing sustainable agriculture practices.

Expansion of Mushroom Species: Currently, research has primarily focused on a limited number of mushroom species with known bioactive properties. In the future, there is an opportunity to explore and study a wider range of mushroom species to discover novel bioactive compounds. This

would enable the development of nanobiopesticides with diverse modes of action, targeting a broader spectrum of pests and diseases.
Development of Targeted Delivery Systems: Researchers can focus on refining the delivery systems of nanobiopesticides to improve their efficiency and target specificity. By designing nanoparticles that can specifically recognize and bind to pest or disease targets, the nanobiopesticides can achieve enhanced precision and efficacy. This would minimize off-target effects and reduce the amount of nanobiopesticides required for effective pest and disease control.
Integration with Other Pest Management Strategies: Nanobiopesticides can be integrated with other pest management strategies, such as biological control and cultural practices, to create integrated pest management (IPM) approaches. This combination of multiple methods can enhance the overall effectiveness of pest and disease control while reducing reliance on chemical pesticides. Synergistic interactions between nanobiopesticides and other control methods can provide more sustainable and long-term solutions.
Field Validation and Commercialization: As nanobiopesticides derived from mushrooms continue to show promise in laboratory and greenhouse studies, future research should focus on field trials to evaluate their performance under real-world conditions. These trials will provide valuable data on efficacy, persistence, and environmental impact, paving the way for commercialization and widespread adoption.
Regulations and Safety Assessments: With the emergence of nanobiopesticides, regulatory frameworks and safety assessments need to be developed and implemented. It is important to ensure that these innovative products meet the necessary safety standards for human health, non-target organisms, and the environment. Establishing clear guidelines and regulations will facilitate the responsible use and commercialization of nanobiopesticides.

References

1. Abdollahdokht, D., Gao, Y., Faramarz, S., Conventional agrochemicals towards nano-biopesticides: an overview on recent advances. *Chem. Biol. Technol. Agric.*, 9, 13, 2022.
2. Aboody, M.S.A. and Mickymaray, S., Anti-Fungal Efficacy and Mechanisms of Flavonoids. *Antibiotics.*, 9, 45, 2020.
3. Abubakar, I., Mann, A., Mathew, J.T., Evaluation of Phytochemical, Antinutritional and Antioxidant Potentials of Flower and Seed Methanol Extracts of *Senna alata* L. Grown in Nigeria. *Am. J. Appl. Chem.*, 3, 93–100, 2015a.

4. Abubakar, I., Mann, A., Mathew, J.T., Phytochemical composition, Antioxidant and Anti-nutritional properties of root-bark and leaf methanol extracts of *Senna alata L.* Grown in Nigeria. *Afr. J. Pure Appl. Chem.*, 9, 91–97, 2015b.
5. Adetunji, C.O., Mathew, J.T., Inobeme, A., Olaniyan, O., Singh, K.R.B., Ogundolie, F.A., Vanya, N., Singh, J., Singh, R.P., Microbial and Plant Cell Biosensors for Environmental Monitoring, in: *Nanobiosensors for Environmental Monitoring*, pp. 175–190, Springer, Cham, 2022a.
6. Adetunji, C.O., Inobeme, A., Olaniyan, O.T., Anani, O.A., Mathew, J.T., Bodunrinde, R.E., Adetunji, J.B., Hefft, D.I., Nanomaterials as Effective Tools for Detection and Microbial Diagnosis of Foodborne Pathogens, in: *The Science of Nanomaterials: Basics and Applications*, Apple Academic Press, Inc. Co-published with CRC Press (Taylor & Francis), Oakville, Ontario, Canada, 2022b.
7. Adetunji, C.O., Anani, O.A., Olaniyan, O.T., Inobeme, A., Bodunrinde, R.E., Mathew, J.T., Adetunji, J.B., Hefft, D.I., Prospect of Utilizing Novel Nanomaterials for Effective Food Security, Safety, Packaging, and Preservation, in: *The Science of Nanomaterials: Basics and Applications*, Apple Academic Press, Inc. Co-published with CRC Press (Taylor & Francis), Oakville, Ontario, Canada, 2022c.
8. Adetunji, C.O., Ogundolie, F.A., Mathew, J.T., Inobeme, A., Olotu, T., Olaniyan, O.T., Ijabadeniyi, O.A., Ajiboye, M.D., Ajayi, O.O., Dauda, W., Graphene-based nanomaterials for targeted drug delivery and tissue engineering, in: *Novel Platforms for Drug Delivery Applications*, pp. 277–288, Elsevier Ltd., Oxford, United Kingdom, 2022d.
9. Adetunji, C.O., Olaniyan, O.T., Shariati, M.A., Olugbenga, S.M., Osemwegie, O.O., Rebezov, M., Ajayi, O.O., Farrukh, M., Bodunrinde, R.E., Adetunji, J.B., *et al.*, Medical Attributes of Breynia disticha J.R. Forst. & G. Forst, in: *Phytochemical Composition and Pharmacy of Medicinal Plants*, vol. 2, T. Pullaiah (Ed.), pp. 531–538, Apple Academic Press, Inc. Co-published with CRC Press (Taylor & Francis), Oakville, Ontario, Canada, 2023a.
10. Adetunji, C.O., Olaniyan, O.T., Shariati, M.A., Olugbenga, S.M., Rebezov, M., Ajayi, O.O., Goncharov, A., Bodunrinde, R.E., Adetunji, J.B., Adeniyi, M.J., *et al.*, Biological Constituents and Medicinal Attributes of Spondias mombin L, in: *Phytochemical Composition and Pharmacy of Medicinal Plants*, vol. 2, T. Pullaiah (Ed.), pp. 517–529, Apple Academic Press, Inc. Co-published with CRC Press (Taylor & Francis), Oakville, Ontario, Canada, 2023b.
11. Adetunji, C.O., Olaniyan, O.T., Shariati, M.A., Inobeme, A., Mathew, J.T., Olugbenga, S.M., Adetunji, J.B., Hefft, D.I., Rebezov, M., Ajayi, O.O., *et al.*, Phytochemical Constituents and Medicinal Effectiveness of Buchholzia coriacea Engl., in: *Phytochemical Composition and Pharmacy of Medicinal Plants*, vol. 2, T. Pullaiah (Ed.), pp. 483–498, Apple Academic Press, Inc. Co-published with CRC Press (Taylor & Francis), Oakville, Ontario, Canada, 2023c.

12. Adetunji, C.O., Olaniyan, O.T., Olugbenga, S.M., Shariati, M.A., Osarenkhoe, O.O., Odoh, U.E., Oyedara, O.O., Rebezov, M., Ajayi, O.O., Babaeva, M., *et al.*, Pharmacology, Photochemical, and Healing Effectiveness of *Chromolaena odorata* (L.) R.M. King and H. Robinson, in: *Phytochemical Composition and Pharmacy of Medicinal Plants*, vol. 2, T. Pullaiah (Ed.), pp. 461–482, Apple Academic Press, Inc. Co-published with CRC Press (Taylor & Francis), Oakville, Ontario, Canada, 2023d.
13. Alengebawy, A., Abdelkhalek, S.T., Qureshi, S.R., Wang, M.Q., Heavy Metals and Pesticides Toxicity in Agricultural Soil and Plants: Ecological Risks and Human Health Implications. *Toxics*, *9*, 3, 42, 2021.
14. Arunachalam, K., Sasidharan, S.P., Yang, X., A concise review of mushrooms antiviral and immunomodulatory properties that may combat against COVID-19. *Food Chem. Adv.*, *1*, 100023, 2022.
15. Ayilara, M.S., Adeleke, B.S., Akinola, S.A., Fayose, C.A., Adeyemi, U.T., Gbadegesin, L.A., Omole, R.K., Johnson, R.M., Uthman, Q.O., Babalola, O.O., Biopesticides as a promising alternative to synthetic pesticides: A case for microbial pesticides, phytopesticides, and nanobiopesticides. *Front. Microbiol.*, *14*, 1040901, 2023.
16. Bains, A., Chawla, P., Kaur, S., Najda, A., Fogarasi, M., Fogarasi, S., Bioactives from Mushroom: Health Attributes and Food Industry Applications. *Materials*, *14*, 7640, 2021.
17. Berestetskiy, A. and Hu, Q., The Chemical Ecology Approach to Reveal Fungal Metabolites for Arthropod Pest Management. *Microorganisms*, *9*, 1379, 2021.
18. Bhambri, A., Srivastava, M., Mahale, V.G., Mahale, S., Karn, S.K., Mushrooms as Potential Sources of Active Metabolites and Medicines. *Front. Microbiol.*, *13*, 837266, 2022.
19. Chaud, M., Souto, E.B., Zielinska, A., Severino, P., Batain, F., Oliveira-Junior, J., Alves, T., Nanopesticides in Agriculture: Benefits and Challenge in Agricultural Productivity, Toxicological Risks to Human Health and Environment. *Toxics*, *9*, 131, 2021.
20. Chun, S., Gopal, J., Muthu, M., Antioxidant Activity of Mushroom Extracts/Polysaccharides-Their Antiviral Properties and Plausible AntiCOVID-19 Properties. *Antioxidants*, 10, 1899, 2021.
21. Damalas, C.A. and Koutroubas, S.D., Farmers' Exposure to Pesticides: Toxicity Types and Ways of Prevention. *Toxics*, *4*, 1, 1, 2016.
22. Deka, B., Babu, A., Baruah, C., Barthakur, M., Nanopesticides: A Systematic Review of Their Prospects With Special Reference to Tea Pest Management. *Front. Nutr.*, *8*, 686131, 2021.
23. Dyussembayev, K., Sambasivam, P., Bar, I., Brownlie, J.C., Shiddiky, M.J.A., Ford, R., Biosensor Technologies for Early Detection and Quantification of Plant Pathogens. *Front. Chem.*, *9*, 636245, 2021.
24. Ekiz, E., Oz, E., Abd El-Aty, A.M., Proestos, C., Brennan, C., Zeng, M., Tomasevic, I., Elobeid, T., Çadırcı, K., Bayrak, M., *et al.*, Exploring the

Potential Medicinal Benefits of Ganoderma lucidum: From Metabolic Disorders to Coronavirus Infections. *Foods*, *12*, 1512, 2023.

25. Ganaie, M.I., Jan, I., Mayer, A.N., Dar, A.A., Mayer, I.A., Ahmed, P., Sofi, J.A., Health Risk Assessment of Pesticide Residues in Drinking Water of Upper Jhelum Region in Kashmir Valley-India by GC-MS/MS. *Int. J. Anal. Chem.*, 1, 6802782, 2023.
26. Inobeme, A., Adetunji, C.O., Maliki, M., Onyeachu, B.I., Kelani, T., Eziukwu, C.A., Olori, E., Mathew, J.T., Bamigboye, M.O., Strategies to synthesize, advantages, and disadvantages of pharmaceutical nanoparticles, in: *Nanotechnology for Drug Delivery and Pharmaceuticals*, pp. 371–402, Academic Press is an imprint of Elsevier, Cambridge, Massachusetts, USA, 2023.
27. Jiang, Q., Zhang, M., Mujumdar, A.S., UV induced conversion during drying of ergosterol to vitamin D in various mushrooms: Effect of different drying conditions. *Trends Food Sci. Tech.*, *105*, 200–210, 2020.
28. Kruk, M.E., Gage, A.D., Arsenault, C., Jordan, K., Leslie, H.H., Roder-DeWan, S., Adeyi, O., Barker, P., Daelmans, B., Doubova, S.V., English, M., García-Elorrio, E., Guanais, F., Gureje, O., Hirschhorn, L.R., Jiang, L., Kelley, E., Lemango, E.T., Liljestrand, J., Malata, A., Pate, M., High-quality health systems in the Sustainable Development Goals era: time for a revolution. *Lancet Global Health*, *6*, e1196–e1252, 2018.
29. Kubiak, A., Wolna-Maruwka, A., Niewiadomska, A., Pilarska, A.A., The Problem of Weed Infestation of Agricultural Plantations vs. the Assumptions of the European Biodiversity Strategy. *Agronomy*, *12*, 1808, 2022.
30. Kumar, A., Singh, S.K., Kant, C., Verma, H., Kumar, D., Singh, P.P., Modi, A., Droby, S., Kesawat, M.S., Alavilli, H., *et al.*, Microbial Biosurfactant: A New Frontier for Sustainable Agriculture and Pharmaceutical Industries. *Antioxidants*, *10*, 1472, 2021.
31. Kumari, S., Goyal, A., Sönmez Gürer, E., Algın Yapar, E., Garg, M., Sood, M., Sindhu, R.K., Bioactive Loaded Novel Nano-Formulations for Targeted Drug Delivery and Their Therapeutic Potential. *Pharmaceutics*, *14*, 1091, 2022.
32. Landi, N., Clemente, A., Pedone, P.V., Ragucci, S., Di Maro, A., An Updated Review of Bioactive Peptides from Mushrooms in a Well-Defined Molecular Weight Range. *Toxins*, *1*, 84, 2022.
33. Manzari, M.T., Shamay, Y., Kiguchi, H., Rosen, N., Scaltriti, M., Heller, D.A., Targeted drug delivery strategies for precision medicines. *Nat. Rev. Mater.*, *6*, 4, 351–370, 2021.
34. Mathew, J.T., Ndamitso, M.M., Shaba, E.Y., Mustapha, S., Muhammed, S.S., Salihu, A.S., Physicochemical and Phytochemical Composition of locust bean tree emperor moth larvae (*Bunaea alcinoe)* from Gurara Local Government Area, Niger state, Nigeria. *Int. J. Eng. Sci. Invent.*, 3, 14–18, 2014.
35. Melanie, M., Miranti, M., Kasmara, H., Malini, D.M., Husodo, T., Panatarani, C., Joni, I.M., Hermawan, W., Nanotechnology-Based Bioactive Antifeedant for Plant Protection. *Nanomat.*, *12*, 630, 2022.

36. Mishra, A., Behura, A., Mawatwal, S., Kumar, A., Naik, L., Mohanty, S.S., Manna, D., Dokania, P., Mishra, A., Patra, S.K., *et al.*, Structure-function and application of plant lectins in disease biology and immunity. *Food Chem. Toxicol.*: an international journal published for the British Industrial Biological Research Association, *134*, 110827, 2019.
37. Mohammad, Z.H., Ahmad, F., Ibrahim, S.A., Application of nanotechnology in different aspects of the food industry. *Discov. Food*, 2, 12, 2022.
38. Mutha, R.E., Tatiya, A.U., Surana, S.J., Flavonoids as natural phenolic compounds and their role in therapeutics: an overview. *Future J. Pharm. Sci.*, 7, 25, 2021.
39. Nongbet, A., Mishra, A.K., Mohanta, Y.K., Mahanta, S., Ray, M.K., Khan, M., Baek, K.-H., Chakrabartty, I., Nanofertilizers: A Smart and Sustainable Attribute to Modern Agriculture. *Plants*, *11*, 2587, 2022.
40. Pateiro, M., Gómez, B., Munekata, P.E.S., Barba, F.J., Putnik, P., Kovačević, D.B., Lorenzo, J.M., Nanoencapsulation of Promising Bioactive Compounds to Improve Their Absorption, Stability, Functionality and the Appearance of the Final Food Products. *Molecules*, *26*, 1547, 2021.
41. Pathak, V.M., Verma, V.K., Rawat, B.S., Kaur, B., Babu, N., Sharma, A., Dewali, S., Yadav, M., Kumari, R., Singh, S., *et al.*, Current status of pesticide effects on environment, human health and it's eco-friendly management as bioremediation: A comprehensive review. *Front. Microbiol.*, *13*, 962619, 2022.
42. Rajput, V.D., Minkina, T., Upadhyay, S.K., Kumari, A., Ranjan, A., Mandzhieva, S., Sushkova, S., Singh, R.K., Verma, K.K., Nanotechnology in the Restoration of Polluted Soil. *Nanomaterials*, *12*, 769, 2022.
43. Rizzo, D.M., Lichtveld, M., Mazet, J.A.K., Togami, E., Miller, S.A., Plant health and its effects on food safety and security in a One Health framework: four case studies. *One Health Outlook*, *3*, 6, 2021.
44. Sánchez-Bayo, F., Indirect Effect of Pesticides on Insects and Other Arthropods. *Toxics*, *9*, 177, 2021.
45. Senapaty, M.K., Ray, A., Padhy, N., IoT-Enabled Soil Nutrient Analysis and Crop Recommendation Model for Precision Agriculture. *Computers*, *12*, 61, 2023.
46. Seo, D.J. and Choi, C., Antiviral Bioactive Compounds of Mushrooms and Their Antiviral Mechanisms: A Review. *Viruses*, *13*, 350, 2021.
47. Shelar, A., Nile, S.H., Singh, A.V., Rothenstein, D., Bill, J., Xiao, J., Chaskar, M., Kai, G., Patil, R., Recent Advances in Nano-Enabled Seed Treatment Strategies for Sustainable Agriculture: Challenges, Risk Assessment, and Future Perspectives. *Nano Micro Lett.*, *15*, 54, 2023.
48. Siddiqui, J.A., Fan, R., Naz, H., Bamisile, B.S., Hafeez, M., Ghani, M.I., Wei, Y., Xu, Y., Chen, X., Insights into insecticide-resistance mechanisms in invasive species: Challenges and control strategies. *Front. Physiol.*, *13*, 1112278, 2023.

49. Sosan, M.B., Adeleye, A.O., Oyekunle, J.A.O., Udah, O., Oloruntunbi, P.M., Daramola, M.O., Saka, W.T., Dietary risk assessment of organochlorine pesticide residues in maize-based complementary breakfast food products in Nigeria. *Heliyon, 6*, e05803, 2020.
50. Šunjka, D. and Mechora, Š., An Alternative Source of Biopesticides and Improvement in Their Formulation-Recent Advances. *Plants, 11*, 3172, 2022.
51. Tsado, A.N., Abdulkadir, A., Shaba, A.M., Mathew, J.T., Umar, A.M., Chirama, D.N., Phytochemical, Antioxidant and Antimicrobial Potentials of Methanol Seed Extracts of *Carica Papaya*. *World Wide J. Multidiscip. Res. Dev.*, 4, 24–28, 2018.
52. Tudi, M., Daniel Ruan, H., Wang, L., Lyu, J., Sadler, R., Connell, D., Chu, C., Phung, D.T., Agriculture Development, Pesticide Application and Its Impact on the Environment. *Int. J. Environ. Res. Public Health, 18*, 1112, 2021.
53. Wang, H., Peng, H., Li, W., Cheng, P., Gong, M., The Toxins of *Beauveria bassiana* and the Strategies to Improve Their Virulence to Insects. *Front. Microb.*, 12, 705343, 2021.

16

Nutraceutical, Mineral, Proximate Constituents from Beneficial Mushrooms

John Tsado Mathew[1*], Abel Inobeme[2], Charles Oluwaseun Adetunji[3], Abdulfatai Aideye Otori[4], Elijah Yanda Shaba[5], Musah Monday[1], Yakubu Azeh[1], Ismail Haruna[1], Amos Mamman[1], Jemkur Maurice[6], Oluwafemi Adebayo Oyewole[7], Eniola K. I. T.[8] and Mohammed Bello Yerima[9]

[1]Department of Chemistry, Ibrahim Badamasi Babangida University, Lapai, Niger State, Nigeria
[2]Department of Chemistry, Edo University, Uzairue, Auchi, Edo State, Nigeria
[3]Applied Microbiology, Biotechnology and Nanotechnology Laboratory, Department of Microbiology, Edo University, Uzairue, Auchi, Edo State, Nigeria
[4]Department of Chemical Engineering, Federal Polytechnic, Bida, Niger State, Nigeria
[5]Department of Chemistry, Federal University of Technology, Minna, Nigeria
[6]Department of Chemistry, Air Force Institute of Technology, Kaduna State, Nigeria
[7]Department of Microbiology, School of Life Sciences, Federal University of Technology, Minna, Nigeria
[8]Department of Biological Sciences, Joseph Ayo Babalola University, Ikeji Arakeji, Osun State, Nigeria
[9]Department of Microbiology, Sokoto State University, Sokoto, Sokoto State, Nigeria

Abstract
Mushrooms, belonging to the fungal kingdom, have garnered significant attention for their potential health benefits and are considered valuable sources of nutraceuticals, minerals, and proximate constituents. Nutraceuticals are bioactive compounds derived from natural sources that provide both nutritional and pharmaceutical benefits. Mushrooms are rich in nutraceutical constituents, with beta-glucans being one of the most notable. Beta-glucans, complex polysaccharides found in species like Reishi (Ganodermalucidum), Shiitake (Lentinulaedodes),

**Corresponding author*: johntsadom@gmail.com; jmathew@ibbu.edu.ng

Charles Oluwaseun Adetunji, Chukwuebuka Egbuna, Oluwatosin Ademola Ijabadeniyi and Samantha C. Karunarathna (eds.) Mushroom Biotechnology for Improved Agriculture and Human Health, (337–358) © 2025 Scrivener Publishing LLC

and Maitake (Grifolafrondosa), have been associated with immune-enhancing properties and potential anti-cancer effects. These bioactive compounds have gained attention for their potential in supporting immune function and preventing various diseases. Mushrooms are low in calories and fat but high in dietary fiber, making them beneficial for maintaining healthy blood sugar levels and promoting feelings of fullness. Additionally, mushrooms provide proteins that contain all the essential amino acids required by the human body, making them a valuable protein source, especially for individuals following vegetarian or vegan diets. Incorporating mushrooms into a balanced diet can provide a natural and nutritious way to support immune function, promote antioxidant defense, and maintain overall well-being. Further research is warranted to explore the full extent of their health-promoting properties and to understand their mechanisms of action.

Keywords: Beta-glucans, mushrooms, nutraceuticals

16.1 Introduction

Mushrooms have long been recognized for their unique flavors and culinary versatility, but they also possess a range of health-promoting properties. In recent years, mushrooms have gained attention as a source of nutraceuticals, minerals, and proximate constituents that offer various potential health benefits. Nutraceuticals are bioactive compounds derived from natural sources, such as plants and fungi that provide both nutritional and pharmaceutical benefits. Mushrooms, as part of the fungal kingdom, are a rich source of nutraceutical constituents that have been studied for their therapeutic properties. One of the prominent nutraceutical constituents found in mushrooms is beta-glucans. These complex polysaccharides are known for their immune-enhancing properties and potential anticancer effects. Mushroom species like Reishi (Ganodermalucidum), Shiitake (Lentinulaedodes), and Maitake (Grifolafrondosa) are particularly rich in beta-glucans. It is important to note that the composition and concentrations of these constituents can vary among different mushroom species. Therefore, the specific health benefits and nutritional profile may differ depending on the type of mushroom consumed [16, 21].

In addition to nutraceuticals, mushrooms are also abundant in various minerals that are essential for optimal health. For example, mushrooms such as Shiitake are a good source of selenium, a trace mineral with antioxidant properties that supports immune function and may have anticancer effects. They also contain potassium, copper, zinc, and other minerals that contribute to various physiological processes in the body. Mushrooms are low in calories and fat but high in dietary fiber, making them a healthy addition to a balanced diet. They also provide proteins that contain all the essential amino acids required by the human body, making them a valuable protein source, especially for individuals following vegetarian or vegan diets [24, 48].

It is important to note that different mushroom species may contain varying levels of nutraceuticals, minerals, and proximate constituents. Therefore, the specific health benefits and nutritional profile may differ among mushroom varieties. In summary, mushrooms are more than just a delicious ingredient in culinary dishes. They offer a range of nutraceutical, mineral, and proximate constituents that contribute to their potential health benefits (Table 16.1). Incorporating beneficial mushrooms into your diet can provide a natural and nutritious way to support overall well-being and promote a healthy lifestyle [8, 22]. The chapter presents the nutraceutical, proximate, and mineral compositions of essential mushrooms (Figure 16.1).

16.2 Nutraceutical Constituents of Mushrooms

Mushrooms are known for their diverse array of nutraceutical constituents (Figure 16.1), which are bioactive compounds that provide both nutritional and pharmaceutical benefits. Most of the nutraceutical constituents found in mushrooms are:

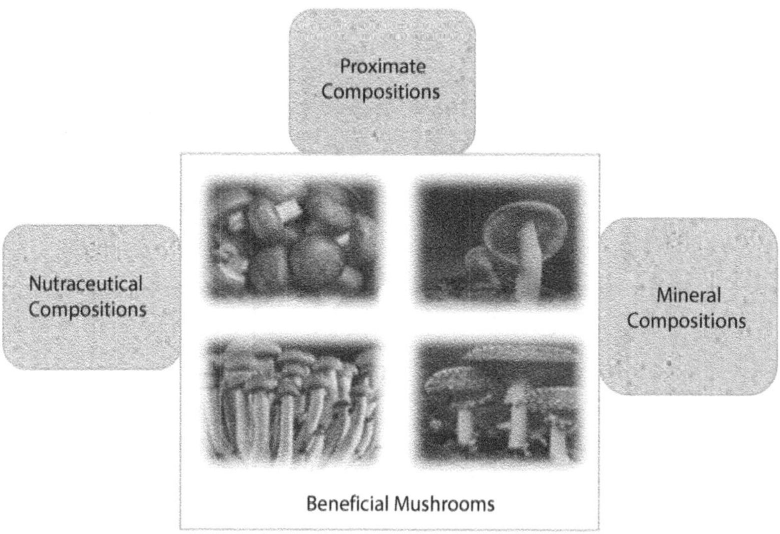

Figure 16.1 The nutraceutical constituents found in mushrooms.

Table 16.1 Beneficial mushroom, nutraceuticals, minerals, and proximate constituents.

Beneficial mushroom	Nutraceuticals	Minerals	Proximate constituents	References
Reishi (Ganodermalucidum)	Polysaccharides, triterpenoids, ganoderic acids	Potassium, magnesium, calcium, phosphorus	Protein, carbohydrates, dietary fiber, fats	[20, 23]
Lion's Mane (Hericiumerinaceus)	Hericenones, erinacines, polysaccharides	Potassium, zinc, copper, selenium	Protein, carbohydrates, dietary fiber, fats	[19, 22]
Chaga (Inonotusobliquus)	Betulinic acid, polysaccharides, melanin	Potassium, calcium, manganese, iron	Protein, carbohydrates, dietary fiber, fats	[14]
Cordyceps (Cordycepssinensis)	Cordycepin, adenosine, polysaccharides	Zinc, iron, magnesium, potassium	Protein, carbohydrates, dietary fiber, fats	[32, 39]
Shiitake (Lentinulaedodes)	Lentinan, eritadenine, polysaccharides	Copper, selenium, manganese, zinc	Protein, carbohydrates, dietary fiber, fats	[36, 56]
Maitake (Grifolafrondosa)	Beta-glucans, ergosterol, polysaccharides	Potassium, phosphorus, calcium, magnesium	Protein, carbohydrates, dietary fiber, fats	[62, 22]

Beta-Glucans: Beta-glucans are complex polysaccharides found in various mushroom species, such as Reishi (Ganodermalucidum), Shiitake (Lentinulaedodes), and Maitake (Grifolafrondosa). These compounds have immune-enhancing properties and are known for their potential anti-cancer effects. Beta-glucans stimulate immune cells, promote the production of cytokines, and enhance the body's defense against infections and tumors [43].

Triterpenes: Triterpenes are bioactive compounds found in specific mushroom species, including Reishi, Turkey Tail (Trametesversicolor), and Chaga (Inonotusobliquus). Triterpenes exhibit anti-inflammatory, antioxidant, and immune-modulating properties. They have been studied for their potential to support liver health, reduce inflammation, and boost immune function [2, 3, 19].

Ergosterol/Vitamin D: Mushrooms have the unique ability to synthesize ergosterol, which can be converted into vitamin D2 when exposed to sunlight or UV radiation. Vitamin D plays a crucial role in immune function, bone health, as well as overall well-being. Certain mushroom varieties, such as those exposed to sunlight during growth, can be a natural source of vitamin D, particularly for individuals with limited sun exposure [35, 52].

Lectins: Lectins are carbohydrate-binding proteins found in mushrooms like Shiitake and Snow Fungus (Tremellafuciformis). These compounds have demonstrated antitumor, immunomodulatory, and antimicrobial properties. Lectins may also have the potential to regulate blood sugar levels and support digestive health [55].

Aromatics and Phenolic Compounds: Many mushrooms contain aromatic compounds and phenolic compounds, including flavonoids and phenolic acids. These compounds possess antioxidant and anti-inflammatory properties and contribute to the overall antioxidant capacity of mushrooms. Examples of mushrooms rich in phenolic compounds include Porcini (Boletus edulis) and Oyster (Pleurotus spp.) mushrooms. It is important to note that the specific composition and concentration of nutraceutical constituents can vary among different mushroom species and even within different parts of the same mushroom. Additionally, the bioavailability and health benefits of these constituents can depend on factors such as cooking methods and processing techniques [1, 12].

16.3 Mineral Constituents of Mushrooms

Mushrooms are not only valued for their nutraceutical constituents but also for their mineral content. While the specific mineral composition can vary among different mushroom species, here are some commonly found mineral constituents in mushrooms:

Selenium: Certain mushroom species, such as Shiitake (Lentinulaedodes), are known for their selenium content. Selenium is a trace mineral with antioxidant properties that plays a crucial role in immune function, thyroid health, and antioxidant defense. It also has potential anti-cancer effects and supports the body's natural detoxification processes. The selenium content in mushrooms is generally expressed in micrograms per 100 grams of the edible portion [25, 50]. Some examples of the approximate selenium content in common mushroom species: White Button Mushrooms (Agaricusbisporus); approximately 4–12 micrograms of selenium per 100 grams. Portobello Mushrooms (Agaricusbisporus, mature); roughly 5–18 micrograms of selenium per 100 grams. Shiitake Mushrooms (Lentinulaedodes); nearly 7–22 micrograms of selenium per 100 grams. Oyster Mushrooms (Pleurotus spp.); in the region of 1–4 micrograms of selenium per 100 grams. While mushrooms may not be a significant source of selenium on their own, incorporating them into a balanced diet that includes other selenium-rich foods can contribute to overall selenium intake. Selenium is an essential trace mineral involved in various physiological processes, including antioxidant defense, thyroid function, immune system support, and reproductive health [34].
Potassium: Mushrooms are a good source of potassium, an essential mineral involved in maintaining proper heart and muscle function, regulating blood pressure, and supporting nerve function. Adequate potassium intake is important for maintaining electrolyte balance and overall cellular health. The potassium content in mushrooms is generally expressed in milligrams per 100 grams of the edible portion [18, 30]. There are some examples of the approximate potassium content in common mushroom species: White Button Mushrooms (Agaricusbisporus); roughly 318–429 mg of potassium per 100 grams. Portobello Mushrooms (Agaricusbisporus, mature); about 364–521 mg of potassium per 100 grams. Shiitake Mushrooms (Lentinulaedodes); in the region of 308–506 mg of potassium per 100 grams. Oyster Mushrooms (Pleurotus spp.); approximately 364–533 mg of potassium per 100 grams. While mushrooms may not be a significant source of potassium on their own, incorporating them into a balanced

diet that includes other potassium-rich foods can contribute to overall potassium intake. Potassium is an essential mineral involved in various physiological processes, including maintaining proper heart and muscle function, regulating blood pressure, and supporting nerve function [51].

Copper: Copper is another essential mineral found in mushrooms, albeit in varying amounts. Copper is involved in the production of red blood cells, energy metabolism, and the formation of connective tissues. It also acts as an antioxidant and contributes to the proper functioning of the immune system. The copper content in mushrooms is generally expressed in milligrams per 100 grams of the edible portion [11]. There are some examples of the approximate copper content in common mushroom species: White Button Mushrooms (Agaricusbisporus); approximately 0.1–0.2 mg of copper per 100 grams. Portobello Mushrooms (Agaricusbisporus, mature); approximately 0.2–0.3 mg of copper per 100 grams. Shiitake Mushrooms (Lentinulaedodes); approximately 0.1–0.3 mg of copper per 100 grams. Oyster Mushrooms (Pleurotus spp.); approximately 0.1–0.3 mg of copper per 100 grams. It is important to note that these values are approximate and can vary depending on factors such as cultivation methods, maturity of the mushrooms, and variations within species. While mushrooms may not be a significant source of copper on their own, incorporating them into a balanced diet that includes other copper-rich foods can contribute to overall copper intake. Copper is an essential mineral involved in various physiological processes, including the formation of red blood cells, energy metabolism, and antioxidant defense [46].

Zinc: Mushrooms can be a source of zinc, an essential mineral involved in numerous enzymatic reactions in the body. Zinc plays a critical role in immune function, wound healing, DNA synthesis, and the maintenance of proper taste and smell sensations. It also acts as an antioxidant and supports overall growth and development. The zinc content in mushrooms is generally expressed in milligrams per 100 grams of the edible portion [5, 45]. Here are some examples of the approximate zinc content in common mushroom species: White Button Mushrooms (Agaricusbisporus); about 0.5–1 mg of zinc per 100 grams. Portobello Mushrooms (Agaricusbisporus, mature); in the region of 0.7–1.1 mg of zinc per 100 grams. Shiitake Mushrooms (Lentinulaedodes); roughly 0.7–1.2 mg of zinc per 100 grams. Oyster Mushrooms (Pleurotus spp.); approximately 0.4–0.6 mg of zinc per 100 grams. It is important to note that these values are approximate and can vary depending on factors such as cultivation methods, maturity of the mushrooms, and variations within species. While mushrooms may not be a significant source of zinc on their own, incorporating them into a balanced diet that includes other zinc-rich foods can contribute to overall zinc intake. Zinc is an essential mineral involved in various physiological processes, including immune function, DNA synthesis, growth and development, and wound healing [33].

Iron: The iron content in mushrooms can vary depending on the species and growing conditions. Generally, mushrooms are not considered a significant source of iron compared to other iron-rich foods like meat or legumes. However, they can still contribute to overall iron intake. The iron content in mushrooms is typically expressed in milligrams per 100 grams of the edible portion [7, 13]. Some examples of iron content in common mushroom species: White Button Mushrooms (Agaricusbisporus); approximately 0.5 mg of iron per 100 grams. Portobello Mushrooms (Agaricusbisporus, mature); nearly 0.5 mg of iron per 100 grams. Shiitake Mushrooms (Lentinulaedodes); approximately 1.2 mg of iron per 100 grams. Oyster Mushrooms (Pleurotus spp.); in the region of 0.5–1 mg of iron per 100 grams. It is important to note that these values are approximate and can vary depending on factors such as cultivation methods, maturity of the mushrooms, and variations within species. While mushrooms may not be a significant source of iron on their own, they can still contribute to overall dietary iron intake as part of a varied and balanced diet. Iron absorption from plant-based sources like mushrooms can be enhanced by consuming them alongside foods rich in vitamin C, which improves non-heme iron absorption [9, 38].

Magnesium: The magnesium content in mushrooms is generally expressed in milligrams per 100 grams of the edible portion. There are some examples of the approximate magnesium content in common mushroom species: White Button Mushrooms (Agaricusbisporus); approximately 9–10 mg of magnesium per 100 grams. Portobello Mushrooms (Agaricusbisporus, mature); approximately 9–10 mg of magnesium per 100 grams. Shiitake Mushrooms (Lentinulaedodes); approximately 34–38 mg of magnesium per 100 grams. Oyster Mushrooms (Pleurotus spp.); approximately 8–10 mg of magnesium per 100 grams. It is important to note that these values are approximate and can vary depending on factors such as cultivation methods, maturity of the mushrooms, and variations within species. While mushrooms may not be a significant source of magnesium on their own, incorporating them into a balanced diet that includes other magnesium-rich foods can contribute to overall magnesium intake. Magnesium is an essential mineral involved in various physiological processes, including energy metabolism, nerve function, muscle contraction, and bone health [58, 63].

Mushrooms may also contain trace amounts of other minerals, including calcium, phosphorus, and manganese. While the concentrations may vary, these minerals are important for various bodily functions, such as oxygen transport, bone health, energy production, and enzyme activity. It is worth noting that the mineral content of mushrooms can be influenced by factors like soil composition, cultivation methods, and the mushroom's growth stage. Additionally, the bioavailability of minerals from mushrooms may vary, and it can be influenced by cooking methods and the presence of certain compounds that may enhance or hinder mineral absorption [57].

16.4 Proximate Constituents of Mushrooms

Mushrooms are complex organisms that contain various proximate constituents, including proteins, carbohydrates, vitamins, fats, fiber as well as minerals [54]. Here is a breakdown of some key constituents found in mushrooms (Figure 16.2):

Carbohydrates: Mushrooms are low in carbohydrates compared to other plant-based foods. The main carbohydrate found in mushrooms is chitin, a complex polysaccharide. Chitin provides structural support to the mushroom cell walls. Mushrooms are predominantly composed of water, which accounts for a significant portion of their weight. The carbohydrate content in mushrooms typically ranges from 1% to 5% of their fresh weight. On average, fresh mushrooms may contain around 2–3 grams of carbohydrates per 100 grams. The main carbohydrate found in mushrooms is

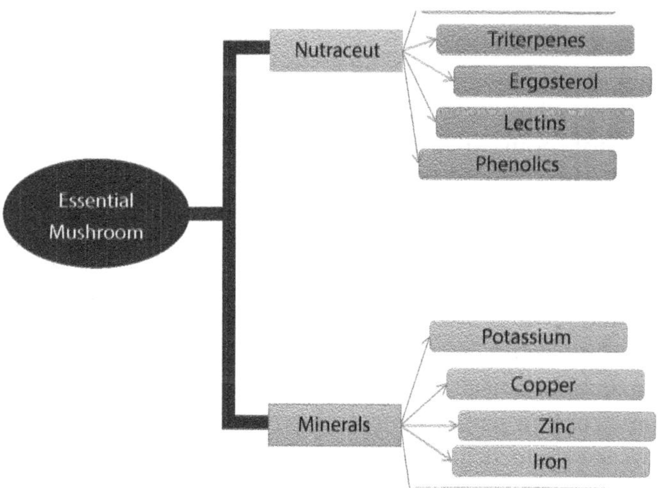

Figure 16.2 Nutraceutical and mineral components of mushroom.

chitin, a complex polysaccharide that provides structural support to the mushroom cell walls. Chitin is not easily digested by humans and is considered a dietary fiber rather than a readily available carbohydrate source [41, 61].

Proteins: Mushrooms are a good source of dietary proteins, although the protein content varies among different mushroom species. The protein content generally ranges from 2% to 5% of the mushroom's dry weight. Mushrooms contain all the essential amino acids required by the human body. Mushrooms are not typically recognized as a significant source of protein compared to animal-based foods or legumes. However, they do contain some amount of protein that can contribute to overall protein intake [42]. The protein content in mushrooms can vary depending on the species and growing conditions. The protein content of mushrooms generally ranges from 2% to 5% of their dry weight. On average, fresh mushrooms may contain around 1–3 grams of protein per 100 grams. The protein content in mushrooms is relatively low compared to other sources like meat, legumes, or dairy products [8].

Fats: Mushrooms are low in fat, with most species containing less than 1% fat. The fat content is mainly composed of polyunsaturated fatty acids, including linolenic acid as well as linoleic acid. However, certain wild mushrooms may contain higher amounts of fat. Most mushroom varieties contain less than 1% fat by weight. The fat content in mushrooms is typically less than 0.5 grams per 100 grams of fresh mushrooms. This low-fat content makes mushrooms a favorable choice for those following low-fat diets or seeking to reduce their overall fat intake. The fat found in mushrooms is primarily in the form of monounsaturated fatty acids (MUFAs) and polyunsaturated fatty acids (PUFAs). Some common types of fatty acids found in mushrooms include linoleic acid and linolenic acid, which are both essential fatty acids [53].

Vitamins: Mushrooms are a natural source of several vitamins, including B vitamins such as riboflavin (B2), thiamine (B1), pantothenic acid (B5), folate (B9) as well as niacin (B3). They are also a good source of vitamin D, particularly when exposed to sunlight or UV light. Mushrooms contain thiamine, which is important for energy metabolism and nervous system function. Riboflavin is involved in energy production and helps support healthy skin and vision. Niacin is essential for energy production and plays a role in maintaining healthy skin, digestion, and nervous system function. Mushrooms contain pantothenic acid, which is involved in various metabolic processes and helps produce energy from food. Folate is important for DNA synthesis and cell division. It is particularly important for pregnant women as it supports the growth and development of the fetus.

Mushrooms can naturally produce vitamin D when exposed to sunlight or UV light. Vitamin D is crucial for bone health, immune function, and various other physiological processes [26, 27].
Fiber: Mushrooms contain varying amounts of dietary fiber, but the exact fiber content can depend on the species and the part of the mushroom being consumed (e.g., cap and stem). On average, mushrooms have a moderate to high fiber content compared to many other vegetables. The fiber content of mushrooms ranges from approximately 1.5% to 3.5% of their fresh weight. This means that mushrooms typically contain around 1.5 to 3.5 grams of fiber per 100 grams of fresh mushrooms. The specific type of fiber found in mushrooms is predominantly non-starch polysaccharides, including β-glucans, which are known for their potential health benefits such as supporting gut health and immune function. It is important to note that cooking mushrooms can slightly affect their fiber content, as some fiber can be lost during the cooking process. However, mushrooms are still considered a good source of dietary fiber regardless of whether they are consumed raw or cooked.

16.5 Variation among Mushroom Species

Indeed, the carbohydrate content can vary among different mushroom species. While mushrooms are generally considered low in carbohydrates, there can be variations in their nutritional composition. Here are a few examples of mushroom species and their approximate carbohydrate content [6].

White Button Mushrooms: White button mushrooms, one of the most commonly consumed mushroom varieties, have relatively low carbohydrate content. They typically contain about 1–2 grams of carbohydrates per 100 grams [59].
Portobello Mushrooms: Portobello mushrooms are known for their larger size and robust flavor. They generally have a slightly higher carbohydrate content compared to white button mushrooms, ranging from 2–3 grams of carbohydrates per 100 grams [15].
Shiitake Mushrooms: Shiitake mushrooms have a distinct savory flavor and are widely used in various cuisines. They typically contain around 2–3 grams of carbohydrates per 100 grams.
Oyster Mushrooms: Oyster mushrooms have a delicate texture and mild flavor. They are known to have a slightly higher carbohydrate content compared to some other mushroom varieties, ranging from 3–4 grams of carbohydrates per 100 grams [21].

Morel Mushrooms: Morel mushrooms are highly sought after for their unique appearance and earthy flavor. They generally have low carbohydrate content, similar to white button mushrooms, with approximately 1–2 grams of carbohydrates per 100 grams. Additionally, the carbohydrate content can change when mushrooms are cooked or prepared in different ways [6].

16.6 Health Implications and Potential Benefits

Mushrooms offer several health implications and potential benefits due to their nutrient content and bioactive compounds.

Nutrient Dense: Mushrooms are low in calories and fat while providing essential nutrients like B vitamins (niacin, riboflavin, folate, pantothenic acid) and minerals (such as potassium, copper, selenium, and zinc). Incorporating mushrooms into your diet can contribute to overall nutrient intake. Incorporating mushrooms into your diet can be beneficial for weight management. They are low in calories and fat, making them a satisfying food choice that can help control calorie intake. The fiber content in mushrooms can also contribute to feelings of fullness and aid in weight loss or weight maintenance efforts. Mushrooms are a good source of various nutrients, including B vitamins (such as niacin, pantothenic acid, as well as riboflavin), minerals (such as potassium, copper, selenium, and zinc), and vitamin D (when exposed to sunlight or UV light). These nutrients play important roles in energy metabolism, nerve function, immune support, and bone health [4].

Antioxidant Properties: Mushrooms contain various antioxidants, such as ergothioneine and selenium, which can help protect cells from oxidative stress and reduce inflammation in the body. Mushrooms are known for their unique nutritional profile and have been consumed for centuries due to their various health benefits. When it comes to antioxidant properties, mushrooms are indeed a good source. As a result of normal bodily processes and exposure to the environment, free radicals damage our cells. Antioxidants protect our cells from damage caused by free radicals [40].

Mushrooms, such as white button mushrooms, contain vitamin C, which is a potent antioxidant. Vitamin C helps neutralize free radicals and protect cells from oxidative stress. Certain mushroom varieties, including shiitake and oyster mushrooms, are rich in selenium. Selenium is an essential trace mineral that acts as an antioxidant, working in conjunction with other antioxidants to protect cells from damage. Mushrooms contain various polyphenols, including phenolic acids and flavonoids, which possess antioxidant properties. These compounds help scavenge free radicals and reduce oxidative stress. Many mushroom species, such as white button mushrooms and oyster mushrooms, are notable for their high content of glutathione. It is thought that glutathione plays a crucial role in preventing oxidative damage to cells as it is a powerful antioxidant. Certain mushrooms, such as porcini and oyster mushrooms, are a rich source of ergothioneine. Ergothioneine is an antioxidant amino acid that has been linked to various health benefits, including anti-inflammatory properties and protection against oxidative stress. Mushrooms contain complex carbohydrates called polysaccharides, such as beta-glucans. These polysaccharides exhibit antioxidant activity and have been associated with immune-enhancing effects [29, 37].

Immune System Support: Certain mushrooms, like shiitake and maitake, contain bioactive compounds that have been shown to enhance immune function by stimulating immune cells and promoting the production of immune-modulating substances [60].

Prebiotic Effects: The fiber content in mushrooms, particularly β-glucans, can act as prebiotics, promoting the growth and activity of beneficial gut bacteria. This can contribute to improved digestive health and enhanced overall well-being. Mushrooms contain a significant amount of dietary fiber, including both soluble and insoluble fibers. These fibers pass through the digestive system mostly intact and serve as a food source for beneficial bacteria in the colon. By promoting the growth of these beneficial bacteria, mushrooms can help support a healthy gut microbiota. Mushrooms, particularly varieties like shiitake, maitake, and oyster mushrooms, are rich in beta-glucans. Beta-glucans are a type of soluble fiber that can act as prebiotics. They have been shown to stimulate the growth and activity of beneficial bacteria in the gut, such as Bifidobacteria and Lactobacilli. Some mushroom species, such as shiitake and oyster mushrooms, contain chitin and chitosan. These compounds are not digested by humans but can serve as substrates for the growth of certain beneficial gut bacteria. They have been found to promote the growth of Bifidobacteria and inhibit the growth of potentially harmful bacteria. Certain mushrooms, including the white button mushroom, contain mannans. Mannans are a type of carbohydrate that can act as prebiotics. They have been shown to stimulate the growth of beneficial bacteria, such as Bifidobacteria, in the gut. In addition to their antioxidant properties, the polyphenols found in mushrooms, such as phenolic acids and flavonoids, may also have prebiotic effects. These compounds can positively influence the gut microbiota by stimulating the growth of beneficial bacteria [17, 44].

Potential Anti-Cancer Properties: Some mushrooms, including turkey tail, maitake, and reishi, contain compounds that have shown potential anticancer properties in preclinical studies. These compounds may have immune-enhancing and tumor-suppressive effects, although more research is needed. Certain mushrooms, such as shiitake, maitake, and reishi mushrooms, contain bioactive compounds that have been shown to modulate the immune system. These compounds, such as beta-glucans and polysaccharides, can enhance the activity of immune cells, including macrophages and natural killer cells, which play a crucial role in recognizing and eliminating cancer cells. Some mushrooms possess compounds that have demonstrated direct antitumor activity in laboratory studies. For example, compounds like triterpenoids found in reishi mushrooms have shown potential anticancer effects by inhibiting tumor growth, suppressing metastasis, and inducing apoptosis (programmed cell death) in cancer cells [47].

Cancer development and progression are associated with chronic inflammation, which can contribute to an increased risk of developing cancer. Mushrooms contain various bioactive compounds, including polysaccharides and phenolic compounds, which exhibit anti-inflammatory properties. By reducing inflammation, mushrooms may help create an environment less conducive to cancer growth. Oxidative stress can contribute to DNA damage and the development of cancer. By scavenging free radicals, mushrooms may help protect cells from damage and potentially inhibit the initiation and progression of cancer. Some mushrooms, including the Agaricusbisporus species (commonly known as button mushrooms or white mushrooms), have been found to possess compounds that can inhibit angiogenesis. Angiogenesis is the process by which new blood vessels are formed, and it is crucial for tumor growth and metastasis. Inhibiting angiogenesis can help restrict the blood supply to tumors, thereby impeding their growth [10].

Cardiovascular Health: Mushrooms are naturally low in sodium and high in potassium, which may contribute to maintaining healthy blood pressure levels. Additionally, some studies suggest that certain mushroom varieties may help reduce cholesterol levels and improve lipid profiles. Some mushrooms, particularly oyster mushrooms and shiitake mushrooms, have been shown to have cholesterol-lowering effects in animal and human studies. Certain compounds in mushrooms, such as beta-glucans, have been found to help reduce total cholesterol and LDL (bad) cholesterol levels, potentially benefiting cardiovascular health. Chronic inflammation is associated with the development of cardiovascular disease. Certain compounds in mushrooms, including polysaccharides and phenolic compounds, possess anti-inflammatory properties. By reducing inflammation, mushrooms may help protect against the development and progression of cardiovascular conditions. Mushrooms are a good source of dietary fiber, which can help promote cardiovascular health. Dietary fiber has been associated with a reduced risk of heart disease by helping to lower cholesterol levels, improve blood lipid profiles, and regulate blood sugar levels. Mushrooms are naturally low in fat and sodium, making them a heart-healthy food choice. A diet low in saturated and trans fats, as well as sodium, is important for maintaining cardiovascular health and reducing the risk of heart disease [9, 21].

Weight Management: Mushrooms are low in calories and can be a healthy substitute for higher-calorie ingredients in various recipes. Their high fiber content can also contribute to a feeling of fullness and help with weight management. Mushrooms have a unique umami flavor profile that can enhance the taste of dishes, making them a satisfying addition to meals. By adding mushrooms to recipes, you can enhance the flavor and increase the overall satisfaction of the meal, potentially reducing the desire for additional calorie-dense ingredients [31]. Mushrooms can be used as a substitute for higher-calorie ingredients in recipes. For example, replacing some or all of the meat in a dish with mushrooms can reduce the overall calorie content while still providing texture and flavor. There are fewer calories per gram of food in mushrooms due to their low energy density, which means that they are a good source of calories. Foods with low energy density can help increase meal volume without dramatically increasing calorie intake, helping to manage weight. Despite being low in calories, mushrooms are nutrient-dense, meaning they provide a good amount of essential nutrients such as vitamins (e.g., vitamin D and B vitamins) and minerals (e.g., selenium and potassium) relative to their calorie content. Consuming nutrient-dense foods is important for overall health and can support weight management by providing necessary nutrients without excess calories. Mushrooms are low in calories and virtually fat-free, making them a great addition to a weight-conscious diet. They can provide bulk and texture to meals without adding excessive calories or fat. Mushrooms contain dietary fiber, which can contribute to a feeling of fullness and help control appetite. Foods high in fiber tend to be more satisfying, as they take longer to digest and promote satiety, potentially reducing overall calorie intake [28, 49].

16.7 Conclusion

Beneficial mushrooms have gained recognition as valuable sources of nutraceuticals, minerals, and proximate constituents, offering a range of potential health benefits. These natural compounds found in mushrooms contribute to their therapeutic properties and make them an important component of a healthy diet. The nutraceutical constituents in mushrooms, such as beta-glucans, have been extensively studied for their immune-enhancing properties and potential anti-cancer effects. They have shown promise in supporting immune function and aiding in disease prevention. Triterpenes found in certain mushroom species exhibit anti-inflammatory and antioxidant effects, contributing to their potential health benefits. Additionally, the synthesis of vitamin D from ergosterol found in mushrooms provides an added advantage for bone health, immune function, and overall well-being. Mushrooms also contain a variety of minerals that are essential for optimal health. Selenium, potassium, copper, zinc, and other minerals found in mushrooms play vital roles in supporting immune function, heart health, and various physiological processes in the body. Including mushrooms in the diet can be a natural way to ensure an adequate intake of these essential minerals. Proximate constituents in mushrooms, such as proteins and dietary fiber, contribute to their nutritional value. Mushrooms provide all the essential amino acids, making them a valuable source of protein, especially for individuals following vegetarian or vegan diets. The high fiber content in mushrooms supports healthy digestion, aids in maintaining blood sugar levels, and promotes satiety. Incorporating mushrooms into a balanced diet can provide a natural and nutritious way to support immune function, promote antioxidant defense, and maintain overall well-being. There is still much work to be done to fully understand these mushrooms' mechanisms of action and explore their potential therapeutic applications.

References

1. Abdelshafy, A.M., Belwal, T., Liang, Z., Wang, L., Li, D., Luo, Z., Li, L., A comprehensive review on phenolic compounds from edible mushrooms: Occurrence, biological activity, application and future prospective. *Crit. Rev. Food Sci. Nutr.*, *62*, 6204–6224, 2022.
2. Abubakar, I., Mann, A., Mathew, J.T., Evaluation of Phytochemical, Anti-nutritional and Antioxidant Potentials of Flower and Seed Methanol Extracts of *Sennaalata* L. Grown in Nigeria. *Am. J. Appl. Chem.*, 3, 93–100, 2015.

3. Abubakar, I., Mann, A., Mathew, J.T., Phytochemical composition, Antioxidant and Anti-nutritional properties ofroot-bark and leaf methanol extracts of *Sennaalata L.* Grown in Nigeria. *Afr. J. Pure Appl. Chem.*, 9, 91–97, 2015.
4. Agarwal, S. and Fulgoni I, V.L., Nutritional impact of adding a serving of mushrooms to USDA Food Patterns - a dietary modeling analysis. *Food Nutr. Res.*, 12, 65, 2021.
5. Ahmed, M.H., Hassan, A., Molnár, J., The Role of Micronutrients to Support Immunity for COVID-19 Prevention. *Rev. Bras. Farmacogn.: Orgaooficial SociedadeBrasileira Farmacogn.*, *31*, 361–374, 2021.
6. Amara, A.A. and El-Baky, N.A., Fungi as a Source of Edible Proteins and Animal Feed. *J. Fungi*, *9*, 73, 2023.
7. Andreani, G., Sogari, G., Marti, A., Froldi, F., Dagevos, H., Martini, D., Plant-Based Meat Alternatives: Technological, Nutritional, Environmental, Market, and Social Challenges and Opportunities. *Nutrients*, *15*, 452, 2023.
8. Assemie, A. and Abaya, G., The Effect of Edible Mushroom on Health and Their Biochemistry. *Int. J. Microbiol.*, 23, 8744788, 2022.
9. Ayimbila, F. and Keawsompong, S., Nutritional Quality and Biological Application of Mushroom Protein as a Novel Protein Alternative. *Curr. Nutr. Rep.*, 14, 1–18, 2023.
10. Bains, A., Chawla, P., Kaur, S., Najda, A., Fogarasi, M., Fogarasi, S., Bioactives from Mushroom: Health Attributes and Food Industry Applications. *Materials*, *14*, 7640, 2021.
11. Bell, V., Silva, C.R.P.G., Guina, J., Fernandes, T.H., Mushrooms as future generation healthy foods. *Front. Nutr.*, *9*, 1050099, 2022.
12. Bhambri, A., Srivastava, M., Mahale, V.G., Mahale, S., Karn, S.K., Mushrooms as Potential Sources of Active Metabolites and Medicines. *Front. Micro*, *13*, 837266, 2022.
13. Budzyńska, S., Siwulski, M., Magdziak, Z., Budka, A., Gąsecka, M., Kalač, P., Rzymski, P., Niedzielski, P., Mleczek, M., Influence of Iron Addition (Alone or with Calcium) to Elements Biofortification and Antioxidants in *Pholiotanameko*. *Plants*, *10*, 2275, 2021.
14. Burmasova, M.A., Utebaeva, A.A., Sysoeva, E.V., Sysoeva, M.A., Melanins of *InonotusObliquus*: Bifidogenic and Antioxidant Properties. *Biomolecules*, *9*, 248, 2019.
15. Castellanos-Reyes, K., Villalobos-Carvajal, R., Beldarrain-Iznaga, T., Fresh Mushroom Preservation Techniques. *Foods*, *10*, 2126, 2021.
16. Cateni, F., Gargano, M.L., Procida, G., Mycochemicals in wild and cultivated mushrooms: nutrition and health. *Phytochem. Rev.*, 21, 339–383, 2022.
17. Chiozzi, V., Eliopoulos, C., Markou, G., Arapoglou, D., Agriopoulou, S., El Enshasy, H.A., Varzakas, T., Biotechnological Addition of β-Glucans from Cereals, Mushrooms and Yeasts in Foods and Animal Feed. *Processes*, *9*, 1889, 2021.

18. Chiu, H.F., Venkatakrishnan, K., Golovinskaia, O., Wang, C.K., Impact of Micronutrients on Hypertension: Evidence from Clinical Trials with a Special Focus on Meta-Analysis. *Nutrients*, *13*, 588, 2021.
19. Chugh, R.M., Mittal, P., Mp, N., Arora, T., Bhattacharya, T., Chopra, H., Cavalu, S., Gautam, R.K., Fungal Mushrooms: A Natural Compound With Therapeutic Applications. *Front. Pharm.*, *13*, 925387, 2022.
20. Cör, D., Knez, Ž., KnezHrnčič, M., Antitumour, Antimicrobial, Antioxidant and Antiacetylcholinesterase Effect of GanodermaLucidumTerpenoids and Polysaccharides: A Review. *Molocules*, 23, 649, 2018.
21. Das, A.K., Nanda, P.K., Dandapat, P., Bandyopadhyay, S., Gullón, P., Sivaraman, G.K., McClements, D.J., Gullón, B., Lorenzo, J.M., Edible Mushrooms as Functional Ingredients for Development of Healthier and More Sustainable Muscle Foods: A Flexitarian Approach. *Molecules*, 26, 2463, 2021.
22. Dimopoulou, M., Kolonas, A., Mourtakos, S., Androutsos, O., Gortzi, O., Nutritional Composition and Biological Properties of Sixteen Edible Mushroom Species. *Appl. Sci.*, *12*, 8074, 2022.
23. El Sheikha, A.F., Nutritional Profile and Health Benefits of *Ganodermalucidum* "Lingzhi, Reishi, or Mannentake" as Functional Foods: Current Scenario and Future Perspectives. *Foods*, 11, 1030, 2022.
24. Gopal, J., Sivanesan, I., Muthu, M., Oh, J.W., Scrutinizing the Nutritional Aspects of Asian Mushrooms, Its Commercialization and Scope for Value-Added Products. *Nutrients*, 14, 3700, 2022.
25. Górska, S., Maksymiuk, A., Turło, J., Selenium-Containing Polysaccharides—Structural Diversity, Biosynthesis, Chemical Modifications and Biological Activity. *Appl. Sci.*, 11, 3717, 2021.
26. Hanna, M., Jaqua, E., Nguyen, V., Clay, J., B Vitamins: Functions and Uses in Medicine. *Perm. J.*, 26, 89–97, 2022.
27. Hrubša, M., Siatka, T., Nejmanová, I., Vopršalová, M., KujovskáKrčmová, L., Matoušová, K., Javorská, L., Macáková, K., Mercolini, L., Remião, F., *et al.*, Biological Properties of Vitamins of the B-Complex, Part 1: Vitamins B_1, B_2, B_3, and B_5. *Nutients*, 14, 484, 2022.
28. Huling, R., Applications of Fungi for Alternative Protein, in: *Fungi and Fungal Products in Human Welfare and Biotechnology*, T. Satyanarayana, and S.K. Deshmukh, (Eds.), Springer, Singapore, 2023.
29. Inobeme, A., Nayak, V., Mathew, J.T., Okonkwo, S., Ekwoba, L., Ajai, A.I., Bernard, E., Inobeme, J., Agbugui, M.M., Singh, K.R.B., Chemometric approach in environmental pollution analysis: A critical review. *J. Environ. Manage.*, 309, 1–18, 2022.
30. Inobeme, A., Adetunji, C.O., Mathew, J.T., Ajai, A.I., Inobeme, J., Oyedolapo, B., Bioremediation of Chlorinated Compounds, in: *Modern Approaches in Waste Bioremediation*, M.P. Shah (Ed.), pp. 101–115, Springer Nature Switzerland AG, Switzerland, 2023.

31. Inobeme, A., Mathew, J.T., Adetunji, C.O., Ajai, A.I., Inobeme, J., Maliki, M., Okonkwo, S., Adekoya, M.A., Bamigboye, M.O., Jacob, J.O., *et al.*, Recent advances in nanotechnology for remediation of heavy metals. *Environ. Monit. Assess.*, 195, 1–24, 2023.
32. Jędrejko, K.J., Lazur, J., Muszyńska, B., Cordycepsmilitaris: An Overview of Its Chemical Constituents in Relation to Biological Activity. *Foods*, 10, 2634, 2021.
33. Jeong, S.C., Jeong, Y.T., Yang, B.K., Islam, R., Koyyalamudi, S.R., Pang, G., Cho, K.Y., Song, C.H., White button mushroom (Agaricusbisporus) lowers blood glucose and cholesterol levels in diabetic and hypercholesterolemic rats. *Nutr. Res.*, 30, 49–56, 2010.
34. Jeong, S.C., Koyyalamudi, S.R., Pang, G., Dietary intake of Agaricusbisporus white button mushroom accelerates salivary immunoglobulin A secretion in healthy volunteers. *Nutrients*, 28, 527–531, 2012.
35. Jiang, Q., Zhang, M., Mujumdar, A.S., UV induced conversion during drying of ergosterol to vitamin D in various mushrooms: Effect of different drying conditions. *Trends Food Sci. Tech.*, *105*, 200–210, 2020.
36. Kaleta, B., Roszczyk, A., Zych, M., Kniotek, M., Zagożdżon, R., Klimaszewska, M., Malinowska, E., Pac, M., Turło, J., Selective Biological Effects of Selenium-Enriched Polysaccharide (Se-Le-30) Isolated from Lentinulaedodes Mycelium on Human Immune Cells. *Biomolecules*, 11, 1777, 2021.
37. Kora, A.J., Nutritional and antioxidant significance of selenium-enriched mushrooms. *Bull. Natl. Res. Cent.*, 44, 34, 2020.
38. Kushairi, N., Phan, C.W., Sabaratnam, V., David, P., Naidu, M., Lion's Mane Mushroom, *Hericiumerinaceus* (Bull.: Fr.) Pers. Suppresses H_2O_2-Induced Oxidative Damage and LPS Induced Inflammation in HT22 Hippocampal Neurons and BV2 Microglia. *Antioxidants*, *8*, 261, 2019.
39. Liu, Y., Wang, J., Wang, W., Zhang, H., Zhang, X., Han, C., The Chemical Constituents and Pharmacological Actions of Cordycepssinensis. *Evid. Based Complement. Alternat. Med.:eCAM*, 9, 575063, 2015.
40. Liuzzi, G.M., Petraglia, T., Latronico, T., Crescenzi, A., Rossano, R., Antioxidant Compounds from Edible Mushrooms as Potential Candidates for Treating Age-Related Neurodegenerative Diseases. *Nutr.*, *15*, 1913, 2023.
41. Mathew, J.T., Dauda, B.E.N., Mann, A., Ndamitso, M.M., Fadipe, A.L., Shaba, E.Y., Assessment of Nutritional Properties of Fermented and Unfermented Seed of *Cissus populnae* from Niger State, Nigeria. *Assum. Univ.-eJournal Interdisc. Res.*, 3, 70–77, 2018.
42. Mathew, J.T., Dauda, B.E.N., Mann, A., Ndamitso, M.M., Etsuyankpa, M.B., Nasirudeen, M.B., Nutrient And Anti-Nutrient Assessment of Fermented and Unfermented Seed of *Haematostaphis barteri* from Niger State, Nigeria. *Int. J. Appl. Biol. Res.*, 9, 55–66, 2018.
43. Mirończuk-Chodakowska, I., Kujawowicz, K., Witkowska, A.M., Beta-Glucans from Fungi: Biological and Health-Promoting Potential in the COVID-19 Pandemic Era. *Nutrients*, *13*, 3960, 2021.

44. Mitsou, E.K., Saxami, G., Stamoulou, E., Kerezoudi, E., Terzi, E., Koutrotsios, G., Bekiaris, G., Zervakis, G.I., Mountzouris, K.C., Pletsa, V., Effects of Rich in B-Glucans Edible Mushrooms on Aging Gut Microbiota Characteristics: An *In Vitro* Study. *Molecules, 25,* 2806, 2020.
45. Name, J.J., Souza, A.C.R., Vasconcelos, A.R., Prado, P.S., Pereira, C.P.M., Zinc, Vitamin D and Vitamin C: Perspectives for COVID-19 With a Focus on Physical Tissue Barrier Integrity. *Front. Nutr., 7,* 606398, 2020.
46. O'Brien, B.J., Physicochemical Properties of Residuals from Anaerobic Digestion of Dairy Manure and Food Waste: Nutrient Cycling Implications and Opportunities for Edible Mushroom Cultivation. Graduate College Dissertations and Theses, p. 1015, 2019.
47. Panda, S.K., Sahoo, G., Swain, S.S., Luyten, W., Anticancer Activities of Mushrooms: A Neglected Source for Drug Discovery. *Pharmaceuticals (Basel, Switzerland), 15,* 176, 2022.
48. Podkowa, A., Kryczyk-Poprawa, A., Opoka, W., Culinary–medicinal mushrooms: a review of organic compounds and bioelements with antioxidant activity. *Eur. Food Res. Technol.*, 247, 513–533, 2021.
49. Ponzo, V., Pellegrini, M., Costelli, P., Vázquez-Araújo, L., Gayoso, L., D'Eusebio, C., Ghigo, E., Bo, S., Strategies for Reducing Salt and Sugar Intakes in Individuals at Increased Cardiometabolic Risk. *Nutrients, 13,* 279, 2021.
50. Rahman, M.A., Abdullah, N., Aminudin, N., *Lentinulaedodes* (shiitake mushroom): An assessment of *in vitro* anti-atherosclerotic bio-functionality. *Saudi J. Biol. Sci., 25,* 1515–1523, 2018.
51. Reis, F.S., Barros, L., Martins, A., Ferreira, I.C., Chemical composition and nutritional value of the most widely appreciated cultivated mushrooms: an inter-species comparative study. *Food Chem. Toxicol.: Int. J. Publ. Brit. Indust. Biol. Res. Ass., 50,* 191–197, 2012.
52. Rondanelli, M., Moroni, A., Zese, M., Gasparri, C., Riva, A., Petrangolini, G., Perna, S., Mazzola, G., Vitamin D from UV-Irradiated Mushrooms as a Way for Vitamin D Supplementation: A Systematic Review on Classic and Nonclassic Effects in Human and Animal Models. *Antioxidants, 12,* 736, 2023.
53. Sande, D., Oliveira, G.P., Moura, M.A.F.E., Martins, B.A., Lima, M.T.N.S., Takahashi, J.A., Edible mushrooms as a ubiquitous source of essential fatty acids. *Food Res. Int., 125,* 108524, 2019.
54. Shaba, E.Y., Ndamitso, M.M., Mathew, J.T., Etsunyakpa, M.B., Tsado, A.N., Muhammad, S.S., Nutritional and anti-nutritional composition of date palm *(Phoenix dactylifera* L.) fruits sold in major markets of Minna Niger State, Nigeria. *Afr. J. Pure Appl. Chem.*, 9, 167–174, 2015.
55. Shamim, M.Z., Mishra, A.K., Kausar, T., Mahanta, S., Sarma, B., Kumar, V., Mishra, P.K., Panda, J., Baek, K.-H., Mohanta, Y.K., Exploring Edible Mushrooms for Diabetes: Unveiling Their Role in Prevention and Treatment. *Molecules, 28,* 2837, 2023.

56. Spim, S.R.V., Castanho, N.R.C.M., Pistila, A.M.H., Jozala, A.F., Oliveira Júnior, J.M., Grotto, D., *Lentinulaedodes* mushroom as an ingredient to enhance the nutritional and functional properties of cereal bars. *J. Food Sci. Tech.*, *58*, 1349–1357, 2021.
57. Tardy, A.-L., Pouteau, E., Marquez, D., Yilmaz, C., Scholey, A., Vitamins and Minerals for Energy, Fatigue and Cognition: A Narrative Review of the Biochemical and Clinical Evidence. *Nutrients*, *12*, 228, 2020.
58. Uffelman, C.N., Chan, N.I., Davis, E.M., Wang, Y., McGowan, B.S., Campbell, W.W., An Assessment of Mushroom Consumption on Cardiometabolic Disease Risk Factors and Morbidities in Humans: A Systematic Review. *Nutrients*, *15*, 1079, 2023.
59. Usman, M., Murtaza, G., Ditta, A., Nutritional, Medicinal, and Cosmetic Value of Bioactive Compounds in Button Mushroom (Agaricusbisporus): A Review. *Appl. Sci.*, *11*, 5943, 2021.
60. Venturella, G., Ferraro, V., Cirlincione, F., Gargano, M.L., Medicinal Mushrooms: Bioactive Compounds, Use, and Clinical Trials. *Int. J. Mol. Sci.*, *22*, 634, 2021.
61. Vetter, J., The Mushroom Glucans: Molecules of High Biological and Medicinal Importance. *Foods*, *12*, 1009, 2023.
62. Wu, J.Y., Siu, K.C., Geng, P., Bioactive Ingredients and Medicinal Values of *Grifolafrondosa* (Maitake). *Foods (Basel, Switzerland)*, *10*, 1, 95, 2021.
63. Zięba, P., Sękara, A., Bernaś, E., Krakowska, A., Sułkowska-Ziaja, K., Kunicki, E., Suchanek, M., Muszyńska, B., Supplementation with Magnesium Salts-A Strategy to Increase Nutraceutical Value of *Pleurotusdjamor* Fruiting Bodies. *Molecules*, *26*, 3273, 2021.

17

Application of Mushrooms in the Promotion of Longevity

Victory Igiku[1], Charles Oluwaseun Adetunji[1]*, Oluwafemi Adebayo Oyewole[2], Eniola K. I. T.[3] and Mohammed Bello Yerima[4]

[1]Department of Microbiology, Edo State University, Uzairue, Edo State, Nigeria
[2]Department of Microbiology, Federal University of Technology, Minna, Nigeria
[3]Department of Biological Sciences, Joseph Ayo Babalola University, Ikeji Arakeji, Osun State, Nigeria
[4]Department of Microbiology, Sokoto State University, Sokoto, Sokoto State, Nigeria

Abstract
There has been increased awareness of the several beneficial importance of mushrooms, due to several studies reporting their several health benefits. Mushrooms are very important storage of several bioactive compounds, both known and novel and they have found several importance as antibiotics, antioxidants, antidiabetes, immunomodulatory, and antitumor. The ability of mushrooms to promote good health and healthy living can be associated with their ability to promote longevity.

Keywords: Mushrooms, longevity, bioactive compounds, antioxidants, immunomodulatory, medicinal properties, antidiabetic, anticancer

17.1 Introduction

Mushrooms have long been regarded as a highly nutritious, medicinal, and tasty food. Mushrooms and fungi are incredibly diverse, occurring in different places around the world. Malty *et al.* [28] described mushroom as a renewable gift from nature to humanity, endowed with special flavor, taste, and medicinal qualities. Only 10% of the estimated 150,000–160,000 species of mushrooms are known to science, and there are about 1.5 million of

**Corresponding author*: adetunjicharles@gmail.com

Charles Oluwaseun Adetunji, Chukwuebuka Egbuna, Oluwatosin Ademola Ijabadeniyi and Samantha C. Karunarathna (eds.) Mushroom Biotechnology for Improved Agriculture and Human Health, (359–370) © 2025 Scrivener Publishing LLC

them in the world. Consequently, our current understanding of the global fungal biota is limited to 1%. Although not all of the 700 pharmacologically significant species are edible, about 2000 are safe.

While some mushrooms are Ascomycetes, the majority of the mushrooms are in the class Basidiomycetes. The Basidiomycota is a monophyletic group with more than 31,000 living species known, approximately one-third of all fungi; however, molecular and genetic studies indicate that there is greater diversity within this group yet to be discovered [39]. The term "mushroom" refers to a "macrofungus," or a fungus that is visible to the naked eye.

There have been increased reports on the number of antinutrients and nutrients present in mushrooms due to the increased interest of researchers in this area. The nutritional value of mushrooms has been related to their richness in proteins, several vitamins, low total fat and high unsaturated fatty acid proportion, and the presence of no cholesterol, chitin, and minerals [13]. However, as a result of a large number of species occurring worldwide and also due to varied cooking methods, our understanding of the nutrient composition of mushrooms is still deficient [10].

Mushrooms are delicious foodstuff and a good flavoring substance. According to Garnango *et al.* [13], this has been associated with the volatile carbon 8-molecules and the availability of water-soluble taste elements such as soluble sugars, polyols, free amino acids, organic acids, and five nucleotides.

Mushrooms are useful sources of a number of beneficial bioactive compounds. Several substances with potent antibacterial properties are found in the mycelia and fruiting bodies of mushrooms [31]. Examples of these bioactive compounds include polysaccharides, alkaloids, enzymes, carotenoids, phenolics, lipids, tocopherols, and terpenes. They have been regarded as very good candidates for naturally occurring antibiotics. The immunomodulatory effect of glucans present on the cell wall of mushrooms has been reported by Stachowiak and Regula [38]. The authors also reported that the secondary metabolite produced by mushroom mycelia inhibits viruses and bacteria. The pharmacological potential of mushrooms has been largely recognized but they are also largely untapped sources of novel pharmaceutical products. The clinical functions of medicinal mushrooms are diverse. Studies have identified some of the functions of mushrooms as antioxidants, immunomodulating, antiviral, antiparasitic, antibacterial, cholesterol-lowering, detoxicative, antiobesity, antidiabetic, neuroprotective, hepatoprotective, anticancer, and so many more functions. Additionally, compounds gotten from medicinal mushrooms have been used as analgetics and painkillers [29]. A very effective use of medicinal mushroom drugs has been in the prevention of immune disorders mostly in immunocompromised patients [43].

Generally, the consumption of mushrooms can greatly impact human health in a lot of beneficial ways as it promotes overall good health and well-being. These benefits have led to an increase in the global consumption of mushrooms, with agaricus, pleurotus, auricularia, porphyra, luminaria, flammulina, and lentinula being the most often consumed species [34]. Different medicinal benefits of mushrooms have been reported in different literature. The present study tries to explore the application of mushrooms in the promotion of longevity.

17.2 Health Benefits of Mushrooms

The increased recognition of the benefits of mushrooms has geared the interest of both consumers and researchers. Several benefits of medicinal mushrooms have been reported in several literature. Although there is limited understanding of the vast medicinal properties of mushrooms, the bioactive compound in medicinal mushrooms possesses antidiabetic, anticancer, immunomodulatory effects, and many more [25].

The antioxidant properties of mushrooms have been reported to be responsible for their ability to neutralize free oxygen radicals since its presence leads to the generation of reactive oxygen which could cause damage in the cells which is a major factor responsible for aging and several other degenerative illnesses [36]. These compounds are usually found in the fruiting bodies, culture broth, and mycelium and they include polysaccharides, phenolics, ascorbic acids, and ergosterol to mention a few. The mycelium and fruiting bodies can be exploited to produce bioactive compounds in a very short time and this is a huge advantage in the extraction of antioxidants from mushrooms. [36]. Mu *et al.* [30] reported the antioxidant properties of crude polysaccharides gotten from inontus oblique mushrooms. Similarly, an investigation by Zhang *et al.* [50] reported the *in vitro* antioxidant properties of polysaccharides gotten from Hericium erinaceus mushrooms on mice model and protected them against liver damage resulting from carbon tetrachloride. Samanta *et al.* [35] have also reported the immunostimulating and antioxidant properties of a fucogalactan obtained from *Macrolepiota dolichaula* mushroom. The antioxidant compounds from mushrooms may be isolated and used as functional additives or incorporated into the food regime, thereby acting as an alternative food source to prevent the damage caused by oxidation in humans [36].

The potency of polysaccharides extracted from mushrooms as antioxidants can be determined using different chemical tests such as reducing power, ferric reducing antioxidant power (FRAP), nitric oxide, 1,1-diphenyl-2-picrylhydrazyl (DPPH) and azinobis-3-ethylbenzthiazoline-6-sulphonate (ABTS) tests and Lipid peroxide [2].

Different human studies and *in vitro* tests in diabetic rodent research have proposed that different medicinal mushrooms, such as *Aspergillus blazei* [21], *Grifola frondosa* [16], *Agaricus bisporus* [46], *Ganoderma lucidum* [27], *Hericium erinaceus* [40], and *Pleurotus spp.* [18, 26], have a very important potential in the normalization of blood glucose levels. The favorable effects of several medicinal mushrooms on blood pressure and blood lipids have also been examined. Yang *et al.* [47] reported that *Lentinus edodes* reduces cholesterol levels in animal blood. A study by Chang *et al.* [5] revealed that the methanol extract of *H. erinaceus* was able to reduce the blood sugar level, total cholesterol, and total glycerides level in streptozotocin-induced high-fat diet rats.

The antidiabetic efficacy of mushrooms is greatly dependent on their polysaccharides such as *ß*- and D-glucan and their vitamin D contents [6]. Lee and Hong [23] reported that polysaccharides gotten from the fruiting bodies of Cordyceps militaris mushrooms were responsible for preventing melanoma from growing *in vivo* in a mouse. Similarly, Huang *et al.* [17] demonstrated that administering polysaccharides extracted from the pleurotus tuber-regium mushroom to high-fat diet rats decreased the levels of fasting blood sugar and glucosylated hemoglobin while also restoring serum insulin, triglycerides, cholesterol, HDL, and LDL. The study also noted the inhibition of the lipid peroxidation index and also the restoration of oxidative enzymes such as glutathione peroxidase and superoxide dismutase functions in the liver. The study further proposed that the antioxidant system reduces oxidative stress and might be enhanced by the antioxidative property of the polysaccharide, which proposes the possible use of the antioxidant in the treatment of hyperglycemia and other forms of diabetes associated with oxidative stress. When diabetic mice were administered a combination of polysaccharides extracted from Catathelasma ventricosum mushrooms for 30 days, blood glucose, triglycerides, LDL, HDL, total, and antioxidant enzyme levels were all significantly reduced [24]. According to Calvo *et al.* [4], the medicinal mushroom Grifola frondosa broth culture and submerged mycelia decreased type 2 diabetes in normal mice by enhancing hyperglycemia in diabetes-induced natural immunity as well as reducing cell-mediated immunity.

Another study showed that the administration of P. linteus polysaccharide for 4 weeks in diabetic rats rapidly reduced fasting blood glucose and improved the tolerance of glucose through the alteration of hepatic phospholipid metabolism and rescuing insulin signaling transduction through forkhead transcription factor (FOXO) signaling pathway [24]. Additionally, Liu *et al.* [24] reported that selenium polysaccharide from C. ventriculosu demonstrated strong antihyperlipidemic, antihyperglycemic, and antioxidant activities. The authors also noted that they prevented injuries resulting from increased blood glucose levels in the kidney, pancreas, and liver of hyperglycemic mice. A study by Reza *et al.* [33] showed that the treatment of obese mice with 70% ethanol extract of Auricularia auricula-judae mushroom resulted in a reduced chance of hepatic steaosis through the modulation of plasma lipids by regulating both lipogenic and adipogenic transcription factors.

An investigation by Wahab *et al.* [44] identified three bioactive proteins (catalase-like protein, glyceraldehyde-3-phosphate dehydrogenase-like protein, and profilin-like protein), isolated from Pleurotus pulmonarius basidiocarp protein with the ability to inhibit α- glucosidase and α-amylase and are therefore very good potential antidiabetics. These two enzymes are usually involved in carbohydrate digestion but their inhibition can significantly reduce blood glucose, hence they are very good candidates for the treatment and management of type 2 diabetes mellitus [48]. Significant progress has been observed in testing the antidiabetic potential of mushrooms in humans. A study by Calvo *et al.*, 4, found that the consumption of *Agaricus bisporus* mushrooms reduced the risk associated with diabetes among 37 human participants. A previous study revealed the hypolipidemic and hypoglycemic activity of Agaricus bisporus mushroom in rats. Wang *et al.* [45] found that after a 70 percent partial pancreatectomy, mice subjects treated with glycoproteins derived from Agaricus bisporus showed the regeneration of pancreatic β-cells. The impact of culinary Pleurotus ostreatus and P. cystidiosus on healthy individuals and type 2 diabetic patients was examined by Jayasuriya *et al.* [19]. The results of the study indicated that patients with type 2 diabetes had higher serum insulin levels, lower postprandial serum glucose levels, and lower fasting and postprandial blood glucose levels. It is also crucial to remember that data from a small number of trials contradict the use of Ganoderma lucidum mushrooms to treat cardiovascular disease risk factors in individuals with type 2 diabetes [22].

The mechanism of action of mushrooms as antidiabetics includes the reduction of plasma glucagon and increase of plasma insulin, improving insulin sensitivity through AMPK activation, a decrease of insulin resistance through the induction of PPAR (peroxisome proliferator–activated receptor) gamma, further reduction of insulin resistance and increased insulin sensitivity through the reduction of gut lipopolysaccharides and gut dysbiosis, increased hepatic glycogen, improved glucose utilization from peripheral tissues, quenching reactive oxygen species, α-glucosidase enzyme inhibition, short-chain fatty acid production, and increased glucose utilization and glucose transporter 4 transloocation through the activation of protein kinase B and insulin receptor substrate 2 pathway [2].

More than half of the anticancer drugs have been linked to natural products but none so far has been developed from mushrooms [7]. Current anticancer therapies have been reported to show several undesirable side effects in patients as well as inducing resistance [32]. Several studies have found that mushrooms have anticancer properties and minimal unfavorable side effects, such as anemia, bone marrow suppression, nausea, and insomnia. They have also been shown to reduce the resistance to drugs following radiation and chemotherapy [18].

The components responsible for the anticancer activity of mushrooms include antroquimomol, hidpolon, krestin, cordycepin, lectin, lentinan, sulfated polysaccharide, maitake D fraction, and polysaccharides [3]. Among these components, polysaccharides present the most potent group with both immunomodulatory and antitumor activity [32]. As a non-self molecule, β-glucan, a polysaccharide, stimulates the immune system by altering the synthesis of cytokines that activate phagocytes and leukocytes [3]. Both lectins and lentinan exhibit cytotoxic activity on breast cancer cells. Mushroom polysaccharides increase immune response through the stimulation of B cells, T cells, natural killer cells, and macrophages [15]. Cordycepin (3-deoxyadenosine) has been identified as a very important anticancer compound found in Cordyceps species. The compounds exhibit an apoptotic activity by attaching to the location where the nucleic acid is to be attached, resulting in dysregulated polyadenylation and the cessation of DNA or RNA elongation [49].

An investigation by Tsai *et al.* [42] demonstrated that patients with advanced adenocarcinoma receiving *Antrodia cinnamomea* mushroom treatment experienced less severe gastrointestinal symptoms, such as diarrhea and abdominal pain, compared to the placebo group.

Also, *Agaricus blazei Murill* lowered prostate-specific antigen levels and controlled recurrent prostate cancer by lowering immunosuppressive factors [43]. Ahn *et al.* [1] found that patients with gynecological cancer receiving chemotherapy in addition to Agaricus blazei Murill treatment experienced less adverse effects, such as baldness, general weakness, and appetite loss, as compared to the placebo group. Fortes *et al.* [11] reported a decreased glycaemia level in post-surgery cancer patients administered

Agaricus silvaticus. Similarly, Gao *et al.* [12] reported a significant improvement in cancer in comparison to the placebo group (11–43%), 3–84% of patients with advanced lung cancer had significantly better cancer-related signs (like cough, fever, sweating, weakness, and sleeplessness).

Over time, patients who received six courses of chemotherapy in addition to *Agaricus sylvaticus* experienced a decrease in appetite; however, most patients in the placebo group experienced gastrointestinal symptoms in addition to appetite loss, such as nausea, vomiting, diarrhea, and constipation [20]. A randomised controlled study (RCT) by Zhao *et al.* [51] found that patients with breast cancer who received *Ganoderma lucidum* treatment experienced less cancer-related fatigue compared to those who received endocrine therapy. In phase I/II trial of breast cancer survivors, *Grifola frondosa* extract functioned as an immunomodulator and increased the production of interleukin 2, interleukin 10, tumour necrosis factor alpha, and interferon gamma by specific T cells [8]. Hackman *et al.* [9] reported that using *lentigula endodes* alone to treat patients with prostate cancer was ineffective. The anticancer properties of a semisynthetic compound illudin S isolated from *Omphalotus illudens* are attributed to the alkylation of DNA, RNA, and proteins. However, its application in clinical settings is limited because of its significant retinal toxicity and poor therapeutic index [37]. In patients with immunocompromised breast cancer, *Trametes versicolor* improved their immunological state in a phase I trial [41]. *Trametes versicolor* polysaccharides have been shown to have an immunostimulatory impact and to be directly toxic to cancer cells, indicating that their use is not limited to adjuvant therapy [14].

17.3 Conclusion

Mushrooms are excellent sources of various biologically active compounds with applications in medicine, agriculture, biotechnology, pharmaceuticals, and lots more. The role of mushroom in the promotion of longevity can be attributed to its ability to promote human health. Mushroom polysaccharides which have been described as one of the most potent compounds in mushrooms are a very functional compound which has found applications as antioxidants, antidiabetes, immunomodulatory, anti inflammatory, and even antitumor agent. Mushrooms could be the solution to several life-threatening diseases being faced in healthcare currently, and there should be increased awareness on the several benefits of mushroom and the potential significant effects of their bioactive compounds.

The consumption of mushrooms should be encouraged, and there is a need to explore more mushroom species as they are very important reservoirs of novel bioactive compounds.

References

1. Ahn, W.-S., Kim, D.-J., Chae, G.-T., Lee, J.-M., Bae, S.-M., Sin, J.-I., Kim, Y.-W., Namkoong, S.-E., Lee, I.P., Natural killer cell activity and quality of life were improved by consumption of a mushroom extract, Agaricus blazei Murill Kyowa, in gynecological cancer patients undergoing chemotherapy. *Int. J. Gynecol. Cancer*, *14*, 4, 589–594, 2004, https://doi.org/10.1136/ijgc-00009577-200407000-00003.
2. Arunachalam, K., Sreeja, P.S., Yang, X., The Antioxidant Properties of Mushroom Polysaccharides can Potentially Mitigate Oxidative Stress, Beta-Cell Dysfunction and Insulin Resistance. *Front. Pharmacol.*, *13*, 874474, 2022, https://doi.org/10.3389/fphar.2022.874474.
3. Ayeka, P.A., Potential of Mushroom Compounds as Immunomodulators in Cancer Immunotherapy: A Review. *Evid.-Based Complementary Altern. Med.*, *2018*, e7271509, 2018, https://doi.org/10.1155/2018/7271509.
4. Calvo, M.S., Mehrotra, A., Beelman, R.B., Nadkarni, G., Wang, L., Cai, W., Goh, B.C., Kalaras, M.D., Uribarri, J., A Retrospective Study in Adults with Metabolic Syndrome: Diabetic Risk Factor Response to Daily Consumption of Agaricus bisporus (White Button Mushrooms). *Plant Foods Hum. Nutr.*, *71*, 3, 245–251, 2016, https://doi.org/10.1007/s11130-016-0552-7.
5. Chang, C.-H., Chen, Y., Yew, X.-X., Chen, H.-X., Kim, J.-X., Chang, C.-C., Peng, C.-C., Peng, R.Y., Improvement of erinacine A productivity in Hericium erinaceus mycelia and its neuroprotective bioactivity against the glutamate-insulted apoptosis. *LWT Food Sci. Technol.*, *65*, 1100–1108, 2016, https://doi.org/10.1016/j.lwt.2015.08.014.
6. Das, A., Chen, C.-M., Mu, S.-C., Yang, S.-H., Ju, Y.-M., Li, S.-C., Medicinal Components in Edible Mushrooms on Diabetes Mellitus Treatment. *Pharmaceutics*, *14*, 2, 436, 2022, https://doi.org/10.3390/pharmaceutics14020436.
7. Demain, A.L. and Vaishnav, P., Natural products for cancer chemotherapy. *Microb. Biotechnol.*, *4*, 6, 687–699, 2011, https://doi.org/10.1111/j.1751-7915.2010.00221.x.
8. Deng, G., Lin, H., Seidman, A., Fornier, M., D'Andrea, G., Wesa, K., Yeung, S., Cunningham-Rundles, S., Vickers, A.J., Cassileth, B., A phase I/II trial of a polysaccharide extract from Grifola frondosa (Maitake mushroom) in breast cancer patients: Immunological effects. *J. Cancer Res. Clin. Oncol.*, *135*, 9, 1215–1221, 2009, https://doi.org/10.1007/s00432-009-0562-z.
9. deVere White, R.W., Hackman, R.M., Soares, S.E., Beckett, L.A., Sun, B., Effects of a mushroom mycelium extract on the treatment of

prostate cancer. *Urology*, *60*, 4, 640–644, 2002, https://doi.org/10.1016/S0090-4295(02)01856-3.

10. Falandysz, J., Nutritional and Other Trace Elements and Their Associations in Raw King Bolete Mushrooms, Boletus edulis. *Int. J. Environ. Res. Public Health*, *19*, 1, Article 1, 417, 2022, https://doi.org/10.3390/ijerph19010417.
11. Fortes, R.C., Recôva, V.L., Melo, A.L., Novaes, M.R.C.G., Effects of dietary supplementation with medicinal fungus in fasting glycemia levels of patients with colorectal cancer: A randomized, double-blind, placebo-controlled clinical study. *Nutr. Hosp.*, *23*, 6, 591–598, 2008.
12. Gao, Y., Dai, X., Chen, G., Ye, J., Zhou, S., A Randomized, Placebo-Controlled, Multicenter Study of *Ganoderma lucidum* (W.Curt.:Fr.) Lloyd (Aphyllophoromycetideae) Polysaccharides (Ganopoly®) in Patients with Advanced Lung Cancer. *Int. J. Med. Mushrooms*, *5*, 4, 14, 2003, https://doi.org/10.1615/InterJMedicMush.v5.i4.40.
13. Gargano, M.L., van Griensven, L.J.L.D., Isikhuemhen, O.S., Lindequist, U., Venturella, G., Wasser, S.P., Zervakis, G.I., Medicinal mushrooms: Valuable biological resources of high exploitation potential. *Plant Biosyst. - Int. J. Dealing Aspects Plant Biosyst.*, *151*, 3, 548–565, 2017, https://doi.org/10.1080/11263504.2017.1301590.
14. Habtemariam, S., The Chemistry, Pharmacology and Therapeutic Potential of the Edible Mushroom Dictyophora indusiata (Vent ex. Pers.) Fischer (Synn. Phallus indusiatus). *Biomedicines*, *7*, 4, 98, 2019, https://doi.org/10.3390/biomedicines7040098.
15. Hamza, A., Ghanekar, S., Santhosh Kumar, D., Current trends in health-promoting potential and biomaterial applications of edible mushrooms for human wellness. *Food Biosci.*, *51*, 102290, 2023, https://doi.org/10.1016/j.fbio.2022.102290.
16. Hong, L., Xun, M., Wutong, W., Anti-diabetic effect of an α-glucan from fruit body of maitake (Grifola frondosa) on KK-Ay mice. *J. Pharm. Pharmacol.*, *59*, 4, 575–582, 2007, https://doi.org/10.1211/jpp.59.4.0013.
17. Huang, H. Y., Korivi, M., Chaing, Y.-Y., Chien, T.-Y., Tsai, Y.-C., *Pleurotus tuber-regium* Polysaccharides Attenuate Hyperglycemia and Oxidative Stress in Experimental Diabetic Rats. *Evid.-Based Complementary Altern. Med.*, *2012*, e856381, 2012, https://doi.org/10.1155/2012/856381.
18. Jayachandran, P., Battaglin, F., Strelez, C., Lenz, A., Algaze, S., Soni, S., Lo, J.H., Yang, Y., Millstein, J., Zhang, W., Shih, J.C., Lu, J., Mumenthaler, S.M., Spicer, D., Neman, J., Roussos Torres, E.T., Lenz, H.-J., Breast cancer and neurotransmitters: Emerging insights on mechanisms and therapeutic directions. *Oncogene*, *42*, 9, Article 9, 627–637, 2023, https://doi.org/10.1038/s41388-022-02584-4.
19. Jayasuriya, W.J.A.B.N., Wanigatunge, C.A., Fernando, G.H., Abeytunga, D.T.U., Suresh, T.S., Hypoglycaemic Activity of Culinary Pleurotus ostreatus and P. cystidiosus Mushrooms in Healthy Volunteers and Type 2 Diabetic

Patients on Diet Control and the Possible Mechanisms of Action. *Phytother. Res.*, *29*, 2, 303–309, 2015, https://doi.org/10.1002/ptr.5255.

20. Jeitler, M., Michalsen, A., Frings, D., Hübner, M., Fischer, M., Koppold-Liebscher, D.A., Murthy, V., Kessler, C.S., Significance of Medicinal Mushrooms in Integrative Oncology: A Narrative Review. *Front. Pharmacol.*, *11*, 580656, 2020, https://www.frontiersin.org/articles/10.3389/fphar.2020.580656.
21. Kim, Y.-W., Kim, K.-H., Choi, H.-J., Lee, D.-S., Anti-diabetic activity of beta-glucans and their enzymatically hydrolyzed oligosaccharides from Agaricus blazei. *Biotechnol. Lett.*, *27*, 7, 483–487, 2005, https://doi.org/10.1007/s10529-005-2225-8.
22. Klupp, N.L., Chang, D., Hawke, F., Kiat, H., Cao, H., Grant, S.J., Bensoussan, A., Ganoderma lucidum mushroom for the treatment of cardiovascular risk factors. *Cochrane Database Syst. Rev.*, *2015*, 2, CD007259, 2015, https://doi.org/10.1002/14651858.CD007259.pub2.
23. Lee, J.S. and Hong, E.K., Immunostimulating activity of the polysaccharides isolated from Cordyceps militaris. *Int. Immunopharmacol.*, *11*, 9, 1226–1233, 2011, https://doi.org/10.1016/j.intimp.2011.04.001.
24. Liu, X., Luo, D., Guan, J., Chen, J., Xu, X., Mushroom polysaccharides with potential in anti-diabetes: Biological mechanisms, extraction, and future perspectives: A review. *Front. Nutr.*, *9*, 1087826, 2022, https://www.frontiersin.org/articles/10.3389/fnut.2022.1087826.
25. Łysakowska, P., Sobota, A., Wirkijowska, A., Medicinal Mushrooms: Their Bioactive Components, Nutritional Value and Application in Functional Food Production—A Review. *Molecules*, *28*, 14, Article 14, 5393, 2023, https://doi.org/10.3390/molecules28145393.
26. M, P., M, P., R, S., R, K., EVALUATION OF PHYTOCHEMICALS AND *IN VITRO* ANTIINFLAMMATORY, ANTI-DIABETIC ACTIVITY OF THE WHITE OYSTER MUSHROOM, PLEUROTUS FLORIDA. *Int. Res. J. Pharm. Appl. Sci.*, *5*, 1, Article 1, 2015.
27. Ma, H.-T., Hsieh, J.-F., Chen, S.-T., Anti-diabetic effects of Ganoderma lucidum. *Phytochemistry*, *114*, 109–113, 2015, https://doi.org/10.1016/j.phytochem.2015.02.017.
28. Maity, P., Sen, I.K., Chakraborty, I., Mondal, S., Bar, H., Bhanja, S.K., Mandal, S., Maity, G.N., Biologically active polysaccharide from edible mushrooms: A review. *Int. J. Biol. Macromol.*, *172*, 408–417, 2021, https://doi.org/10.1016/j.ijbiomac.2021.01.081.
29. Moazzem Hossen, S.M., Hossain, M.S., Akbar, S., Tahmida, U., Mawa, J., Uddin Emon, N., Wild mushrooms showed analgesic and cytotoxic properties along with phytoconstituent's binding affinity to COX-1, COX-2 and cytochrome P450 2C9. *Heliyon*, *7*, 9, e07997, 2021, https://doi.org/10.1016/j.heliyon.2021.e07997.
30. Mu, D., Li, C., Shi, L., Zhang, X., Ren, A., Zhao, M., Bioinformatic Identification of Potential MicroRNAs and Their Targets in the Lingzhi or

Reishi Medicinal Mushroom *Ganoderma lucidum* (Higher Basidiomycetes). *Int. J. Med. Mushrooms*, *17*, 8, 783–797, 2015, https://doi.org/10.1615/IntJMedMushrooms.v17.i8.80.

31. Mwangi, B., Ebmeier, K.P., Matthews, K., Douglas Steele, J., Multi-centre diagnostic classification of individual structural neuroimaging scans from patients with major depressive disorder. *Brain*, *135*, 5, 1508–1521, 2012, https://doi.org/10.1093/brain/aws084.
32. Park, H.-J., Current Uses of Mushrooms in Cancer Treatment and Their Anticancer Mechanisms. *Int. J. Mol. Sci.*, *23*, 18, 10502, 2022, https://doi.org/10.3390/ijms231810502.
33. Reza, M.A., Hossain, M.A., Damte, D., Jo, W.-S., Hsu, W.H., Park, S.-C., Hypolipidemic and Hepatic Steatosis Preventing Activities of the Wood Ear Medicinal Mushroom Auricularia auricula-judae (Higher Basidiomycetes) Ethanol Extract *In Vivo* and *In Vitro*. *Int. J. Med. Mushrooms*, *17*, 8, 723–734, 2015, https://doi.org/10.1615/intjmedmushrooms.v17.i8.30.
34. Royse, D.J., Baars, J., Tan, Q., Current Overview of Mushroom Production in the World, in: *Edible and Medicinal Mushrooms*, pp. 5–13, John Wiley & Sons, Ltd., 2017, https://doi.org/10.1002/9781119149446.ch2.
35. Samanta, S., Nandi, A.K., Sen, I.K., Maity, P., Pattanayak, M., Devi, K.S.P., Khatua, S., Maiti, T.K., Acharya, K., Islam, S.S., Studies on antioxidative and immunostimulating fucogalactan of the edible mushroom Macrolepiota dolichaula. *Carbohydr. Res.*, *413*, 22–29, 2015, https://doi.org/10.1016/j.carres.2015.05.006.
36. Sánchez, C., Reactive oxygen species and antioxidant properties from mushrooms. *Synth. Syst. Biotechnol.*, *2*, 1, 13–22, 2017, https://doi.org/10.1016/j.synbio.2016.12.001.
37. Seiden, M.V., Gordon, A.N., Bodurka, D.C., Matulonis, U.A., Penson, R.T., Reed, E., Alberts, D.S., Weems, G., Cullen, M., McGuire, W.P., A phase II study of irofulven in women with recurrent and heavily pretreated ovarian cancer. *Gynecol. Oncol.*, *101*, 1, 55–61, 2006, https://doi.org/10.1016/j.ygyno.2005.09.036.
38. Stachowiak, B. and Reguła, J., Health-promoting potential of edible macromycetes under special consideration of polysaccharides: A review. *Eur. Food Res. Technol.*, *234*, 3, 369–380, 2012, https://doi.org/10.1007/s00217-011-1656-9.
39. Taylor, T.N., Krings, M., Taylor, E.L., 9—Basidiomycota, in: *Fossil Fungi*, T.N. Taylor, M. Krings, E.L. Taylor (Eds.), pp. 173–199, Academic Press, Cambridge, MA., 2015, https://doi.org/10.1016/B978-0-12-387731-4.00009-8.
40. Thongbai, B., Rapior, S., Hyde, K.D., Wittstein, K., Stadler, M., Hericium erinaceus, an amazing medicinal mushroom. *Mycol. Prog.*, *14*, 10, 91, 2015, https://doi.org/10.1007/s11557-015-1105-4.
41. Torkelson, C.J., Sweet, E., Martzen, M.R., Sasagawa, M., Wenner, C.A., Gay, J., Putiri, A., Standish, L.J., Phase 1 Clinical Trial of Trametes versicolor in Women with Breast Cancer. *ISRN Oncol.*, *2012*, 251632, 2012, https://doi.org/10.5402/2012/251632.

42. Tsai, M.-Y., Hung, Y.-C., Chen, Y.-H., Chen, Y.-H., Huang, Y.-C., Kao, C.-W., Su, Y.-L., Chiu, H.-H.E., Rau, K.-M., A preliminary randomised controlled study of short-term Antrodia cinnamomea treatment combined with chemotherapy for patients with advanced cancer. *BMC Complementary Altern. Med.*, *16*, 1, 322, 2016, https://doi.org/10.1186/s12906-016-1312-9.
43. Twardowski, P., Kanaya, N., Frankel, P., Synold, T., Ruel, C., Pal, S.K., Junqueira, M., Prajapati, M., Moore, T., Tryon, P., Chen, S., A phase I trial of mushroom powder in patients with biochemically recurrent prostate cancer: Roles of cytokines and myeloid-derived suppressor cells for Agaricus bisporus–induced prostate-specific antigen responses. *Cancer*, *121*, 17, 2942–2950, 2015, https://doi.org/10.1002/cncr.29421.
44. Wahab, N.A.A., Abdullah, N., Aminudin, N., Characterisation of potential antidiabetic-related proteins from Pleurotus pulmonarius (Fr.) Quél. (Grey oyster mushroom) by MALDI-TOF/TOF mass spectrometry. *Biomed Res. Int.*, *2014*, 131607, 2014, https://doi.org/10.1155/2014/131607.
45. Wang, Y., Liu, Y., Wang, H., Li, C., Qi, P., Bao, J., Agaricus bisporus lectins mediates islet β-cell proliferation through regulation of cell cycle proteins. *Exp. Biol. Med. (Maywood, N.J.)*, *237*, 3, 287–296, 2012, https://doi.org/10.1258/ebm.2011.011251.
46. Yamac, M., Kanbak, G., Zeytinoglu, M., Senturk, H., Bayramoglu, G., Dokumacioglu, A., Van Griensven, L., Pancreas Protective Effect of Button MushroomAgaricus bisporus (J.E. Lange) Imbach(Agaricomycetidae) Extract on Rats withStreptozotocin-Induced Diabetes. *Int. J. Med. Mushrooms*, *12*, 379–389, 2010, https://doi.org/10.1615/IntJMedMushr.v12.i4.50.
47. Yang, H., Hwang, I., Kim, S., Hong, E.-J., Jeung, E.-B., Lentinus edodes promotes fat removal in hypercholesterolemic mice. *Exp. Ther. Med.*, *6*, 6, 1409–1413, 2013, https://doi.org/10.3892/etm.2013.1333.
48. Yi, X., Dong, M., Guo, N., Tian, J., Lei, P., Wang, S., Yang, Y., Shi, Y., Flavonoids improve type 2 diabetes mellitus and its complications: A review. *Front. Nutr.*, *10*, 1192131, 2023, https://www.frontiersin.org/articles/10.3389/fnut.2023.1192131.
49. Yoon, S.Y., Park, S.J., Park, Y.J., The Anticancer Properties of Cordycepin and Their Underlying Mechanisms. *Int. J. Mol. Sci.*, *19*, 10, Article 10, 3027, 2018, https://doi.org/10.3390/ijms19103027.
50. Zhang, Z., Lv, G., Pan, H., Pandey, A., He, W., Fan, L., Antioxidant and hepatoprotective potential of endo-polysaccharides from Hericium erinaceus grown on tofu whey. *Int. J. Biol. Macromol.*, *51*, 5, 1140–1146, 2012, https://doi.org/10.1016/j.ijbiomac.2012.09.002.
51. Zhao, H., Zhang, Q., Zhao, L., Huang, X., Wang, J., Kang, X., Spore Powder of *Ganoderma lucidum* Improves Cancer-Related Fatigue in Breast Cancer Patients Undergoing Endocrine Therapy: A Pilot Clinical Trial. *Evid.-Based Complementary Altern. Med.*, *2012*, e809614, 2011, https://doi.org/10.1155/2012/809614.

Index

Also of Interest

Handbook of Agricultural Biotechnology

Volume I Nanopesticides
Edited by Charles Oluwaseun Adetunji and Julius Kola Oloke
Published 2024. ISBN 978-1-119-83614-8

Volume II Nanobioherbicides
Edited by Charles Oluwaseun Adetunji and Julius Kola Oloke
Published 2024. ISBN 978-1-119-83615-5

Volume III Nanofungicides
Edited by Charles Oluwaseun Adetunji and Julius Kola Oloke
Published 2024. ISBN 978-1-119-83616-2

Volume IV Nanoinsecticides
Edited by Charles Oluwaseun Adetunji and Julius Kola Oloke
Published 2024. ISBN 978-1-119-83617-9

Volume V Nanobiofertilizers
Edited by Charles Oluwaseun Adetunji, Chukwuebuka Egbuna, Anton Ficai and Oluwatosin Ademola Ijabadeniyi
Published 2024. ISBN 978-1-394-21149-4

Next-Generation Algae, Volume I Applications in Agriculture, Food and Environment
Edited by Charles Oluwaseun Adetunji, Julius Kola Oloke, Naveen Dwivedi, Sabeela Beevi Ummalyma, Shubha Dwivedi, Juliana Bunmi Adetunji and Daniel Ingo Hefft
Published 2023. ISBN 978-1-119-85727-3

Next-Generation Algae, Volume II Applications in Medicine and the Pharmaceutical Industry
Charles Oluwaseun Adetunji, Julius Kola Oloke, Naveen Dwivedi, Sabeela Beevi Ummalyma, Shubha Dwivedi, Juliana Bunmi Adetunji and Daniel Ingo Hefft
Published 2023. ISBN 978-1-119-85728-0

Printed and bound by CPI Group (UK) Ltd, Croydon, CR0 4YY
17/09/2026
14964280-0004